COST PLANNING OF BUILDINGS

Cost Planning of Buildings

DOUGLAS J. FERRY
PhD, FRICS

PETER S. BRANDON
MSc, FRICS

SIXTH EDITION

OXFORD
BSP PROFESSIONAL BOOKS
LONDON EDINBURGH BOSTON
MELBOURNE PARIS BERLIN VIENNA

BSP Professional Books
A division of Blackwell Scientific
  Publications Ltd
Editorial offices:
Osney Mead, Oxford OX2 0EL
25 John Street, London WC1N 2BL
23 Ainslie Place, Edinburgh EH3 6AJ
3 Cambridge Center, Cambridge
  MA 02142, USA
54 University Street, Carlton
  Victoria 3053, Australia

First published in Great Britain by Crosby
  Lockwood & Sons Ltd 1964
Second edition (metric) 1970
Third edition 1972
Reprinted 1974
Fourth edition 1980 by Granada Publishing
Reprinted 1981, 1983
Fifth edition 1984 by Granada Publishing
Reprinted by Collins Professional and Technical
  Books 1986, 1987
Reprinted by BSP Professional Books 1988
Sixth edition 1991 by
  BSP Professional Books

Typeset by DP Photosetting, Aylesbury, Bucks
Printed and bound in Great Britain by
Billing & Sons, Worcester

DISTRIBUTORS
  Marston Book Services Ltd
  PO Box 87
  Oxford OX2 0DT
  (*Orders*: Tel: 0865 791155
         Fax: 0865 791927
         Telex: 837515)

USA
  Blackwell Scientific Publications, Inc.
  3 Cambridge Center
  Cambridge, MA 02142
  (*Orders*: Tel: (800) 759-6102)

Canada
  Oxford University Press
  70 Wynford Drive
  Don Mills
  Ontario M3C 1J9
  (*Orders*: Tel: (416) 441-2941)

Australia
  Blackwell Scientific Publications
  (Australia) Pty Ltd
  54 University Street
  Carlton, Victoria 3053
  (*Orders*: Tel: (03) 347-0300)

British Library
Cataloguing in Publication Data

Ferry, Douglas J. (Douglas John)
  Cost planning of buildings.
  1. Buildings. Construction. Estimating
  I. Title II. Brandon, Peter S.
  692.5

ISBN 0–632–02367–8

Contents

Preface to the Sixth Edition

A major change since the last edition was published has been the switch of the bulk of the industry's clientele from the public sector (in which formal cost planning techniques first developed) to the private sector, together with the much greater incidence of refurbishment work. These changes have increased the need for a thorough understanding of the principles behind planning and controlling the cost of a project, whilst perhaps reducing the role of highly formalised systems.

The material in the Fifth edition has been largely retained and updated, but it has been re-arranged and developed into a different sequence which is more in keeping with the present-day emphasis of the subject.

The book is therefore now divided into three sections – Setting the Scene, Cost and Design, and Cost Planning Techniques. The last of these sections, which at one time formed the major thrust of the work, might now be seen as perhaps less vital than the strategic matters discussed in the first two sections.

Techniques are constantly changing in the light of office practice, new technologies, client emphasis, and procurement methods, and towards the end of this section some of the most advanced approaches are examined. However, it is useful to be familiar with classic cost planning methods and for examination purposes it is essential. Nevertheless, as their practical application is more commonly associated with the buildings of the 1960s and 1970s it has not been thought necessary to revise some of the examples although their costs have been updated.

Finally, to repeat what has been written in previous prefaces, the book seeks to provide an essential groundwork in all aspects of the discipline whilst containing ample material as a basis for seminar work, discussion, and class development. However, it cannot claim to give full coverage to each of the many topics which are dealt with in principle, and a reader who wishes to explore any of these in greater depth will need to refer to specialist works.

Douglas J. Ferry
Peter S. Brandon

Publisher's note

Some of the techniques described in this book have been programmed for a microcomputer and can be found in *Computer Programs for Building Cost Appraisal* by Peter S. Brandon, R. Geoffrey Moore and Phillip Main.

Part 1
Setting the Scene

Chapter 1
Introduction

From the earliest times people have needed some idea of what a new building was going to cost before they started work on it. The New English Bible says 'Would any of you think of building a tower without first sitting down and calculating the cost, to see whether he could afford to finish it? Otherwise, if he has laid its foundations and then is not able to complete it, all the onlookers will laugh at him. "There is the man" they will say "who started to build and could not finish".' (St Luke, Ch 14.)

Forecasting the cost of a building, however, is not the same thing as planning the cost, any more than a weather forecaster on television can be said to be planning the weather – in both cases things may turn out very differently from what was expected for reasons quite outside the forecaster's control. Nevertheless, until the early nineteenth century rough-and-ready forecasting satisfied society's needs fairly adequately. Most major building was undertaken either as an act of religious faith or by the very rich for their own pleasure and gratification, and in both cases the necessary resources were likely to be forthcoming in the end. The building process itself comprised a series of fairly autonomous craft activities, of which the costs had become established and known over a lengthy period of time. Even so miscalculations occurred – the building of Blenheim Palace almost bankrupted the Duke of Marlborough, and it was not at all unknown for prospective owners of buildings to suffer the fate of the man in the Bible. Readers of Robert Louis Stevenson's *Kidnapped* will remember the unfinished House of Shaws which provided such an exciting start to the tale.

With the advent of the Industrial Revolution, however, something better was clearly needed for three reasons.

First, the people commissioning large building projects were increasingly cost-conscious, being either industrialists concerned with profitability, government bodies concerned with accountability, or joint stock companies concerned with both. Second, the projects themselves were of an increasing technological complexity. Third, the traditional settled economic and social order was turning into something more sophisticated and dynamic.

In order to deal with the needs of this new situation the price-in-advance system was developed, where responsibility for the execution of the whole project was handed over to a 'general contractor' at a previously agreed price. The building owner and his architect had thus got rid of many of the problems of organisation, and secured a firm cost commitment before starting work, by this single ingenious move.

For over a century this system worked very well indeed, and it is still in widespread use today although much of the simplicity of the original concept has been lost in recent years. However, it did place one additional burden upon the architect – complete drawings and specification of the work had to be prepared before prices could be sought.

Prices were normally obtained by competitive tender; the documents would be sent to a number of general contractors who would submit sealed offers to be opened by the architect at a time and date set in advance. The job would then usually be given to the firm whose tender was the lowest.

In order for contractors to prepare their tenders it was necessary for them to calculate the quantities of labour and material which would be required to build the project. This was done by measurement off the drawings, and since the figures should have been the same for all the firms tendering (unless a mistake were made) it was obviously more economic for this task to be undertaken by one person on behalf of them all. This was the origin of the quantity surveyor, whose task was the preparation of a 'Bill of Quantities' setting out the labour and material requirements expressed in terms of the quantity of finished work to be produced. His fee was paid by the successful tenderer.

A difficulty immediately arose. The success of the tendering system depended upon there being true competition, and there was always a danger that the tenderers might get together secretly in order to agree on an inflated price and a suitable division of the proceeds. The architectural profession was worried that if the tenderers had to meet in order to agree the appointment of a quantity surveyor they might seize the opportunity to make some other less desirable arrangements at the same time. The custom soon arose of the architect sending out a bill of quantities with the drawings and specifications, so depriving the tenderers of an excuse to consort together, although the successful builder was still at that time responsible for paying the quantity surveyor whom the architect had appointed. From this point, quantity surveying developed in the United Kingdom (and many of its overseas dependencies) into a fully recognised profession.

Although the system of firm price competitive tendering gave the prospective building owner a definite budget prior to committing him to actual expenditure on building, he had already committed himself to a good deal of other expenditure by this stage, which could not be recovered if he found he was unable to go ahead because costs were too high. The architect and other design consultants would have completed most of their work and would require a substantial proportion of their fees, as would the quantity surveyor, irrespective of whether the job went ahead or not. (The only people who would have worked without charge would have been the builders, who would not have expected to be able to recover their tendering costs – a sore point with many of them.) In addition, the client would probably have bought the land for the project and this money would now be locked up in an unproductive asset, in addition to which he might have made plans for the use of the building which would have to be abandoned with consequent loss. An alternative to abandonment, of course, would be complete redesign on a more modest scale and the seeking of fresh tenders, but this would involve a further loss of time and would not relieve the client of the abortive expenditure on the first scheme. It was therefore important that the client should have a good idea of what the building was likely to cost before he had committed himself very far, and this was done by means of an 'approximate estimate'.

This was an attempt to forecast the cost of a building at some stage prior to the completion of the bill of quantities or the obtaining of a tender. Once the bill of quantities was available, of course, quite an accurate estimate could be prepared by the quantity surveyor, pricing it out in detail as though he were a builder. However, this would be far too late in the process to be of much use and would involve an additional fee. (It was sometimes done for a quite different reason, namely as a check on the reasonableness of the builders' prices.)

Because the client was anxious to have some idea of his commitments as early as possible, and because the architect wanted to know whether his design was financially possible before he had gone too far with it, the approximate estimate would be prepared at a fairly early stage when even the basic drawings would be rudimentary, and there would be little or no supporting documentation.

Although the usual terms of appointment of the architect required him to prepare this estimate, it was in practice usually done by the quantity surveyor on his behalf, as the latter had both a greater expertise and a wider range of available cost data. It was not a lengthy task and was normally done free of charge as a piece of goodwill to the architect, who was usually responsible for nominating the quantity surveyor to the client.

The techniques for preparing an estimate of cost in these circumstances will in fact be dealt with later in this book, since they are still valid today in some circumstances as part of a comprehensive system of cost planning and control. These traditional methods were basically 'single-price-rate' estimates; the size of the building was calculated in square or cubic measure and the result multiplied by a single price-rate to give the estimated cost.

However well such an estimate was prepared (and this called for a good deal of experience, skill and judgement) it suffered from a vital defect; it was not possible to relate the cost of the work which was shown on the detailed drawings to the estimate during the progress of the design. The quantity surveyor might feel, as he prepared his bills of quantities, that these drawings were showing something rather grander and more extravagant than he had in mind when he produced the estimate, but it was not possible to prove this contention to the architect or client except by preparing and pricing 'approximate quantities'. These were a simplified version of a formal bill of quantities; major items were grouped and measured together, while less important items were ignored and dealt with by loading the prices of major items.

The disadvantages of such a course are obvious. The task would require extra time and the cost would be too great to be covered by 'goodwill' and would involve an additional fee, the drawings had to be fairly complete and the whole of the building had to be dealt with. Even if the approximate quantities when prepared showed a far higher total than the original estimate, the discrepancy could not be easily traced. It still could not be seen whether any particular part of the building was extravagantly designed and there was nothing to show where the cost could be cut most economically. Even if it were possible to identify the problem areas, the architect would obviously resist an expensive major redesign at this late stage.

All that could usually be done in the circumstances was to cut the more high-class parts of the specification and substitute lower quality, which was not at all the same thing as producing an economically efficient design.

As a rule, therefore, approximate quantities would not be prepared but the job sent out to tender by the architect, who would just hope for the best. If the tenders came in too high the same course would have to be followed of cutting the cost where it was most convenient rather than most efficient to do so (without having to alter too many drawings), and the result was likely to be a lot of wasted work and an unsatisfactory and unbalanced design.

During the unsettled years following the Second World War the art of accurate single-price-rate estimating became increasingly difficult to practice. Both for this reason, and because of a demand by public authorities in particular for value in building, a system of cost planning and control during design stage came into vogue. This system was called 'elemental cost planning' and enabled the cost of the scheme to be monitored during design development, as well as facilitating the rational allocation of costs amongst the various elements of the building structure, finishes and services. The technique is still used for these purposes today, and two chapters of this book are devoted to it exclusively, but more importantly it formed the basis upon which a whole range of cost planning practices was developed.

Unfortunately, no sooner had success been achieved in rationalising the economic approach to the obtaining of tenders than the price-in-advance system itself began to deteriorate. There were two reasons for this, the first being that the British system of tendering upon sophisticated bills of quantities enabled tenders to be obtained upon a basis which was theoretically that of a completed and worked-out design. In practice this was little more than a hypothetical construct, so that an enormous number of changes both great and small would occur during the construction stage. The contractor would thus find it difficult to organise the work properly, but would have ample excuse to charge a price which differed substantially from his original offer. The second reason, very much compounded by the first, was the changed social, economic, and moral climate which made it very difficult for a general contractor to quote a firm price for a major project in competition, which would adequately allow for all the risks involved. Such symptoms as rampant inflation, industrial action, social legislation, and financial irresponsibility showed clearly that society was very different from that which had fostered the Victorian concept of price-in-advance, and the consequent abrogation of management responsibility by the client and architect.

In these circumstances alternative methods of building procurement have come into favour, in some of which there is no place for formal bills of quantities, and in which there may be no such concept as a 'tender price'. Quantity surveyors, by virtue of their cost planning experience in the traditional situation, have been well placed to act as building economists in a wider sense. In the UK the quantity surveyor has become the acknowledged source of expertise and advice on costs and their manipulation, and is often engaged for this purpose on many projects where bills of quantities are not required. However, the importance of cost planning and control is such that members of the architecture, engineering and building professions need to have a working knowledge of the principles, and some of them may come to specialise in this aspect of their work and see themselves as cost planners, as do some of their colleagues on the continent of Europe. The development of integrated computer systems which manipulate cost as one of a number of design parameters will tend to assist this process, as will the undertaking of work elsewhere in the European Community and overseas.

Two further factors in developing more flexible cost planning techniques in the UK have been the increased proportion of 'refurbishment' projects in the 1980s compared to the new work for which the original systems were designed, and the general move from a public to a private sector clientele, with less emphasis on the accountability of a construction programme and more emphasis on the profitability of individual projects.

Today's cost planner therefore needs to be able to do much more than merely use techniques to obtain an acceptable tender price, important though this task may often be. He needs to work with his client from the very inception of a scheme (or even earlier in the devising of an investment or development programme of which the project may one day form part), to the time when the scheme is eventually completed and handed over. To do this a flexibility of approach is essential – the client will need to know the best way of going about the project in the particular socio-economic situation that exists at the time and in the place concerned. There is thus no single, tailor-made set of standard procedures that can be learnt and applied without much thought; what is required is a thorough understanding of the issues involved and the solutions available. We have tended to avoid using the terms 'optimum' or 'optimisation' in this book; anybody who achieves satisfactory results in practice will have done quite well enough.

The role of the cost planner is now so important that he must be prepared to take his full share of social responsibility in the advice which he gives and the work which he does. Whilst there may be occasions where the image of the ruthless cost-cutter is the right one, he must be able to see when this is so, and when it is not. In particular, he must be able to understand the needs and points of view of the client, the architect, and the community in relation to the project, and balance them as far as is possible within his brief. His concern should be value rather than minimum cost, and value often goes far beyond those things that can be easily quantified in money terms. A broad education and a wide range of social and cultural interests are necessary for him to do his job really well, in addition to common sense and a feeling for practical probabilities.

The advent of silicon-chip technology is going to provide not only cost planners but also designers, clients, and builders with ever-increasing opportunities for data collection and manipulation. In this situation it will be most important that the cost planner's special expertise in interpreting and using this material should be recognised and sought; the blind processing of figures, by people who neither understand where they have come from nor exactly what they represent, will be a great source of danger in this situation.

This book is arranged in a manner which should help towards a wide understanding of the subject. The first five chapters set the scene in which building development is carried out, whilst the next six chapters explain the design process and the basic tools available to the cost planner. When all this has been assimilated, the techniques to be used in various circumstances are dealt with at greater length.

In conclusion, a word of apology is due to women readers. In the interests of brevity and because the majority of cost planners are still male, we have sometimes used the male pronoun in referring to cost planners rather than adopting some form of circumlocution. We do hope that this device will not prevent women readers from identifying themselves with the person described.

Chapter 2
The Starting Point – Developer's Motivation and Needs

The carrying out of building work is often termed 'development' and the person or organisation responsible is the 'developer'. More often in the building industry he is called the 'client' or 'building owner', or in some parts of the world the 'proprietor'. In popular usage the term 'developer' is restricted to those who build for profit, but this is not correct and those who build for their own use or for social purposes are equally developers.

All development arises from a consumer demand. It may be a direct demand for housing, for offices, for a town hall or library, or may be an indirect demand; that is a demand for something else which will require building development in order to satisfy it (such as a demand for TV sets which will necessitate a factory to produce them). But the demand must always be an economic one. There must be someone willing to foot the bill or who can be induced to do so. Where the development is likely to be profitable such a person should not be hard to find.

The client, whether an individual or a corporate body, will be building either for direct profit or for use; in the first case his approach to cost will be governed by the terms in which his income will be received and in the second by the units of accommodation which he requires.

A factor which makes building development quite different from other entrepreneurial activities is that there can be no building without a plot of available land. This has all sorts of economic consequences, because the supply of land is largely fixed ('Buy now!' said a recent estate agent's advertisement, 'They have stopped making it!').

No developer can start work until he has acquired an interest in the land on which his building is to stand.

Total building cost

When any type of building development, private or public, takes place the total expenditure on the development will be much higher than the net cost of the building fabric, and will comprise:

1. Cost of land.
2. Legal, etc, costs of acquiring and preparing the site, and obtaining all necessary approvals.
3. Demolition or other physical preparation of the site.
4. Building cost.
5. Professional fees in connection with last.
6. Furnishings, fittings, machinery, etc.
7. Costs in connection with disposal (sale, letting, etc) where the building is to be disposed of at completion.

8 Value-added Tax on above items where chargeable.

9 Cost of financing the project (this principally represents interest, etc, on the expenditure before any return is obtained either by way of income or of use).

10 Cost of management, running, and maintenance where the building is to be retained by the client for use or only partially disposed of, or where the building is let to a tenant but the owner has made himself responsible for some or all of these costs.

It is important to realise that 'building cost' is only one of ten items of total cost which is incurred by the client, and that in the end it is the total which counts. It may be thought, therefore, that many cost planning exercises, including some in this book, are unduly obsessed with this one item rather than with optimising the whole package. There is certainly some truth in this allegation, which probably has its roots in the fact that so many of the techniques were developed in the public service where total costs were rarely considered. We must always be on our guard to avoid this trap unless it is forced on us, and even then we should be ready to query its validity.

For instance, the cost of the land may be such an important factor in the costing exercise that everything else has to be fitted round it and the problem then is to get the maximum amount of permissible accommodation almost regardless of building cost. The effect of land costs on the developer's budget is further considered in Chapter 3.

It sometimes happens that the building is only a small part of the total project, as in the setting up of a main TV transmitter station where the mast, electronic equipment and cable-laying dominate the project and the transmitter building is of minor importance. In such circumstances it would be a mistake to consider the cost of the building in isolation – it is much more important that is should phase in with the general programme and that its completion should not be a source of worry to the engineer in charge. It may not be necessary to indulge in cost planning of the building at all.

Cost targets for profit development and social development

In considering cost targets we may differentiate between the private and public sectors of the economy, and also between buildings which are to be economically viable (i.e. self-supporting in the sense that receipts from the use of the buildings will more than cover costs) and those which are not.

Economic viability is not exclusive to private development (local authorities or government boards often undertake profitable development) neither is private development always aimed at making a profit (for instance, churches or charitable buildings). Some development may be partially self-supporting, such as local authority housing schemes paid for partly by rents and partly by subsidies.

Because the calculations involved in budgeting for both the above types of development are similar it is easy to forget that the basic situations are fundamentally different.

1 Profit development

The setting of budget targets for profit development (eg office blocks for letting, housing for letting or sale) must be related to free-enterprise economics; the sole purpose of the development is to make a profit, and setting a target is straightforward. A calculation by the developer will soon show how much he can afford to spend under the heading of 'building costs' for his scheme to be profitable after the other factors in the list have been taken into account. There is a certain amount of flexibility here because the standard of building will partly determine the rents or selling price that can be asked, although this income is often dictated within fairly close limits by the environment and by the cost of the land. It would obviously be unwise to erect a high-cost luxury building for sale or rent in an unpopular neighbourhood, just as it would not be economic to try a low-cost, low-income development in a fashionable district where land costs are high and there is a good income potential.

Once a final decision has been made on income levels the cost will have been determined and must not be exceeded; from this point onwards tight cost planning and control will be vital and if any item goes up in cost the money must be saved elsewhere. Otherwise the anticipated profit will not be made by the developer, and from his point of view the project will be a failure, however pleasant or useful the resulting buildings may be. There is no room for 'hard luck stories', allowances for abnormals, or anything of this kind, but on the other hand there is the challenge of working in a situation where, for instance, time means money. Early completion of a department store may involve substantial extra profits, late completion (e.g. after Christmas instead of before Christmas) can be financially disastrous.

In the early stages of considering alternative ways of developing a site the probable receipts must be compared with the probable costs, and a misjudgment of either figure will have exactly the same effect on the profitability of the scheme; neither is more important than the other, and no amount of detailed cost planning is going to rescue a scheme based on unsound investment assumptions.

The valuation surveyor is an expert on the value of buildings for investment and the income which various types of development can be expected to produce, but as a rule his knowledge of construction costs (and particularly of the sort of factors which we will be discussing in this book) is far more limited. Just as it would be unwise for the client to ask his quantity surveyor to estimate the likely income from a development in order to arrive at a conclusion on feasibility, so it would be foolish for him to rely upon his valuer's ideas of relative construction costs.

In conclusion, however complicated the situation may become, with grants, subsidies, taxation reliefs and so on to be taken into account, the basic exercise remains delightfully straightforward because profit (the main reason for the building) is quantifiable. Even if some non-quantifiable element, like preserving the environment, has to be taken into account it will have to be assessed against its effect on profit. If an adequate profit cannot be foreseen the building will not be built.

2 Social development

In social development, however, the main object of the exercise is the actual provision of the buildings, and it becomes very difficult to set realistic cost targets since there is no definite limit at which the building ceases to be possible. The only limit is what can be afforded, or what can be raised, and while this may be a very real factor in some private non-profit building, it can hardly apply in most public development where it is usually possible (though not always expedient) to raise whatever sum of money may be required for the purpose. Any constraint is usually at a higher level than the individual project, for example, the proportion of the national budget which is allocated to education may determine the amount to be spent on school building as a whole.

Therefore, in order to determine a reasonable cost for this type of development it used to be common to set artificial limits, based upon the cost of similar buildings erected elsewhere. The gross floor area was too crude as a basis for the purposes of this comparison and so various targets based upon user requirements were set up, often by the ministries responsible so as to ensure a nation-wide standard. In the case of schools the unit of cost was the number of 'cost places' (a fictitious number of pupils calculated from the teaching space), while for hospitals the number of beds was once the basic yardstick.

These cost targets were usually determined by a set of artificial standards which had to be adhered to; these although tight in some ways usually made exceptions for technical difficulties associated with a particular project. But here we are not in the world of simple profit economics; the early completion of a school would usually be a financial burden to the Education Committee, however welcome it might be in other ways.

The cost planning of social development projects, therefore, may resemble the playing of a board-game, where one tries to win according to a set of rules which have little validity in the real world outside. Sometimes the rules may be very crude indeed, such as a rule that money cannot be transferred from one fund to another (so that it is useless trying to save money on furnishings or running costs by spending extra on the building); or that money cannot be spent outside the financial year in which it is allocated; or that no major contract can be let except by competitive tendering on a firm bill of quantities with no allowance for cost fluctuations.

In these circumstances the skill in cost planning may consist of 'loophole' designing, to take advantage of the regulations in the same way that a clever accountant takes advantage of the tax laws. An example of this, from outside the UK, occurred when a national system for cost control of flat-building at one time gave a greater cost allowance for balconies than the actual cost of providing them. The blocks of flats built during this period can be identified by their lavish provision of unnecessary balconies.

It is recognised by the government that unrealistic cost rules lead to bad design, and many far-sighted efforts have been made to do away with the cruder kinds of inconsistency, by allowing money saved in one direction to be spent in another, by trying to bring the assessment of running costs into the cost comparisons, or by working out very complex cost criteria for such buildings as hospitals instead of 'so much per bed'. This is a very commendable attempt to get nearer to reality, but while it has undoubtedly led to some improvement it also tends to make the 'game' more complicated, and the loophole-finding more of a challenge to the experts. It becomes increasingly difficult to avoid the 'balcony' type of inconsistency mentioned above.

The consequence of all this is that everybody can become so preoccupied with trying to meet tight cost targets by clever application of the rules, that they lose sight of the social purpose of the whole exercise. The cost is not a clear-cut measurement of the effectiveness of the project (as it is with profit development); the benefits of a hospital, clinic, school, or police-station are largely unquantifiable.

Although cost planning is essential to meeting cost targets, and cost analysis of past projects will have played its part in setting the targets, nobody really knows whether they ought to be spending twice as much money (or half as much) on buildings of this kind, and no amount of cost planning is going to give them the answer. The most it can do is to help them to use the total allocated funds more effectively within the current framework of rules, and accept that the basic values will be decided for political reasons.

In fact it has been alleged, not without some justification, that the continual paring of public sector building cost targets over the years (without any consumer check on satisfactory standards such as sales or rents) played a large part in creating the massive maintenance programmes with which many public authorities are now faced.

3 Private user development

This third type of development covers such projects as a private house, or an office building for an insurance company's own occupation. This is a much less important sector than the previous two in terms of value, and incorporates some features of both profit and social developments. Because the cost situation is not automatically established, as it is in the two previous cases, it is all the more important to find out what the client's real cost priorities are and to get them defined.

Cost-benefit analysis (CBA) and option appraisal

In order to overcome some of the difficulties previously outlined, and to justify the spending of money on public projects, the techniques of cost-benefit analysis were developed. Such analyses attempted to quantify all factors including the various social benefits and disadvantages, and were widely used in connection with traffic and airport schemes and with hospital building. Cost-benefit analysis might be applied,for instance, to a proposal to carry out works to remove a sharp curve and speed restriction from an inter-city railway. The cost of the work could be set against the saving in fuel, and the wear and tear on equipment caused by braking and re-acceleration; however, it is possible that on these grounds alone it might not quite be worthwhile. On the other hand, if the curve is removed it might save two minutes each on a million passenger journeys a year – thirty thousand man-hours are worth something, and if they are priced and included among the benefits the scheme might now be justifiable in relation to the national economy.

This is all very well, but one is soon faced with problems of attaching money values to things which cannot be quantified. As an illustration of the problem, what value should one attach to a human life? One could work this out on the basis of the contribution which the individual person makes to the economic life of the community measured in terms of salary, so that a surgeon might be worth hundreds of thousands of pounds while an unemployed labourer might have a negative value. Even on practical grounds this is obviously unacceptable; if you attempted a cost-benefit analysis of a geriatric hospital on this basis you would find that it would be cheaper to let people die and build a mortuary instead! Or, as another less extreme example, you could justify a ring road which saved a few minutes for 'important' people while wasting the time of humble pedestrians.

It was therefore customary to take a notional figure representing the worth of an 'average' person. There was little wrong with this except that such figures, being notional, were conjured out of thin air and could be used in practice to 'prove' that a politically desired result was the right one. As an example, suppose a traffic improvement scheme was going to cost £100,000 a year and was estimated to reduce road accident deaths by three per year. If you cost a life at £50,000 the scheme was obviously worthwhile; if you cost a life at £20,000 it wasn't.

As a real-life example, in 1988 the UK Department of Transport for political purposes arbitrarily doubled the value for a human life used in its calculations – this simple stroke of the pen increased the benefit expected from its road schemes by an average of 4.5%, although nothing in fact had changed.

If you remember that the reduced number of deaths is a guess anyhow, you can see that this sort of exercise is not really worth very much, and the more complicated it gets and the more social benefits that are quantified, the more questionable is the result.

A further difficulty is that the values assigned to non-quantifiables tend to be those of the cultured middle class; the relative values of preserving the environment as against providing local employment, for instance, might not be those which a working family living in the area would choose.

Cost-benefit analysis in its extreme form is now largely discredited, but its successor in the public domain, option appraisal, draws upon its techniques. Option appraisal can be applied to any proposal for public investment, and involves the appraisal of all possible options (including the 'do-nothing' option). CBA is used to evaluate those aspects which have a clear money value, both of a capital and recurrent nature, but intangibles are merely assessed and shown separately. It is left to the administrators to make the subjective decisions, knowing the financial outcome of the more tangible parts of each option.

The client's needs

Every building client will have a set of needs, some of which are more important to him than others. We have to look at these rather carefully, because it is common to be told for instance, that a low cost is required, or that time is important, without these very basic requirements being defined in more detail – and it is the detail that decides the best way of tackling the project. We can now look at the main time and money requirements which the client may have, and the different forms which each may take, remembering that some clients may have more than one of the requirements under each heading. These time and cost requirements are considered in relation to particular contract types, in Chapter 7.

Time requirements

1 *No critical time requirements.* This is quite common, especially in social development.

2 *Shortest overall time, from the inception of the idea to 'turning the key'.* This is likely to be the requirement on a simple profit development.

3 *Shortest contract period from the time the contractor is appointed.* This by itself is not often relevant to the client's needs, but it is surprising how often it is asked for.

4 *Shortest contract period from the time the contractor starts on site.* This is a reasonable requirement where, for example, there is a delay in acquiring the property, or where people have to be moved out of property on the site before demolition can take place, or where the building work will cause inconvenience or disruption.

5 *Early start on site.* This may be required where the payment of a grant or subsidy depends upon work having been started by a certain date. It is also asked for sometimes by a dilatory architect to give the client the impression that something is happening at last.

6 *Reliable guaranteed completion date stated by contractor.* This may be wanted so that firm arrangements can be made well in advance for commissioning the building.

7 *Reliable guaranteed completion date stated by client.* This may apply where the client is under notice to quit existing premises or where there is a particular event that the building must be open for, such as the beginning of the summer season for a hotel.

8 *Early completion unwelcome.* It is often assumed that if the client says that time is critical then he will be even more pleased if he gets the building early. This is not necessarily the case; if, as in (6) or (7), he is making all his arrangements on the basis of a particular date, early completion simply means that he has got to waste money watching and maintaining an empty building and also that he has to pay for the building earlier than he need, so upsetting his cash flow position.

9 *Phased programme to fit in with plant installation.* This is especially important in the case of sophisticated projects such as TV transmitters or chemical works where the actual building work is only a small part of the total scheme.

10 *Handing over in sections.* This is often very important in alteration works or in rebuilding, where people or processes from one section have to be rehoused elsewhere before work can be carried out in that section.

Cost requirements

11 *No critical cost requirements.* Today this is not very common.

12 *Low total cost of whole project.* By contrast, this is usually said to be the main priority. However, we also need to see whether any of the following points are also important.

13 *Low cost in relation to units of accommodation.* This is a most usual cost requirement, in both the profit and social sectors.

14 *Good budgetary control of whole project.* In many cases it is important that all the initial cost forecasts should be accurate, even if this does not mean the lowest cost that the market might possibly produce at the time the orders are placed.

15 *Low total cost of building contract.* This is not unlike (12), but the concern here is to keep the building cost to the minimum, even if this does not minimise administration and supervisory costs, or costs of furnishing and maintenance. It is often required on public projects where these things come out of different funds.

16 *Good budgetary control of building contract.* Similar to (14), in the circumstances outlined in (15).

17 *Good forecast of cost at contractual commitment.* This is required by a client who wants an accurate forecast of final cost before committing himself contractually.

18 *Best combination of capital and maintenance costs.* One would like to think that this was more common than it is – there are all sorts of reasons like taxation, cost yardsticks, etc, which tend to prevent these two things from being weighted equally.

19 *Low capital cost.* The more usual requirement, especially if the building is going to be sold, or if running costs come out of a different fund.

20 *Low maintenance cost.* Less usual, but may be required if maintenance is going to be inconvenient, for example by putting the building out of commission while it is going on.

21 *Timing of cash flow.* This may be required in order to optimise the cost on a Discounted Cash FLow (DCF) basis (see Chapter 3) or else to phase in with the availability of the client's funds.

22 *Minimum capital commitment.* This would be required if the client wanted the contractor to bear most of the cost until the building was handed over.

23 *Share in risk of development.* A variation on the last, where the contractor is paid by a share in the profits; this has been used on large speculative developments.

The client, of course, may wish to combine three or four or more requirements from the above list; as a result there are many different sets of possibilities, and the way in which he goes about the project should reflect his individual combination of needs. He should not adopt a standard solution simply because it is the usual one.

Chapter 3
Framing the Developer's Budget

Money, time and investment

If you were asked 'Would you rather be given £1,000 now, or in five years time?' you would almost certainly say 'Now', particularly if inflation was rampant, but even if you were living in a time of zero inflation you would still be right to prefer to have the money right away. After all, you might be dead in five years' time, or there might be a revolution or world war before you had the opportunity to collect. Alternatively, the person who was going to pay you might have died, or disappeared, or gone bankrupt, or forgotten their promise. Even if you were not going to need the money for five years and were not worried about risks you would still do better to have it now and place it in a savings bank where it could accumulate interest.

We can therefore see that a sum of money in the future will always be worth less than the same amount of money today, and the difference will depend upon the length of time involved and the probable interest rate. In doing the calculations one might assume an interest rate that would reflect likely inflation and any special risks, rather than a rate which might actually be obtainable at the time.

Just as a future lump sum is worth less than its equivalent today so are future recurrent expenses or receipts. If you had to put a sum of money aside to pay somebody £1,000 a year for ten years the amount required would be much less than £10,000, because the unexpended balance of the sum would be earning interest each year and in the early years in particular this would be quite a lot of money. (In fact if interest rates were as high as 10% the £10,000 would provide £1,000 a year for ever and ever, not just for ten years.) The actual sum which would be needed to provide £1,000 a year for 10 years is an amount which, with all its interest earnings, would be exactly used up at the end of ten years when the last £1,000 had been paid out. Its actual worth would depend upon current interest rates.

Conversely, of course, a sum of money in the past is worth more than the same sum today. £1,000 given to somebody ten years ago and invested would be now be worth two or three times that amount.

Until mid-century it was common practice for cost planners to consider the cost of the building as a lump sum of money, and when interest rates were only 2% or 3%, and construction times were quick, this was possibly good enough.

However, it seems likely that interest rates will continue to be high for some years (especially during expansionary periods when large programmes of building work are being undertaken), and with the increasing complexity of large buildings it is not uncommon for some years to elapse from the start of expenditure on the project to the time when an income is produced. Under these circumstances it is necessary to consider the phasing of the project finance very carefully; an apparently low total cost may not be much of a bargain if it involves a long-drawn-out contract with high early expenditure on which interest charges are piling up. An exceptional example of the effect of financing charges concerns a very expensive office block in Sydney which met with many difficulties and delays in construction, and whose financing charges alone came to almost as much as the *total cost* of the notorious Sydney Opera House!

It may be asked 'Do financing charges apply to all development, or only where profit is involved?' Basically they apply to all development, because the money is either being borrowed (and attracting interest charges), or else it is the developer's own money, which being spent on the development is not available for investing elsewhere (and is not accumulating interest).

Strangely enough, although most cost planners are aware of this factor, some clients' accounting systems do not recognise it and so in practice one is not always able to demonstrate any benefit from optimising payment in relation to time. However, private developers in particular are certainly aware of the importance of the timing of income and expenditure, and building contractors have always recognised its significance even if they have not always been able to formalise their methods of dealing with it.

Methodology

In considering development finance we have three kinds of expenditure/income which we need to compare with each other:

1 Lump sums today.
2 Lump sums in the future.
3 Sums of money occurring at regular intervals during the period under consideration (wages, rents, etc.).

We cannot compare these, one with the other, unless we modify them in some way in order to put them on a common basis. There are two basic methods, and as usual they are just different ways of expressing the same thing:

1 Present-day value.
2 Annual equivalent.

1 Present-day value

All expenditure is expressed in terms of the capital sum which would have to be set aside today in order to meet present commitments and also to provide for future lump sums and future regular payments. These future payments are 'discounted' to allow for accumulation of interest. Income is similarly treated; future income is discounted to the present day in the same way.

2 Annual equivalent

All items of lump sum expenditure are expressed in terms of the annual payments which would be required to pay them off over the term of years in question (rather like paying off house purchase through a building society). These payments are usually shown in two parts:

1 The interest on the capital cost.
2 A 'sinking fund' – an amount of money put away annually to repay the capital cost at the end of the period.

To these amounts are added the amounts of any regular annual payments, such as wages, rents, etc. Income is similarly treated.

Which of these two methods is used is largely a matter of convenience and depends upon whether one is thinking mainly in terms of capital finance or in terms of annual income and expenditure.

Calculations

It is usual these days to undertake the necessary calculations for these comparisons with the aid of a computer or even a pocket calculator rather than by looking up the values in compendious books of tables. The formulae are set out in Appendix D, but are repeated here; short tables showing a few principal rates of interest are also included in Appendix D to assist those who find them more convenient than doing the calculations.

In the following formulae n represents the number of periods and i the interest rate expressed as a decimal fraction of the principal, eg 5% = 0.05.

Calculations are often required that involve monthly, or even weekly, payments and in such circumstances annual interest rates are not of much use. Such calculations should use an equivalent rate for the period, dividing the annual rate by 12 or 52. Strictly speaking this is not correct, as it ignores the effect of weekly or monthly compounding. The exact annual equivalent of a monthly interest rate i is not $12i$ but $(1+i)^{12}-1$. The yearly equivalent of 1% per month is therefore 12.68%, and conversely the monthly equivalent of 12% per annum is 0.94888%. Where comparisons between two or more alternatives are being considered this difference is rarely of much significance, but it should be allowed for if specific annual interest rates are an important factor.

Formula 1 – Compound interest $(1+i)^n$

If a sum of money is invested for a number of years it will have earned some interest by the end of the first year. Compound interest assumes that this earned money is immediately added to the principal and re-invested on the same terms, this process being repeated annually. Many forms of actual investment provide for doing this automatically, but it is in any case the correct method to use when making investment calculations, since it should not be assumed that money would be allowed to lie idle.

Example

What will be the value of £5,500 invested at 9% compound interest for five years?
Formula $(1+i)^n = (1.09)^5$ and shows that £1 so invested will grow to £1.54.
£5,500 will therefore grow to 5,500 × £1.54 = £8,470.
Note that this formula is also useful for extrapolating inflation rates over a period of years.

Formula 2 – Future value of £1 invested at regular intervals $[(1+i)^n - 1]/i$

Instead of a single lump sum being invested we might put away a regular annual amount on the same basis. At the end of each year the total in the fund would comprise all previous investments, plus the compound interest earned on them, and to this would then be added the next annual contribution.

Example

If £150 is invested annually at a rate of 11% compound interest for six consecutive years what would the fund be worth at the end of the sixth year?

Formula $[(1.11)^6 - 1]/0.11$ shows that £1 annually so invested will grow to £7.91, assuming that the investment is made at the *end* of each year.

If the investment is regularly made at an earlier part of the year then 11% pa interest will require to be added to £7.91 for the additional part of the year involved. If the investment was always made at the *beginning* of each year then a whole year's interest would have to be added, that is, 11% of £7.91 = £0.87. The total per £1 at the end of the sixth year would therefore be:

$$£7.91 + £0.87 = £8.78$$

The example concerned a figure of £150 annually, and the result for £1 must therefore be multiplied by 150 in each case.

The answer is therefore £1,186.50 if the money is deposited at the end of the year and £1,317 if it is deposited at the beginning.

Formula 3 – present value of £1 $1/(1+i)^n$

In the compound interest example it will be remembered that £5,500 invested for 5 years at 9% compound interest grew to £8,470. The converse of this is that the *present* value of £8,470 in five years time at 9% interest is £5,500, i.e. that is the amount which would grow to that sum at the end of five years.

To avoid having to work out present values backwards in this fashion we use the reciprocal of the compound interest formula. One pound in five years time at 9% will therefore have a present value of $1/(1.09)^5 = 65.0$p, so £8,470 × 65p = £5,505. (The small error of £5 is due to the value being taken to the nearest tenth of a penny only. We shall discover later that the calculations for which these formulae are used are so conjectural that such an 'error' is of no real significance.)

Another example

What is the present value of £1,200 in 35 years time discounted at 15% per annum?

By use of the formula the present value of £1 in such circumstances is 0.8p. The present value of £1,200 is therefore 1,200 × 0.8p = £9.60. We can understand why there is little point in having 60 year tables when interest rates are high.

Formula 4 – present value of £1 payable at regular intervals $[(1+i)^n - 1]/[i(1+i)^n]$

Just as the previous formula showed the present value of future lump sums this shows the present value of future regular periodic payments or receipts over a limited term of 40 years (or other periods). It is, therefore, very useful for assessing the capital equivalent of things like running costs, wages, or rents.

Example

What is the present value of £1,200 payable annually for ten years assuming an interest rate of 8% per annum?

We see from the formula that the present value of £1 paid annually in such circumstances is £6.71. The present value of £1,200 annually is therefore $1{,}200 \times £6.71 = £8{,}052$. This is the sum which would have to be invested today at 8% compound interest in order to discharge such an obligation and leave nothing over at the end.

It is interesting to see what difference would result if the money were to be paid in quarterly instalments of £300 instead of at the end of each year. Eight per cent per annum is equivalent to 2% per quarter, and (according to the formula) the present value of £1 over $10 \times 4 = 40$ periods at 2% is £27.36. The present value of £300 paid quarterly for ten years is therefore:

$$300 \times £27.36 = £8{,}208$$

compared to £8,052 for the yearly payments of £1,200.

The present value of £1,200 payable for ten years is sometimes referred to as 'Ten year's purchase of £1,200'. There is little point in using this old-fashioned land agents' jargon, but you might come across it somewhere.

Formula 5 – Annuity purchased by £1 $[i(1+i)^n]/[(1+i)^n - 1]$

This is the reciprocal of the previous formula, and gives the annuity (or regular annual payment) purchased by a lump sum payment of £1. At the end of the given number of years the money will be exhausted. It is therefore useful for calculating the annual equivalent of a given present-day lump sum (whereas Formula 4 calculated the present-day lump-sum equivalent of a given annual amount).

Example

What annual saving in maintenance costs over a period of ten years would justify an increase in capital costs of £90,000, assuming an interest rate of 8%?

The annual equivalent of £1 by the formula in these circumstances is 14.9p. The annual equivalent of £90,000 is therefore $90{,}000 \times 14.9p = £13{,}410$; if the saving in maintenance costs exceeds this amount then the additional capital investment of £90,000 would be justified.

Formula 6 – Sinking fund $i/[(1+i)^n - 1]$

It will be remembered that Formula 2 showed the lump sum which would result in the future from investing a fixed sum of money each year (or other periodic interval). However, sometimes you want the same information a different way round; you want to know how much you must invest each year to accumulate a certain sum in a certain number of years (for example to provide for replacing some worn-out piece of equipment). The formula for calculating this is the reciprocal of Formula 2 (Future value of £1).

Example

How much must be invested at the end of each year at 7% pe annum to amount to £20,000 at the end of 12 years?
    We see from the formula that 5.6p has to be invested each year to produce £1 in such circumstances. For £20,000 the figure is therefore:

$$20,000 \times 5.6p = £1,120$$

In practice, money is rarely put away into a sinking fund. However, it is a valuable economic concept.

A worked example

We can now proceed to work out a very simple cost-benefit analysis to demonstrate the use of the formulae. As the example involves a comparison between recurrent expenditure and capital costs, both the 'present-day value' and the 'annual equivalent' method are equally appropriate, and it is worked out both ways to show that the two methods obtain the same result.

Example

Comparison of cost of providing a mechanical stoker for a boiler with the cost of employing a fireman, assuming the life of the installation is twenty years and the interest rate is 8% per annum.

1 *Present day value*
    (a) Fireman's wages etc. say £12,000 p.a.
         3 men (three shifts per day) = £36,000 p.a.
         £1 for twenty years at 8% = £9.82
         (Table 4)   353,520
                                               36,000 × £9.82   £235,680

    (b) Cost of mechanical firing equipment,
         together with consequent alterations to
         boiler house.   £300,000
         Total saving by using mechanical firing   £53,520

2 *Annual equivalent*
    (a) Firemen's wages etc. (as above)   £36,000
    (b) 8% interest on cost of mechanical firing
         equipment, etc (£300,000)   £24,000

(c) Sinking fund to repay capital cost of the equipment at the end of 20 years. (Taken at 8% although it is common practice to adopt a lower interest rate for this purpose) £300,000 at 2.2p (Table 6)

	6,600	
	30,600	£30,600
Annual saving by use of mechanical firing		£5,400

3 *Check*

£5,400 for 20 years at 8% (Table 4)   £53,028
approximately the same as the capital
saving of £53,520

Note that in order to keep the example simple, various extra factors (like maintenance, different type of fuel, etc.) have been ignored.

It is important that the principle of these techniques should be understood, as they have to be used extensively in cost studies and are referred to and developed throughout the rest of the book.

Development budgets

A feature of all cost planning and control procedures is the setting of a budget, and the subsequent budgetary control which is exercised to make sure that the client's financial resources are not overspent. At the root of any budget for a specific design is the allowance made by the developer for the building in the overall costs of the project, which as we have already seen will include such items as the cost of land, legal fees, external works, loan charges, interest on money expended during development, and advertising costs as well as the cost of the building itself. In commercial development there is a very close relationship between how much the project will produce in revenue and its commercial value, the latter being largely the capitalised value of the former, and the total costs of the scheme must no exceed this commercial value. In social development this relationship may be expressed in cost-benefit terms, by quantifying what are considered to be social benefits, but the issue is never so clear-cut and other criteria may be more important in assessing the amount that can be spent on the scheme. As we have seen, one of the most important items on any developer's balance sheet is the cost of the land and we must now consider this aspect.

Land values and development

This is really the specialist province of the valuation surveyor, but anybody who is involved in cost planning of buildings needs to have some knowledge of the factors affecting the costs of land for development.

Like the prices of other commodities, land values are affected by the laws of supply and demand. However, unlike, say, sugar, the amount of land available in any one area is finite and cannot be increased at times of high demand by manufacturing at a higher rate, or bringing some in from outside. This means that if there is a high demand in a particular area the prices of land in the locality will rise very steeply, even though similar land may still be cheap fifty miles away.

Building development of any kind requires a plot of land on which it can take place, and which, once used, will no longer be available for any other development, unless the first one is either demolished or converted. The 'development value' of a piece of land is the difference between the cost of erecting or converting buildings on it and the market price of the finished development (including the land). Nobody can commission building operations on a piece of land unless they have an interest in it; if possible they would wish to own it, so as to obtain the full benefit of the 'development value' although in some areas where land is in very short supply they may have to lease it.

The development value of land will be determined by:

1 Its position – both its geographical region and its local position in that region in regard to amenities, communications, etc.
2 Restrictions on its use, either imposed by the vendors in the form of covenants, or by the community in the form of planning restrictions.
3 Any 'easements' which go with the land, such as right of way across it.
4 The physical state of the land – whether it is level or very hilly, and whether there are buildings on it which need to be demolished.
5 The current state of the economy.

To illustrate how the value may be determined by position, we can do a simple cost-benefit analysis showing the cost effect of a family house in an inner London suburb compared to a similar house in the outer commuter area, from the point of view of somebody employed in London.

Annual cost	£	£
London house, building society repayments		
(after tax rebate)	7,000	
Annual season ticket to City	500	
Travelling time say 1½ hours per day		
200 days per annum = 300 hours at £10	3,000	
	£10,500	10,500
Similar outer area house, building society		
repayments (after tax rebate)	3,000	
Annual season ticket to City	1,000	
Travelling time say 3 hours per day 200 days		
per annum = 600 hours at £10	6,000	
	£10,000	10,000
Annual saving of outer area house		£500

The annual costs of the two alternatives are fairly equal, and it is this calculation which helps to explain the different prices of two quite similar houses whose actual building costs would be identical.

Effect of building cost on land value

We can express the value of land for development as an equation:

Value of development = Price of undeveloped land + Building Costs + Profit

As we have seen, the value of the development in a free market is its worth to the consumer (as compared to other alternatives), and this in turn largely fixes the price of the undeveloped land – the equation can therefore be better expressed as:

Price of undeveloped land = Value of development – (Building Costs + Profit)

The people selling the undeveloped land should be just as capable of doing this calculation as the developer is, and will fix their selling price accordingly. It is therefore not really correct to say that the cost of land is pushing up the price of housing or other accommodation; in fact it is the other way around – the market price of development pulls up the cost of land.

In conditions where there is a shortage of suitable building land and a constant rise in development values, as in the south-east of England, fluctuations in building costs will have little effect on land prices. However, if the situation were reversed and development values remained constant or even fell, then an increase in building costs would have the effect of reducing undeveloped land values. It is usually the price of 'undeveloped land' which is the result of the equation, when the other figures have been filled in.

Social considerations

Most people agree that a completely free market in land for development is socially undesirable, and the setting of limits on the scale and nature of development under Town and Country Planning legislation brings a measure of public control to the process. Even so, there is a strong body of opinion that wants to see the public obtain much of the benefit of any increased development value of private land, and this would certainly make it much easier to undertake social development in urban areas.

The more radical members of this group dislike the whole idea of anybody making money which they have not actually earned, but the moderate elements base their arguments upon:

1 the loss of green land or other pleasant environment to the community as a whole when development is undertaken for profit
2 the fact that public expenditure on infrastructure (roads, services, transport systems, etc.) has often contributed to the rise in value
3 the way in which decisions taken by the planning authorities are handing windfalls to some lucky land-owners and withholding them from others, thus providing strong incentives for undesirable (or even improper) influence upon those decisions.

Several unsuccessful attempts have been made by British governments to deal with the problem, including the Town and Country Planning Act of 1947 and the Community Land Act of 1975. These Acts tended to inhibit development of any kind, as landowners had little incentive to allow their land to be built on, and preferred to wait until a fresh government repealed the legislation. However, so that the community does not actually lose money over private development, the developer may often be asked to contribute towards infrastructure costs as a condition of the granting of planning approval. The agreement of these costs is obviously an area where the cost advisers on both sides should be involved. An interesting technique has been used in recent years in Hong Kong, where the new underground railway system has been largely financed from the increase in development values in the areas around its suburban stations.

Effect of land values on cost planning techniques

It is possible to try and optimise building costs by looking at such factors as low wall/ floor ratios, the avoidance of multi-storey construction, the centralisation of services, and so on, in isolation from land costs. Where land prices are relatively low, or in those parts of the public sector where land costs and building costs are not considered together, this may be a perfectly valid way of looking at the problem.

However, where development values and land prices are high the picture alters somewhat. Suppose that the current price of flats in a fashionable urban area is 20% building cost and 80% land cost (and profit) and a piece of land has been bought at a price which assumes that 20 flats can be built on it. If it should prove possible to build more than 20 flats on the same piece of land, the building cost of these extra flats would be only 20% of their market value, leaving 80% profit. It would therefore be worthwhile to accept a less efficient building configuration to provide these extra flats – even if this increased the building cost slightly there would still be a handsome profit.

In these circumstances cost planning may develop into an exercise to secure the maximum number of accommodation units on a given site, at the same time optimising the design from the point of view of grants and subsidies (if there are any of these to be had). Unfortunately for the developer, a scheme which squeezes the utmost in hard cash out of the site is rarely acceptable to the planning authorities, who have more regard for the amenities and appearance of the neighbourhood. Their veto on a scheme is final, apart from the possibility of a lengthy and time-consuming appeal to the central government (which has the same general motivation).

On such projects the costing and income appraisal of many alternative schemes will be required, and this will be the main cost planning contribution. Once a scheme has been approved by the authorities speed may be vital (because of the need to recover a massive investment as soon as possible), and the cost planner will be required to advise on a suitable contractual system to achieve speed with an appropriate measure of cost control.

However, not all private development is of this speculative kind. The client may be wishing to erect a building for his own use – as offices, warehouse, store, etc. In this case the effect of high land costs on the cost planning exercise should be to cause an examination of the economics of possible alternatives – perhaps carrying out the development in a less expensive area, or possibly changing the client's requirements for a building by solving his problem in another way. An example of the latter might be to avoid the cost of branch warehouses by a system of daily distribution from the factory. It might be thought that there would be an advantage in building in a high cost area because the firm would always have the value of the site on its books, but there are many disadvantages in having too many of a firm's assets tied up in site values.

Grants, subsidies, and taxation concessions

Whilst action under Town and Country Planning legislation can prevent undesirable development it cannot cause socially desirable development to be undertaken. It can certainly earmark a certain area for a particular type of development, but unless the development itself is to be carried out by a public agency, or unless it is obviously going to be highly profitable, the matter will rest there. In order to make such development attractive to a private investor the government may therefore offer special financial inducements. They are usually offered on a sector basis (e.g. for hotel or other tourist building) or on a regional basis, or both (e.g. for factories in an area of high unemployment). These incentives will need to be taken into account in preparing budgets for the type of development concerned, and may be of crucial importance in deciding between alternative sites. The extent to which incentives of this kind may be nothing more than compensation for straightforward commercial disadvantages, such as high transport costs, will certainly emerge from such calculations.

The incentives may be of several different kinds:

1. *Low rent or rates.* Land (or even completed buildings) may be made available by a public body at a very low rent or with substantial relief from local authority charges, or both. This concession usually runs for a limited period of years, after which more normal conditions will apply, so that some kind of Discounted Cash Flow analysis (see chapter 5) will probably be necessary.

2. *Grant towards capital costs.* The Department of Industry used to provide grants towards the cost of factory building and plant located in 'areas of expansion', and an increasing number of authorities provide partial funding for new development, rehabilitation, or improvement in the fields of tourism, leisure, conservation, etc. The situation is always changing over the years as different political priorities come into play. A most important factor in assessing the value of such a grant is its timing in relation to the developer's expenditure, particularly whether it is a reimbursement or an advance.

3. *Taxation relief.* Taxation relief may be given under some circumstances on development and plant costs. The main disadvantage of taxation relief compared to a grant is that there must be a tax liability against which the relief can be set. It is therefore necessary to make a profit before the benefit can be obtained.

It should be noted that other forms of financial encouragement, such as a subsidy paid for each person employed, may need to be taken into account in a development budget even though strictly speaking it is not a grant towards the development itself. It is also worth pointing out that it is possible for the government to impose financial disincentives for types of development of which it does not approve.

The lack of grants for industrial building in some regions can be seen as such; a more extreme example has been the long-term discouragement of private rental housing development by fixing rents at uneconomic levels. Although the UK government currently wishes to reverse this trend prospective developers need to be able to look ahead for more than the five-year term of a government, and the treatment of housing (and other development matters) as a political football by the two main British political parties has discouraged a stable property rental market, as can be seen by comparison with other European countries.

Social development

In the past the costs of site acquisition have usually been ignored in cost planning social development, at any rate as far as the cost planner is concerned. One cannot prepare a profit-investment appraisal for such projects, and the cost of the land is just one of the many items on the expenditure side with which no comparison with income can be made. However, there is an increasing tendency to use the profit-development type of calculation on social projects, such as housing, where there is a real or hypothetical income.

Cost of financing the project

This is the intangible component of total building cost; all the other things like buying land, paying professional fees, etc, involve paying money to somebody else in exchange for property or services.

As soon as this tangible expenditure starts, however, the developer will be laying out his money, and there will be nothing to show for it in terms of income (or use) until the building is ready for occupation or can be disposed of. Hence the cost of lending this money to the development for a period of possibly several years becomes part of the cost of the development itself. When interest rates are high this financing cost can be considerable, as has already been pointed out, so the project should be planned to avoid unnecessarily early expenditure on any part of it.

Where the money comes from

The capital for the development may come from a number of alternative sources.

Private development

1 *Bank overdraft.* This is rarely available as a means of financing a whole development, but may be useful for short term bridging purposes; the money is lent on the security of the development.
2 *Loan from merchant bank or insurance company.* This is similar to the last, except that it is possible to obtain longer term finance from these sources. Interest rates are usually somewhat higher.
3 *Capital account.* Where the development is being undertaken for a firm's own use – such as factory, warehouse, etc. – the capital for the development can be regarded as part of the general capital expenditure of the firm.
4 *Trading funds.* Property companies will have funds for carrying on their business.
5 *Shares in the development.* Speculative development is sometimes financed by paying the builder with shares instead of cash, or by allowing him to develop part of the site for himself. Alternatively a 'joint venture' may be undertaken with another developer.
6 *Finance by the intended occupier.* It may be possible to persuade the intended occupier to pay part of the sale price during the erection of the building – this is often the case with houses which are built for sale.

Public (social) development

7 *Government funds.* These will be available either as direct finance for government building (e.g. defence works) or as a grant or subsidy for local authority building.

8 *Loans from government funds.* These are usually available at a lower rate of interest than money raised on the open market, but are very restricted as to eligibility and amount.

9 *Issue of stock.* Local authorities and public boards are empowered to issue fixed-interest loan stock on the Stock Exchange. Full market rates of interest have to be paid, and there is a long term commitment to these which can prove very expensive if interest rates subsequently fall.

10 *Local authority mortgages.* These are a useful way of raising money, as there is a firm commitment by both sides, but for a short period of three to five years only. Interest rates are usually a little higher than current yield on gilt-edged securities.

11 *Loans from other funds.* A public authority can borrow from its other funds (if it has any) for development purposes.

12 *General income.* It is common for a local authority to raise money for development from its charges and rates – this has the advantage that the money does not have to be paid back!

13 *Private developer.* In large-scale urban redevelopment a private developer can be required to undertake some social development in exchange for being allowed to pursue his scheme; alternatively a developer can be offered some profitable role within a social development if he contributes to the cost of the public part.

The cost planner and the budget

Many years ago a quantity surveyor was often not appointed until the architect had prepared his working drawings of the final scheme, or, if he were appointed earlier, he would play no part in things until this stage was reached and he could start work on the bill of quantities. Things have changed since those days, and now the quantity surveyor, or other cost planner, is sometimes appointed before any other professional adviser and takes over total responsibility for the client's financial interests in the scheme. However, most appointments fall between these two extremes.

When the cost planner is appointed there are a number of questions which will have a bearing on his concern with the budget, and which need to be answered:

1 *Who is responsible for his appointment?* If he has been appointed on the recommendation of the client's architect his relations with the client will tend to be conducted through the architect, whereas if his appointment has been made by the client as a result of previous experience or outside recommendation his relationship will usually be more direct. Apart from any other factor, a client who decides to appoint a cost planner directly will probably have an above-average interest in costs, which will therefore play a predominant part in the scheme design.

2 *What is his role to be?* This may well be established by the nature of the appointment, as above, but it is important to establish whether the cost planner

will be concerned with the total budgeting of the project or whether he will be limited to particular areas, such as capital expenditure, or building and furnishing costs, or net building costs, or indeed (if he is a quantity surveyor) whether he will be expected to do anything more than merely prepare a bill of quantities or other contract documentation.

3 *Who else has been appointed?* If other appointments have not been made, the cost planner could assist the cost-oriented client in setting up his team of advisers, and this would obviously be of assistance in the context of total cost control.

4 *What decisions have been made, or steps taken?* Any decisions which have already been taken, or implemented, will obviously constrain any cost optimisation programme which the cost planner may produce. However, he should be very careful about questioning any of these unless his advice is definitely being sought or unless he feels that they are so fundamentally wrong as to make it impossible for him to carry out the task for which he has been appointed. In particular it would only be in the most extraordinary circumstances, and where he was in a very strong position, that he would cast doubts on specialist professional advice which the client had already received from valuers, accountants, lawyers, etc.

5 *Is cost control to continue until the completion of the project?* In the past it was not unusual for the cost planner's cost control role to finish at the point where a contract was signed with a contractor. In view of the many major problems which can occur subsequently this is a very short-sighted policy, as will be explained in Chapter 18.

It will therefore be seen that the extent to which the cost planner will be able to use the various techniques described in this book depends not only upon the priorities of his client but also upon the terms and circumstances of his appointment. One point which must be remembered is that where the cost planner has been entrusted with the overall cost management of a project he must resist the temptation to make firm proposals to his client based upon his own elementary knowledge of property values, investment, taxation, etc. He is not an expert in these fields (although a large quantity surveying firm might possibly employ such experts) and the purpose of his education in this area is to make him aware of circumstances in which these matters may be important, and to help him in briefing specialists and assessing their advice.

Let us now look at two simple budget examples. These are in no sense 'typical' in that every project has its own particular problems and priorities, but they give an idea of the type of basic calculation which is involved. The method of arriving at the various estimated costs has not been shown, as the concern here is how to manipulate the answers.

The interest rates used for most of the calculations are based on a 'criterion rate of return' – the return which the developer requires in order to make a reasonable profit.

A budget for a profit-development for sale (24 town houses)

	£	£
1 Cost of land	1,000,000	
2 Legal and professional cost of acquiring land, obtaining planning permission, etc.	100,000	
3 Demolition of existing property	20,000	

4 Building cost and professional fees in connection	960,000	
5 Site layout, ditto	40,000	
6 Agent's charges for selling	20,000	
	£2,140,000	2,140,000
7 Cost of finance (2% per month compound interest – criterion rate of return)		
Items (a), (b), (c) £1,120,000 for 12 months	= 302,400	
Item (d), (e), £1,000,000 for 4 months (av)	= 80,000	
Item (f) is paid after sale, no finance charge	= —	
	£382,400	£382,400
		£2,522,400

Minimum economic selling price = £2,522,400 ÷ 24 = £105,100 say £106,000 per house. Anything less than this will not produce the criterion rate of return, and if there is any doubt about obtaining such a figure then the development should not go ahead in this form.

It will be seen that instead of discounting all expenditure and receipts to the commencing date of the scheme, they have been carried forward to the end of the scheme and compound interest added. This has the same effect – it is easier to do it this way in this instance as it is the future selling price which we are interested in and not its present day value.

A budget for profit-development (for rental offices)	£	£
1 Cost of land	500,000	
2 Legal and professional cost of acquiring land, obtaining planning permission, compensation, etc.	170,000	
3 Demolition	20,000	
4 Building cost and professional fees in connection	700,000	
	£1,390,000	1,390,000
5 Cost of finance (2% per month compound interest – criterion rate of return)		
Items (a), (b) £670,000 for 36 months	= 696,800	
Item (c) £20,000 for 18 months	= 8,600	
Item (d) £700,000 for 10 months (av)	= 154,000	
	859,400	859,400
Total capital cost at completion		£2,249,400
		say £2,250,000
Annual income from rents (figure given by valuer)		£600,000
Less		
Allowance for vacant tenancies (voids)	50,000	

Maintenance, repairs and redecorations (excluding tenant's responsibilities)	30,000	
Cleaning, heating, lighting and rates on public part of building (staircases, entrance halls)	50,000	
Management expenses, including caretaker, arranging lettings, and collecting rents	18,000	
Sinking fund to replace capital at end of 40 years at 2½% pa £2,250,000 less cost of land (which will still be there when the building is demolished) = £1,750,000 at 1.5p =	26,250	
	174,250	174,250
Net annual income		425,750

This gives a return of 18.9% on the capital of £2,250,000. An alternative way of setting out this budget would be the following, which starts with the current rental value of office accommodation as a criterion.

	£
Estimated annual income from rents	600,000
Less expenses (excluding sinking fund	148,000
	452,000

Note: The cost of the sinking fund cannot be accurately assessed at this stage, as the capital value of the buildings, etc., is now known. It could be approximated thus:

Capitalised value of income at 18½% is approximately £2,443,000. Say 25% of this represents residual land value.

Sinking fund to replace £1,832,000 at 40 years at 2½% (at 1.5p)	27,483
Estimated net income	£424,517
40 years purchase of £424,517 at 18% (criterion rate of return) at £5.40 =	2,292,400

This represents the capitalised value of income at the date of completion in 36 months time. Discounted to present value at 2% per month at 49.0p = £1,123,270.

	£	£
Present value of income		£1,123,270
Estimated legal and other costs of acquiring land	170,000	
Estimated cost of demolition of existing buildings (PV of £20,000 in 18 months time at 2% per month = 70.0p per £)	14,000	
Estimated cost of building and professional fees (PV of £700,000 in (average) 26 months time at 2% per month = 59.8p per £)	418,600	
	£602,600	£602,600
		£520,670
The maximum amount which can be paid for the land at the present time to give required rate of return, say		£520,000

At this stage the calculations would have to be based on the minimum amount of accommodation for which the client could expect to get planning permission, unless an outline scheme had already been approved. It should be noted that taxation provisions, which are constantly changing, have been ignored in these calculations but cannot be ignored in practice.

If budgets of this type are prepared by computer it is possible to examine the effect of changing the values of the variables either singly or in combination, and thus to identify those that are crucial to the calculations. Very often in profit development the overall time for designing and building the project is of supreme importance, and simulations might well be carried out to assess the effect on construction costs and financing charges of different time scales and fast-track methods of procurement.

Chapter 4
Life-cycle Costs

When talking about building costs we have been thinking principally of the capital cost of providing a new building, but of course the expenditure does not stop there. Once the building has been handed over and paid for it will continue to cost money – it will need to be maintained, decorated, heated, cleaned, repaired, and so on, and from time to time during its life quite expensive refitting may be required. Also each year the interest on the capital cost will have to be found, and perhaps a sinking fund paid into.

The terms 'costs-in-use' or 'life-cycle costing' have been coined to describe a form of modelling technique to cope with this mixture of capital and running costs, the latter being now the preferred term.

It has been felt by many people that cost targets have for too long been expressed purely in terms of capital cost, and that in order to meet these targets materials or constructions have been used which are low in first cost but which will require constant expenditure on maintenance and repair during the life of the building. If only, it is argued, the architect were allowed to spend more money initially on better materials these would pay for themselves in the long run.

As a simple example of this let us consider the walls of a lavatory in a good class building. To plaster the walls and paint three coats of gloss paint would cost about £6.00 per square metre, whereas glazed tiling would cost £20.00. As the tiling would last fifty years with little attention, and the plaster walls would need to be painted every five years or so at a cost of £2.00, the total cost at the end of fifty years would be £24.00 for the plaster walls against £20.00 for the tiling. Apparently therefore the better and more hygienic finish would cost less than the inferior work.

Similarly with roofs; a sound lead roof will last up to a hundred years and will then have quite a considerable scrap value whereas a built-up felt roof might have to be renewed three or four times during this period. The difference of 400% in the initial cost would appear to cancel out by the end of the century, and the trouble and inconvenience of having the roof stripped and relaid at intervals would be avoided by using the more durable material.

Of course, such examples are over-simplified, because we know that in comparing current and future expenditure we have to consider not the actual amounts to be spent in future years but these amounts discounted to present day value – that is, the sum of money which would have to be invested today in order to accumulate to these amounts by the time they are needed. As money invested at as little as 5% compound interest doubles in just under fifteen years there is obviously a considerable difference between the two figures.

At this juncture we can look at a simple example (the present day values are calculated according to the formulae in Appendix D).

Example

It is desired to compare the life-cycle costs of two types of cladding to a factory building. Cladding A will cost £1,000,000, will require redecorating every four years at a cost of £120,000, and will require renewing after twenty years at a cost of £1,400,000. The life of the building is assumed to be 40 years. The alternative Cladding B will cost £1,800,000, and will last the life of the building without any maintenance, although a sum of £300,000 is to allowed for general repairs after 20 years. The rate of interest allowed is 3%, per annum compound.

Life-cycle costs of Cladding A

			£
Initial cost			£1,000,000
Present value at 3% of:			
Redecoration after 4 years	£120,000 at 88.8p	=	106,560
Redecoration after 4 years	£120,000 at 78.9p	=	94,680
Redecoration after 12 years	£120,000 at 70.1p	=	84,120
Redecoration after 16 years	£120,000 at 62.3p	=	74,760
Renewal after 20 years	£1,400,000 at 55.4p	=	775,600
Redecorating after 24 years	£120,000 at 49.2p	=	59,040
Redecorating after 28 years	£120,000 at 43.7p	=	52,440
Redecorating after 32 years	£120,000 at 38.8p	=	46,560
Redecorating after 36 years	£120,000 at 34.5p	=	41,400
			£2,335,160

Life-cycle costs of Cladding B

	£	
Initial cost	1,800,000	
Present value at 3% of:		
Repairs after 20 years £300,000 at 55.4p =	166,200	
	£1,966,200	1,966,200
Saving by using Cladding B		£368,960

It would therefore appear to be justifiable to use the initially more expensive Cladding B, as this will prove much the cheaper in the long run. Note that the 'present value' method of discounting has been used, as the 'annual equivalent' method (see Chapter 3) would have proved far more complicated in such an instance, where none of the payments or receipts are annual. There is no need to bother with sinking funds, as this is a comparison only and both claddings will be worthless at the end of the 40 years.

It is worth noting that the present value of Cladding A could have been evaluated even more quickly, with a slight loss of accuracy (because of averaging out the decoration costs to a yearly figure), as follows:

Life-cycle costs of Cladding A

	£
Initial costs	1,000,000
Redecoration £120,000 every 4 years =	
£30,000 p.a. 36 years purchase of £30,000	
at 3% (see Appendix D Table 4) at £21.83 =	654,900
Replacement at 20 years	
£1,400,000 less saving in decoration	
£120,000 = £1,280,000	
Present value (3%) at 55.4p =	709,120
	£2,364,020

The slight loss of accuracy by this method is of no practical significance, because the assumptions which have been made are just that – assumptions.

In order to demonstrate the effect of quite small misjudgments of the future, the example relating to Claddings A and B is now worked out again, assuming some slight differences to the assumptions previously made. Cladding A is redecorated every 5 years instead of 4 years at a cost of £100,000 and lasts for 25 years instead of 20, costing £1,200,000 to renew. Cladding B has to be repaired after 15 years and again at 30 years, and interest rates average out at 4% instead of 3%. All these differences lie well within the range of error to be expected with careful estimates in normal times, and exclude either inflation or anything going badly wrong.

Revised life-cycle costs of Cladding A

	£	£
Initial cost	£1,000,000	
Present value at 4% of:		
Redecoration after 5 years £100,000 at 82.2p =	82,200	
Redecoration after 10 years £100,000 at 67.6p =	67,600	
Redecoration after 15 years £100,000 at 55.5p =	55,500	
Redecoration after 20 years £100,000 at 45.6p =	45,600	
Renewal after 25 years £1,200,000 at 37.5p =	450,000	
Redecoration after 30 years £100,000 at 30.8p =	30,800	
Redecoration after 35 years £100,000 at 25.3p =	25,300	
	£1,757,000	

Revised life-cycle costs of Cladding B

	£	£
Initial cost	1,800,000	
Present value at 4% of:		
Repairs after 15 years £300,000 at 55.5p =	166,500	
Repairs after 30 years £150,000 at 30.8p =	61,600	
	£2,028,100	2,028,100
Saving by using Cladding A		£271,100

This calculation gives a completely different result to the original – the initially cheaper cladding proves to be much cheaper in the long run also, and would still have life left in it at the end of 40 years if it were decided to keep the building in commission for a bit longer. The lesson it teaches is that calculations of this kind should not be used as the sole basis for decision unless the cost advantage of one of the alternatives is really massive – not the 15–20% of these examples.

A mathematical technique ('Monte Carlo Simulation') for assessing the effect of probabilities in a situation such as this is given in Chapter 20 and this might assist decision-making in some circumstances. Alternatively, the problem can be solved by computer using a number of different values for the variables; this may enable the important ones to be identified.

Inflation

The last calculation did not take inflation into account at all, but some authorities suggest that it is this factor which really makes complete nonsense of any attempt at long term forecasting. This is to take too gloomy a view; inflation may be no worse than the other factors which we have discussed, because interest rates tend to rise during a period of rapid inflation. If, therefore, the original 'Cladding A' example were worked out on the basis of 10% annual inflation and 13% interest we would get much the same result as with no inflation and interest at 3%. (N.B. The compound interest table has been used for calculating the effect of inflation.)

Revised life-cycle costs of Cladding A (10% annual inflation)

		£
Initial cost		£1,000,000

Present value at 13% of:

Redecoration after 4 years	£175,200 at 61.3p =	107,398
Redecoration after 8 years	£256,800 at 37.6p =	96,556
Redecoration after 12 years	£376,800 at 23.1p =	87,040
Redecoration after 16 years	£550,800 at 14.1p =	77,662
Renewal after 20 years	£9,422,000 at 8.7p =	819,714
Redecoration after 24 years	£1,182,000 at 5.3p =	62,646
Redecoration after 28 years	£1,730,400 at 3.3p =	57,103
Redecoration after 32 years	£2,533,200 at 2.0p =	50,664
Redecoration after 36 years	£3,709,200 at 1.2p =	44,510
		£2,403,293

This is very similar to the figure which we had originally, in spite of the vast increase in cost. However, we must remember that building costs do not necessarily increase in step with general inflation; in particular repair costs, which have a high site labour content and sometimes involve obsolescent materials, may well increase more than a general building cost index would indicate. At the same time it should be realised that income will tend to rise at roughly the same rate as inflation (otherwise building owners would go bankrupt) and this increase should offset the increased running costs. This is a further argument for ignoring the effect of inflation in life-cycle cost studies. However, it may be possible to take into account some aspects of differential inflation, where the price of one type of commodity or factor increases at a different rate to the alternative, although this is of course very difficult to predict. It may be influenced very much by the relative scarcity of different raw materials and other resources in the future.

However, whatever attitude is taken it is important to avoid double-counting inflation through using a discount rate based on current gross interest rates (which allow for inflation), and then adding a separate inflation allowance as well.

Disadvantages of 'life-cycle costs' assessment

The advantages of this method of comparing costs are self-evident; it enables us to consider the long term implications of a decision, and to provide a way of showing the cost consequences of short-sighted economies. Unfortunately there are a number of fundamental disadvantages, which explain why this technique for comparing the cost of the alternative materials and constructions has been seen more often in the examination room than in real life. These disadvantages may be dealt with under two headings.

'Initial and running costs cannot really be equated'

1 Where the building is to be disposed of by sale the maintenance charges will fall upon the purchaser, and so are of little importance to the vendor, who is responsible for the construction costs.

2 Where the building is to be let, or used commercially, the initial cost comes out of capital, while the repairs and maintenance are deducted from the commercial receipts in calculating profit for the year. These maintenance charges are therefore paid out of untaxed income, whereas money spent at construction stage has to be raised as part of the capital cost and is eventually repayable. Repayment of, or interest on, capital expenditure is not normally deductible from receipts for the purposes of tax calculations. While this factor can to some extent be allowed for in the rate chosen for discounting, this does not fully deal with the fact that money for capital expenditure is normally more difficult to find and is subject to more constraints than money for current expenditure.

3 Even with publicly-owned buildings it is an advantage to pay maintenance out of running costs instead of incurring a heavier debt due to high construction costs. In some instances, such as schools, the bulk of the construction costs may be paid by one authority while another authority will be responsible for running costs, so that there will be little incentive to provide an unduly high standard of building with a view to subsequent saving.

4 Although built-in obsolescence is not yet a regular feature of the construction industry, undoubtedly circumstances change and although a building may be perfectly sound after the passage of years it may be too old-fashioned in design and accommodation to do the job that is required of it in modern conditions. Similarly with durable finishes, which although still in good condition may give a very old fashioned appearance to a building. Most readers will be familiar with instances where old joinery, tiling, and other finishes or fittings have been ripped out long before they are life-expired for this reason. This factor particularly applies in the competitive world of commerce; in a fashionable

shopping street it is not unusual for a shop front or interior to be replaced after five or six years because it has begun to look out of date, or does not fit in with current merchandising ideas.

5 In comparing figures of increased present-day expenditure against future costs of repair and renewal it must be remembered that once the money has actually been spent it is not possible to amend the decision in the light of future developments. As an example of this, supposing that very expensive finishes are chosen for a house to save repainting. If after a few years the owner finds himself short of money he still has to repay the annual interest on his expensive house, whereas if he had opted for a cheaper house he could have deferred the repainting for a year or two. (He could also redecorate his house in the latest fashion each time, if he wanted to.) It is questionable practice to restrict the actions of future generations by committing them to high interest and repayment costs; this is just as bad as the other extreme of committing them to inflated running costs and maintenance costs by unduly low standards of design and construction.

6 The previous arguments have suggested that life-cycle cost techniques over-emphasise the importance of running costs, but it can also be argued that they are unrealistic in an opposite sense, because it is most unusual to actually set money aside in this way to meet future expenditure. If, for instance, you had to meet repair bills this year for £90,000 it would be little compensation to learn that somebody discounted this to £20,000 in the year 1970 – you would still have to find the £90,000.

'The future cannot really be forecast'

7 While the capital costs can be estimated quite accurately the cost of maintenance is a pure guess. The amount of money spent on decoration and upkeep of a building is determined far more by the current policy of the body responsible for the maintenance than by any quality inherit in the materials. Some owners will redecorate every few years, mend or replace worn or damaged work immediately, and continuously carry out a policy of minor improvements; others will spend the very minimum necessary to keep the building in operation.

8 Major expenditure on repairs is normally caused by failure of detailing, faulty material, or bad workmanship, rather than by overall ageing, and so is almost impossible to forecast. A well designed and maintained piece of cheap construction might well last much longer than its theoretical life, while some quite expensive work could require early renewal because of, say, entry of water at a badly designed joint.

9 Interest rates [which in general tend to reflect the current MLR (Minimum Lending Rate)] cannot be forecast with any certainty, particularly over long periods of twenty years or more; remember that net interest rates are also affected by changes in taxation. Between 1988 and 1990, although interest rates had been forecast by the Chancellor to remain static or fall, the MLR actually increased from 8% to 15%, so throwing many carefully thought-out property

calculations into confusion. And that was just a period of two to three years – would you like to guess what the Chancellor will do in the year 2025 – and whether he will be Tory, Labour, Marxist, or Anarchist?

10 During the last forty years there has been almost continuous economic inflation, and any building or maintenance work is likely to cost several times what would have been estimated, say, twenty years ago.

Practical use of life-cycle costing techniques

Although life-cycle costing techniques have been available for over twenty years their use in practice has, as suggested above, been very limited. Some of the reasons for this have already been given, but the most powerful cause is that while the present capital costs are real enough the future maintenance and renewal expenditure is usually just 'funny money'. It is not generally the practice to set funds aside for this purpose at the time the capital decisions are made, and when these costs actually have to be faced it is no good turning back to ancient projections of expenditure for a solution. Remembering that it is easy to make assumptions that will give the preferred answer (as demonstrated above) and that the long-term projections are not going to come home to roost, it is little wonder that life-cycle costing has often been viewed with suspicion.

Even where designers genuinely want to base their decisions on life-cycle costs they are often beaten by the pressures of meeting real deadlines and cost targets, and conjectural calculations get relegated to second place.

On the other hand, the works organisations of some public and recently-privatised organisations are now being faced with the situation where for the first time they have to consider interest on capital as a factor, and their governing boards are tending to require life-cycle cost appraisals of projects to help them deal with this unfamiliar situation.

In general, however, life-cycle costing is at its least effective when dealing with long-term static structures such as buildings, and in fact it can give dangerously misleading answers in really long-term situations. If two alternative materials are being considered for a new main sewer – a traditional material with a life exceeding a hundred years and a cheaper material with an estimated life of sixty years – even a modest discount rate will discount both renewal costs to almost zero over such long periods, and the issue will be decided on capital cost and any short-term maintenance. But if you were now involved with a sewer that had been laid in the 1930s the difference between sixty and a hundred years life would be very real!

However, there are two sorts of application where life-cycle costing techniques work very well, and are increasingly being used. One is in dealing with shorter-life assets, such as mechanical or electrical equipment, where foreseeable energy consumption and maintenance and renewal programmes generate much of the future costs, and the other is where both the present and future costs are equally real, for instance in a rolling maintenance programme for a major installation where the money is coming from the same fund and policy can be planned accordingly. If both these conditions are fulfilled then life-cycle costing is a must!

In conclusion, as this is a textbook it is necessary to point out that examiners may not always share the above rather practical view of the benefits of life-cycle costing.

Reasons for accepting higher capital costs

The rather obvious link between low first cost and high maintenance costs was mentioned earlier in this chapter, and could easily be taken for granted. It is only right to mention, therefore, that in spite of quite considerable efforts no evidence has been found from general user experience to suggest that increased capital expenditure on better materials and finishes necessarily produces lower maintenance costs.

Nevertheless there are still good reasons why a more expensive alternative may be preferred at design stage:

1 For prestige reasons. The client may consider that his building is too important to have cheap and inferior materials and workmanship.

2 As well as being more durable, expensive materials are often more pleasant to look at or to use.

3 Replacement or repair may be inconvenient. The re-covering of a roof, or relaying of a floor, will upset the users of the building and could have severe commercial effects (customers prefer not to go into a shop or hotel where building works are in progress).

4 Replacement or repair may be difficult, and therefore expensive. Repairs of the covering of an external cornice might involve scaffolding the whole building, for instance.

5 The saving of money on a specific item may involve repairs out of all proportion to the saving. The use of iron instead of copper for slate nails would eventually necessitate the stripping and re-slating of the whole roof, for the sake of saving a few pence per square metre. Similarly the omission of galvanising on metal windows might involve the replacement of all the windows after some years.

6 Obsolescence may not be a factor of any importance. Such buildings as churches, or especially cathedrals, employ the best materials in the hope that they will endure for hundreds of years.

In designing a new building therefore it is important to consider what its economic life is likely to be, and what changes it is likely to undergo during that life, before coming to a decision on construction and finishes. Obviously this can only be done to a limited extent as it is impossible to look too far into the future, but some types of building are inherently more subject to change than others, and this can be taken into account. Theoretically it might be possible to design a building so that all its components would begin to fail simultaneously at the end of its economic life; in practice this is not possible although the worst incongruities can be avoided. Expensive materials should not be chosen for the sake of their durability if they are likely to outlast the rest of the building, or if they will have to be replaced due to the failure of some other and inferior component.

Life-cycle costs of mechanical and electrical installations

As previously mentioned, this is a rewarding field for the application of life-cycle cost techniques, and as the proportion of building cost represented by mechanical and electrical (M&E) work is continually increasing we may expect the cost planner to become more involved with life-cycle costing.

The conflict between capital outlay and running costs occurs in a pronounced form in the evaluation of energy-consuming systems, because considerable economies in daily fuel and staffing costs can be achieved by additional capital expenditure.

The disadvantages of life-cycle costing calculations which were set out earlier in this chapter will not apply so severely here as in the case of the building fabric. This is because:

1 The running costs of energy-consuming systems form a high proportion of their total life-cycle costs (for this reason it is usually convenient to use the 'annual equivalent' method for the life-cycle cost study rather than the 'present day value').

2 The fuel consumption and staffing requirements (e.g. boiler-man, lift attendant) of different kinds of systems can be calculated quite accurately, and cannot be so easily varied by client policy as can general building maintenance and decoration.

3 Mechanical and electrical installations should be written off over a much shorter period of time than a general building fabric, both because of the more limited life of their components and because they become obsolete much sooner. Not many fifty-year-old buildings still have their original heating and electrical systems. Since one is forecasting over a shorter period any assumptions about cost, interest rates, and taxation are more likely to be valid.

The most simple forms of calculation are those related to capital/fuel costs of different heating systems.

1 Automatic gas-fire heating system:

	£	£
Annual fuel costs	10,000	
5% interest on capital cost of system (including building work) £60,000	3,000	
'Sinking fund' to repay capital cost after 20 years at 3% £60,000 at 3.7p	2,220	
	£15,220	15,220

2 Electric storage heating system

	£	£
Annual fuel costs	13,600	
5% interest on capital cost of system £30,000	1,500	
'Sinking fund' to repay capital cost after 5 years at 3% £15,000 at 5.4p	1,620	
	£16,720	16,720
Annual saving on costs-in-use of gas-fired system		£1,500

A more complex calculation is used to take into account the cost of additional thermal insulation of the building, for example double glazing. Such a calculation might have two different objects:

1 To justify the capital expenditure on the insulation itself, because of savings in fuel cost.

2 To also justify the use of a low-capital/high-fuel-cost heating system, which would become more economic if the required heat loading could be reduced.

Very complex calculations are involved where the scope of the mechanical services are dependent upon the configuration of the building – for example, heat extraction systems might be necessary for a multi-storey building which has a very low wall/floor area ratio and would have to be taken into account in carrying out structural cost comparisons, and relative lift costs would be a factor in trying out alternative arrangements for a tall building.

Finally, in all these comparisons, it is likely that in the end the client's own (possibly subjective) views on acceptable levels of capital cost and running cost would be more important than the exact result of comparing them on an economically equal basis. It is important, however, that he should be given the facts (including these comparisons) before arriving at a decision.

Life-cycle costs of the total project

In arriving at a decision on what kind of building to erect a forecast of the life-cycle cost of the total project may be important for three reasons:

1 To fix rentals of a development, such as an office block or service flats, where the owner will remain responsible for some of the running costs.
2 To see whether a client can afford to run the building when it has been erected – as in the case of deciding the optimum size of a new village hall.
3 To see whether increased expenditure on a building which is to be occupied by the client, such as a hospital, could reduce the general level of running costs by making it possible to operate with fewer, or less skilled, staff.

In all these cases the problem is usually a short term one – involving, say, five-year leases in the first instance and current staffing difficulties in the last, so that what is required is a fairly accurate assessment of expenditure in a typical year in the very near future rather than a 'plus' or 'minus' comparison with capital building cost. In the first of these cases there is little point in equating annual costs with increases or decreases in the capital cost of the development, as long as the rents will bear the likely charges. In the second case, since capital and running costs are almost certain to be met from different sources (perhaps an appeal plus grant for the first, and income from lettings for the running costs), much the same applies. It is only in the last of these three cases that such comparisons are likely to mean anything, although even here the main motivation is likely to be staffing problems rather than costs as such.

Information on maintenance costs

Without the existence of a large quantity of published data it would have been difficult for cost planning to take root, and one of the factors making costs-in-use studies so risky has been the lack of any similar body of information on maintenance costs. In order to remedy this situation a Building Maintenance Cost Information Service (BMCIS) has been set up by the Royal Institution of Chartered Surveyors (RICS), to operate on a fairly similar basis to the earlier established Building Cost Information Service (BCIS). The new service was able to get off to a flying start as all the information which the government had accumulated on its own building maintenance costs has been made available to the information service. For continued success, however, it will be essential for a large number of major property owners to contribute to the store of information, and a major difficulty is that cost records are often not kept in enough detail to enable work to individual components and the like to be identified.

There is however a school of thought which holds that simulated or estimated figures for expenditure are at least as useful as recorded data. This rather surprising view relies on three arguments:

1 Data relates to the past whereas simulation relates to the future, which is when the expenditure is actually going to be incurred. Maintenance and servicing costs of a fleet of cars, for instance, recorded over the last ten years would not be much guide to the next ten years, because of increases in reliability and corrosion protection, and the increasing use of electronic monitoring and controls.

2 One is very fortunate if historic cost data has been kept in exactly the form that is wanted – if you need information on specific items of boiler-house equipment you may well find that costs have only been recorded at the level of boiler-house generally.

3 The basis on which data is gathered is always suspect – e.g. figures filled in on time sheets at random on a Friday afternoon, or the use of wrong job numbers, or the possibility that similarly-named cost centres from different sources will have differently defined parameters. On the other hand the assumptions made in simulation are known.

Taxation and grants

One major assumptions which has to be made in costs-in-use calculation is the degree to which taxation will affect the final results. Taxation is a fluctuating variable, like other assumptions such as building life, component life, inflation, and interest rate. Different governments will have different policies towards the amount and type of taxation to be levied and this will directly or indirectly affect the amount the client pays for his maintenance. It is possible for a company to write off the cost of most operational and maintenance costs against profit before arriving at the net figure on which Corporation Tax is payable. In effect this means that the government is giving a very substantial discount on the cost of these items (25% – 35% in 1989–90). In practice, the benefit is probably not quite as great as the tax percentage, because the relief from tax does not usually occur until the next financial year after the costs have been incurred. The effect of this 'tax lag' is to reduce the saving by the discounted value for the time elapsed between payment for the work, and the time at which tax payment is actually due. Hence with a discount rate of 10% and a tax lag of one year, the real benefit becomes:

$$£0.35 \times 0.909 = £0.318 \text{ in every pound}$$

This position becomes more complex when tax relief is allowed on initial costs. For example, to encourage investment in industrial buildings the government has allowed 4% of the construction cost to be written off against tax in the first year, and 4% per year thereafter until the whole of the capital cost has been accounted for. Each of these savings would need to be discounted using the present value table, bearing in mind once again that there is usually some form of tax lag. In addition to tax savings on initial cost it is also possible to receive direct grants from the government for building in specially designed development areas.

These grants reduce the effective initial capital cost to the client. All these factors should be taken into account when undertaking a costs-in-use exercise. (However it should not be forgotten that the taxation relief is only available on profit and not every firm makes a profit every year.)

One last point to bear in mind is the influence of Value-added Tax, now payable on all construction work except qualifying residential accommodation. Until 1989 VAT was only payable on maintenance and repair work, new construction being zero rated.

Sensitivity analysis

We have already seen that there are a large number of assumptions made in any costs-in-use calculation, and that it is not always possible to assess the effect of changes in these assumptions realistically. One method of testing whether the results achieved by costs-in-use studies are satisfactory for decision-making purposes is to repeat the calculations in a methodical way, changing the value of a single variable (i.e. assumption) each time. It is then possible to see how sensitive the results are to changes in the variable under consideration. If the results are plotted on a graph then a visual check can be made to see when, for example, one component becomes more attractive than another.

Figure 4.1 shows the results of such a sensitivity analysis between two components

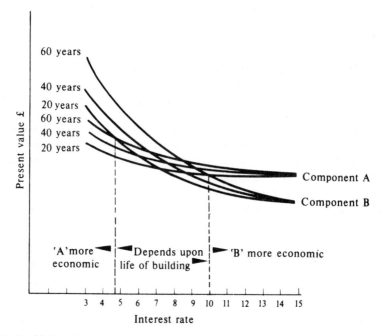

Fig. 4.1. Sensitivity analysis.

where the calculations have been changed by varying the life of a building and the interest rate. These computations have been plotted in the form of a graph with present value on the Y axis and interest rate on the X axis. From a graph of this nature, the interest rate and building life at which one component becomes a better proposition can be clearly seen. The area at the crossover of the graphs for each alternative is not clear and, therefore, at these interest rates there is obviously very little to choose between the two. While this technique is of some advantage in testing the effect of changes in the assumptions it still assumes that the variables are static during the building life. If it is possible to assess the probable distributions of the likely occurrence of values for each variable then the Monte Carlo technique should be used to sample from these distributions. This technique is described in detail in Chapter 20.

Summing up

The forecasting of running costs of a building will often be useful, but this does not always mean that a costs-in-use study (equating capital costs and running costs) will have much meaning for the client. Such studies are most likely to be justifiable where one, or preferably both, of the following conditions are fulfilled.

    1 Both sides of the calculation represent 'real' money, and the figures for maintenance and renewal will actually be used for programming, controlling, and monitoring the costs of such work.

    2 The study relates to energy-consuming equipment with a well-defined servicing programme.

Even so, the following points should be remembered:

    1 The shorter the period involved the more accurate the forecast is likely to be. As an alternative the use of an artificially high rate for discounting will have much the same effect by attaching little value to anything which may happen in the very distant future.

    2 While the results of the study will be of use as one of the factors to bear in mind in arriving at a decision, it would be most unwise to come to a decision regarding alternatives on a costs-in-use basis alone, unless the calculated advantage of one of them is very substantial.

But life-cycle costing of energy-consuming installations and equipment are likely to be much more useful.

Chapter 5
Cash Flow

Project planning needs to take into account not just the total lump sum involved but the timing of the various payments and receipts. This is the concept of 'cash flow'; there is a positive flow when money is received into an organisation and a negative flow when money is paid out.

Cash flow is a valuable concept because it enables us to look at the financing of a project in a different light. We shall first look at its uses in connection with the financing of a project by a building contractor, because it is here that its influence is most dramatic, although it is equally applicable to the client's finances, as we shall see in due course.

Cash flow and the building contractor

As an example a simple contract for a £410,000 building, to be erected in 30 weeks, has been chosen. The prime cost to the contractor is £400,000, leaving £10,000 profit. This represents $2\frac{1}{2}\%$ on turnover, which is fairly usual, but the real profit percentage may be very different to this miserable figure which is so often quoted as an example of the poor financial returns of the building industry.

In Table 1 is shown the weekly outlay on labour, plant hire and overheads, the weekly value of material deliveries, and the subcontractors' accounts which are received monthly. The table also shows the total prime cost at the end of each month together with the quantity surveyor's valuation.

In this particular example it is assumed that 10% of the value of the work is withheld by the client each month as 'retention' until the total of the retention fund amounts to £25,000, after which remaining work is paid for in full. This arrangement would be less generous to the contractor than the current requirements of the British JCT Standard Form of Building Contract, and if this Form was being used the contractor would have rather more cash in hand than is shown in the examples.

It will be seen from Table 1 that as usual the job gets into its stride slowly, following the normal 'S-curve' of expenditure plotted against time (Fig. 5.1), and because much of the early expenditure is in any case related to setting-up rather than producing finished work the first valuation by the quantity surveyor does not meet the full cost. By the time the job is halfway through it is showing a handsome profit (perhaps the 'loading' of Bill rates for the early trades may have helped) although the finishing-off, again as usual, is not very profitable.

Table 2 illustrates the contractor's cash flow, on the assumption that everything goes as it should. The client pays the amounts of the valuations about a fortnight after they are made, and the contractor pays the nominated sub-contractors as soon as he receives the money. Materials are paid for at the end of the month following delivery (in practice

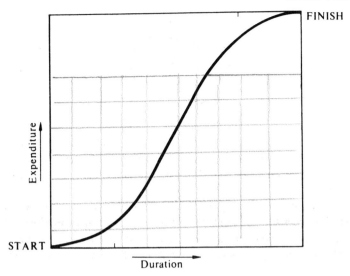

Fig. 5.1. Typical S-curve of project expenditure plotted against duration.

the sub-contractors and materials suppliers would allow discounts for such prompt payment, but these have been ignored in the example in order to keep it reasonably simple). Labour, etc, has to be paid for weekly as the costs are incurred. At the conclusion of the project the client pays half of the retention sum and pays the balance six months later. The cash flow in Table 2 is also shown graphically in Fig 5.2.

What is the first thing we notice about the cash flow as shown in Table 2. Although is a £410,000 contract, and although the contractor is certainly bearing the risk (and undertaking the organisation) of a project of this size, such a figure has nothing to do with his financing of the job. Except for one or two weeks he never has more than £25,000 sunk in the contract and in fact he often has no money invested in it at all – for example, by the middle of June he has received £43,000 more than he has paid out. If we therefore say that he could carry out the job on a working capital of £25,000, his £10,000 profit is not 2½% but 40%, a very different figure! In fact the true situation is even better than this, because his average working capital (setting positive cash flows in some weeks against negative in others) is much less than £25,000, as can be seen from Fig 5.2. This is therefore not a £410,000 job, as far as the contractor's budgeting is concerned.

Before we become too excited at the prospect of making 100% (or more) per annum profit as a matter of course, we should look at Table 3 and the accompanying Fig. 5.3. This represents the cash flow on the same project where the client is not being quite so helpful. The monthly payments are being delayed for a further two weeks, and the quantity surveyor is being 'prudent' with his valuations, finally under-certifying to the extent of £10,000 (possibly because of variations which have not been agreed) although the full amount is eventually paid over.

This is still the same project with the same costs and the same profit, but as far as the contractor is concerned it is on a completely different scale. He now has an almost permanently negative cash flow, often amounting to £30,000–£40,000; this job will

Table 1. Weekly outlays for a 30-week, £410,000 building

	Week No.	Wages, plant hire and overheads	Materials delivered	Subcontractors accounts received	Total prime cost and overheads	QS valuation	Valuation less retention
March	1	1,000	2,000				
	2	1,500	1,000				
	3	1,500	1,000				
	4	2,000	3,000		13,000	10,000	9,000
April	5	3,000	10,000				
	6	3,000	10,000				
	7	3,000	6,000				
	8	4,000	6,000	10,000	68,000	70,000	63,000
May	9	4,000	10,000				
	10	4,000	3,000				
	11	5,000	20,000				
	12	5,000	10,000				
	13	5,000	10,000	10,000	154,000	170,000	153,000
June	14	5,000	17,000				
	15	6,000	15,000				
	16	6,000	10,000				
	17	5,000	5,000	10,000	233,000	250,000	225,000*
July	18	5,000	5,000				
	19	5,000	10,000				
	20	4,000	5,000				
	21	3,000	3,000	30,000	303,000	325,000	300,000
Aug	22	3,000	2,000				
	23	3,000	—				
	24	2,000	5,000				
	25	2,000	—				
	26	2,000	5,000	30,000	357,000	375,000	350,000
Sept	27	4,000	10,000				
	28	4,000	—				
	29	3,000	—				
	30	2,000	—	20,000	400,000	410,000	385,000
	Total	106,000	184,000	110,000	400,000	410,000	385,000
						retention	25,000
							£410,000

* Maximum retention now withheld

Table 2. Payments and receipts in £s based on Table 1.

	Week No.	Wages etc.	Materials	Sub-contractors	Total	Amounts received	Cumulative cash flow
			Payments				
March	1	1,000			1,000		−1,000
	2	1,500			1,500		−2,500
	3	1,500			1,500		−4,000
	4	2,000			2,000		−6,000
April	5	3,000			3,000		−9,000
	6	3,000			3,000	9,000	−3,000
	7	3,000			3,000		−6,000
	8	4,000	7,000 (March)		11,000		−17,000
May	9	4,000			4,000		−21,000
	10	4,000		9,000	13,000	54,000	+20,000
	11	5,000			5,000		+15,000
	12	5,000			5,000		+10,000
	13	5,000	32,000 (April)		37,000		−27,000
June	14	5,000			5,000		−32,000
	15	6,000		9,000	15,000	90,000	+43,000
	16	6,000			6,000		+37,000
	17	5,000	53,000 (May)		58,000		−21,000
July	18	5,000			5,000		−26,000
	19	5,000		9,000	14,000	72,000	+32,000
	20	4,000			4,000		+28,000
	21	3,000	47,000 (June)		50,000		−22,000
Aug	22	3,000			3,000		−25,000
	23	3,000		27,000	30,000	75,000	+20,000
	24	2,000			2,000		+18,000
	25	2,000			2,000		+16,000
	26	2,000	23,000 (July)		25,000		−9,000
Sept	27	4,000			4,000		−13,000
	28	4,000		27,000	31,000	50,000	+6,000
	29	3,000			3,000		+3,000
	30	2,000	12,000 (August)		14,000		−11,000
Oct	32			18,000	18,000	35,000	
		Release of retention		5,500	5,500	12,500	+13,000
	34		10,000 (September)		10,000		+3,000
April	60	Release of retention		5,500	5,500	12,500	+10,000
	Total	106,000	184,000	110,000	400,000	410,000	(10,000)

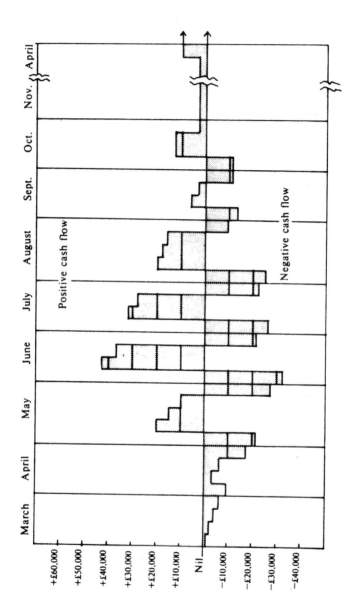

Fig. 5.2. Diagram showing cumulative cash flow as Table 2.

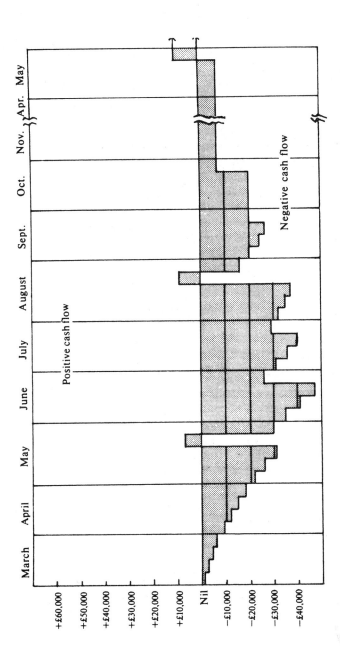

Fig. 5.3. Diagram showing cumulative cash flow as Table 3.

Table 3. As Table 2 but with late and underestimated payment in £s by client.

	Week No.	Wages etc.	Materials	Sub-contractors	Total	Amounts received	Cumulative cash flow
			Payments				
March	1	1,000			1,000		−1,000
	2	1,500			1,500		−2,500
	3	1,500			1,500		−4,000
	4	2,000			2,000		−6,000
April	5	3,000			3,000		−9,000
	6	3,000			3,000		−12,000
	7	3,000			3,000		−15,000
	8	4,000	7,000 (March)		11,000	8,000	−18,000
March	9	4,000			4,000		−22,000
	10	4,000			4,000		−26,000
	11	5,000			5,000		−31,000
	12	5,000		9,000	14,000	52,000	+7,000
	13	5,000	32,000 (April)		37,000		−30,000
June	14	5,000			5,000		−35,000
	15	6,000			6,000		−41,000
	16	6,000			6,000		−47,000
	17	5,000	53,000 (May)	9,000	67,000	88,000	−26,000
July	18	5,000			5,000		−31,000
	19	5,000			5,000		−36,000
	20	4,000			4,000		−40,000
	21	3,000	47,000 (June)	9,000	59,000	70,000	−29,000
Aug	22	3,000			3,000		−32,000
	23	3,000			3,000		−35,000
	24	2,000			2,000		−37,000
	25	2,000		27,000	29,000	75,000	+9,000
	26	2,000	23,000 (July)		25,000		−16,000
Sept	27	4,000			4,000		−20,000
	28	4,000			4,000		−24,000
	29	3,000			3,000		−27,000
	30	2,000	12,000 (Aug)	27,000	41,000	48,000	−20,000
Oct	34		10,000 (Sept)	18,000	28,000	34,000	
		Release of retention					−7,000
				5,500	5,500	12,500	
May	64	Release of retention		5,500	5,500	22,500	+10,000
	Total	106,000	184,000	110,000	400,000	410,000	(10,000)

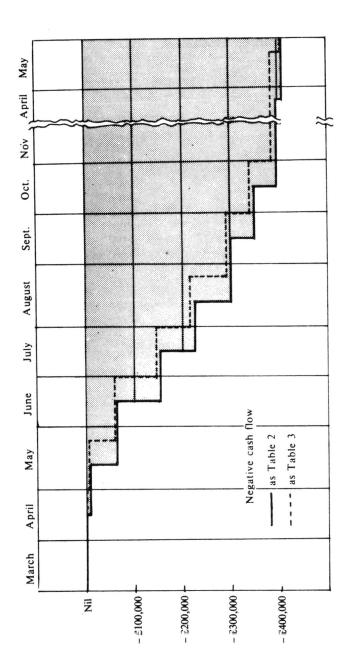

Fig. 5.4. Diagram showing client's cash flow as Table 2 and 3.

Table 4. As Table 3 but with one month's delay in payments (in £s) to materials suppliers.

	Week No.	Wages etc.	Materials	Sub-contractors	Total	Amounts received	Cumulative cash flow
			Payments				
March	1	1,000			1,000		−1,000
	2	1,500			1,500		−2,500
	3	1,500			1,500		−4,000
	4	2,000			2,000		−6,000
April	5	3,000			3,000		−9,000
	6	3,000			3,000		−12,000
	7	3,000			3,000		−15,000
	8	4,000			4,000	8,000	−11,000
March	9	4,000			4,000		−15,000
	10	4,000			4,000		−19,000
	11	5,000			5,000		−24,000
	12	5,000		9,000	14,000	52,000	+14,000
	13	5,000	7,000 (March)		12,000		+2,000
June	14	5,000			5,000		−3,000
	15	6,000			6,000		−9,000
	16	6,000			6,000		−15,000
	17	5,000	32,000 (April)	9,000	46,000	88,000	+27,000
July	18	5,000			5,000		+22,000
	19	5,000			5,000		+17,000
	20	4,000			4,000		+13,000
	21	3,000	53,000 (May)	9,000	65,000	70,000	+18,000
Aug	22	3,000			3,000		+15,000
	23	3,000			3,000		+12,000
	24	2,000			2,000		+10,000
	25	2,000		27,000	29,000	75,000	+56,000
	26	2,000	47,000 (June)		49,000		+7,000
Sept	27	4,000			4,000		+3,000
	28	4,000			4,000		−1,000
	29	3,000			3,000		−4,000
	30	2,000	23,000 (July)	27,000	52,000	48,000	−8,000
Oct	34		12,000 (Aug)	18,000	30,000	34,000	
		Release of retention					+3,000
				5,500	5,500	12,500	
Nov	38		10,000 (Sept)		10,000		−7,000
May	64	Release of retention		5,500	5,500	22,500	+10,000
	Total	106,000	184,000	110,000	400,000	410,000	(10,000)

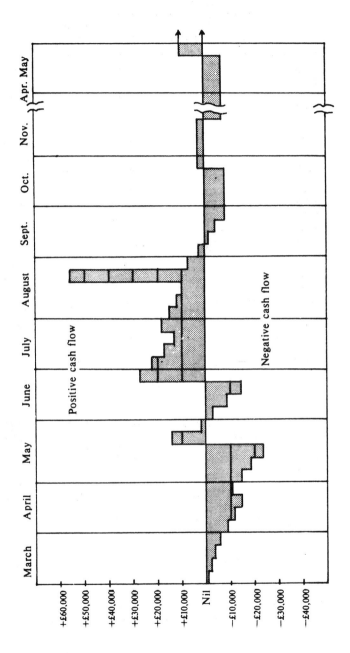

Fig. 5.5. Diagram showing cumulative cash flow as Table 4.

Table 5. As Table 2 but builder's work underestimated by 10%, and one month's delay in payments to materials suppliers. Values in £s.

	Week No.	Wages etc.	Materials	Sub-contractors	Total	Amounts received	Cumulative cash flow
March	1	1,000			1,000		−1,000
	2	1,500			1,500		−2,500
	3	1,500			1,500		−4,000
	4	2,000			2,000		−6,000
April	5	3,000			3,000		−9,000
	6	3,000			3,000	8,100	−3,900
	7	3,000			3,000		−6,900
	8	4,000			4,000		−10,900
March	9	4,000			4,000		−14,900
	10	4,000		9,000	13,000	49,500	+21,600
	11	5,000			5,000		+16,600
	12	5,000			5,000		+11,600
	13	5,000	7,000 (March)		12,000		−400
June	14	5,000			5,000		−5,400
	15	6,000		9,000	15,000	81,900	+61,500
	16	6,000			6,000		+55,500
	17	5,000	32,000 (April)		37,000		+18,500
July	18	5,000			5,000		+13,500
	19	5,000		9,000	14,000	65,700	+65,200
	20	4,000			4,000		+61,200
	21	3,000	53,000 (May)		56,000		+5,200
Aug	22	3,000			3,000		+2,200
	23	3,000		27,000	3,000	70,500	+42,700
	24	2,000			2,000		+40,700
	25	2,000			2,000		+38,700
	26	2,000	47,000 (June)		49,000		−10,300
Sept	27	4,000			4,000		−14,300
	28	4,000		27,000	31,000	48,000	+2,700
	29	3,000			3,000		−300
	30	2,000	23,000 (July)		25,000		−25,300
Oct	32			18,000	18,000	33,500	
		Release of retention					−3,900
	34		12,000 (Aug)		12,000		−15,900
Nov	38		10,000 (Sept)		10,000		−25,900
April	60	Release of retention		5,500	5,500	11,400	−20,000
	Total	106,000	184,000	110,000	400,000	380,000	(−20,000)

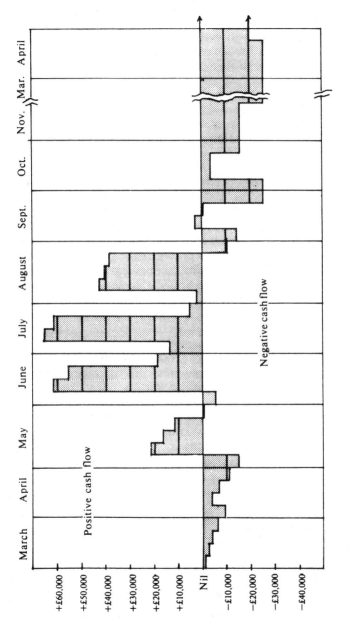

Fig. 5.6. Diagram showing cumulative cash flow as Table 5.

involve two or three times the capital commitment of the previous one and as far as the contractor's financing is concerned it is a project of more than double the size, although the profit is still the same. On the other hand, if the contractor were to complain the client might wonder what all the fuss was about; as Fig. 5.4 shows, the effect of the different payment pattern on his own cash flow is proportionately very small. Many contractors would confirm that this kind of situation is neither unusual nor does it by any means represent the worst that may befall them as regards deferring of payments.

The contractor's remedy is often to defer payment in turn to his own suppliers, which in the above example could more than restore the cash flow figures to their former satisfactory state (Table 4 and Fig. 5.5).

It is possible that delaying payment in this way might cause the contractor to forfeit some of the cash discounts which the materials suppliers give him. These are normally of the order of $2\frac{1}{2}\%$; in the event of all the suppliers refusing to give him the discount he would stand to lose a total of £4,600, although in practice it would be unlikely that more than a few would do so unless the delays became very serious. He might be prepared to accept the loss of a few cash discounts as a small price to pay for the benefit of cash flow so much improved that he could carry out the project without any capital commitment at all after the first 16 weeks.

Finally, however, a dreadful warning: suppose we have an exactly similar job, where the client pays punctually and fully (as in the first example) but where the contractor has underpriced his work by 10% and is therefore not going to make £10,000 profit but £20,000 loss; also, because of financial difficulties, he is paying his suppliers one month late. The resultant cash flow (Table 5 and Fig 5.6) looks eminently satisfactory until the end of the project in September; it is in many ways better than that of the soundly managed contractor in Figs 5.2 and 5.5, and requires a working capital of about £10,000 for a few weeks only. However, instead of a handsome profit this contractor will finish by losing the whole of his working capital twice over. If he has several such jobs going on concurrently, so that the negative cash flows on one coincide with the positive flows on another, he will be able to keep trading for some time before the crash comes, and meanwhile his cash flow figures may look fairly healthy.

It will be seen how difficult it is to distinguish the early symptoms of insolvency when looking at a contractor's accounts, and indeed it is quite possible that a badly managed contracting firm may not itself realise what is happening until it is too late.

When times are bad in the building industry it is common for contractors to submit very low tenders in order to obtain work and so keep their organisation employed, and they can then easily become short of cash. It is also not unknown for contractors who have already got into this position, either for the above reason or through bad management, to continue to quote absurdly low prices. This is because they urgently need the cash flow from new work in order to pay their past debts; if the contractor in the above example is receiving payments on another job by September he will be able to pay his suppliers and keep going (although the new job in turn will be getting into even worse difficulties in due course and will need an even more drastic dose of the same medicine). Quantity surveyors do well to be suspicious of a building company which is expanding its operations rapidly at prices which its competitors cannot match.

However, we can see from the earlier examples that even a soundly managed contractor will find it very tempting to get deeply into debt at the bank if he is able to get returns of more than 50% on money which he is borrowing at less than 15%.

So cash flow assessment of accounts must be used carefully, but these examples do demonstrate that quite small percentage differences in estimating can have a dramatic effect on profitability. This is in fact the principal weakness of the construction industry, that the difference between a substantial profit on capital and a substantial loss can lie inside the normal margin of error in estimating. A greater capital investment in a project on the part of contractors might lead to greater stability, as the required profit on turnover would then have to be much higher than 1% or 2%. If the contractor had to make 10% profit on turnover to get a reasonable return on capital, then estimating errors of 2% or so would have a proportionately smaller effect on his overall profit percentage and would not make the difference between 'boom or bust'. Any move to reduce 'retentions' will have exactly the opposite effect.

Discounted cash flow

In the above examples we have forgotten about the time/money relationship. However, following from what was written earlier we realise that, for example, the £6,000 negative cash flow in the fourth week of Table 2 must be worth more than the £6,000 positive flow in the twenty eighth week. If we therefore discount all the payments and receipts to a common date we shall be able to measure the true profitability of the project. There are basically two ways in which this can be done.

Rate of return

This involves calculating the 'rate of return' (sometimes called 'internal rate of return') on the working capital, and could be used to show the profit as a percentage of average working capital in the kind of study we have been looking at in Tables 2–5. Unfortunately this cannot be done very easily without a computer, because it involves 'trial-and-error' experiments with different discount rates until a rate is found which discounts the total receipts and the total expenditure to exactly the same present-day figure, but it is the only the Discounted Cash Flow (DCF) method which shows the profit as a percentage on capital employed.

As a matter of interest, the rates of return on the contractor's working capital in two of the previous examples are as follows:

Table 2 270% per annum
Table 3   42% per annum

and as proof the figures in Table 3 are shown again in Table 6, discounted at 42% per annum, or 0.80% per week.

Now that we have figures, the enormous difference in return between the Table 2 and Table 3 situation can be seen, caused purely by slightly late and 'conservative' payment by the client.

Very high returns, such as the 270%, would be dependent upon the contractor having other projects in progress on which the surplus funds arising from this contract (e.g. the £43,000 in Week 15) could be profitably but temporarily employed at the same

Table 6. As Table 3, discounted to beginning of week 1 at 42% p.a. (0.8% per week).

	Week No.	Actual cash flow		Discounted cash flow	
		Negative	Positive	Negative	Positive
March	1	1,000		990	
	2	1,500		1,480	
	3	1,500		1,470	
	4	2,000		1,940	
April	5	3,000		2,880	
	6	3,000		2,860	
	7	3,000		2,840	
	8	11,000	8,000	10,320	7,510
May	9	4,000		3,720	
	10	4,000		3,690	
	11	5,000		4,580	
	12	14,000	52,000	12,700	47,260
	13	37,000		33,360	
June	14	5,000		4,470	
	15	6,000		5,320	
	16	6,000		5,280	
	17	67,000	88,000	58,510	76,850
July	18	5,000		4,330	
	19	5,000		4,300	
	20	4,000		3,410	
	21	59,000	70,000	49,910	59,220
August	22	3,000		2,520	
	23	3,000		2,500	
	24	2,000		1,650	
	25	29,000	75,000	23,760	61,450
	26	25,000		20,320	
September	27	4,000		3,230	
	28	4,000		3,200	
	29	3,000		2,380	
	30	41,000	48,000	32,280	37,800
October	34	33,500	46,500	25,550	35,470
May	64	5,500	22,500	3,300	13,510
	Total	400,000	410,000	339,070	339,070

rate. It is unlikely that the cash flows of several projects can be perfectly matched, but much the same result could be obtained by ironing out the worst peaks and troughs in the cash flow diagram, for example by delaying payment to some merchants for a week or two and making up for it by paying them early in a month when funds are plentiful.

In cases such as this (where there are large positive cash flows quite early in the time scale of the project) the 'rate of return' method can be most misleading. It would, for instance, show an annual rate of return of over 200% when applied to Table 5; this money-losing project would therefore appear to be most lucrative provided that the contractor was able to earn 200% on the positive cash flow elsewhere in his business. Such projects would better be dealt with by the 'NPV (Net Present Value) at criterion rate of return' method which is next to be described; alternatively the 'rate of return' method could be used ignoring any short-term positive cash flows on the basis that it would be difficult to obtain any return on these.

Note that the 'rate of return' method cannot be used for calculations where there are no receipts to set against expenditure.

NPV at criterion rate of return

This is a much easier calculation, and will probably be more often used by cost planners in assessing the viability of development. It involves the selection of a 'criterion' rate of return, which would be either the rate at which money can be borrowed to finance the project, or else a minimum acceptable profit, depending on the circumstances of the case. The income and expenditure are both discounted at this rate and the difference found; this is the NPV or Net Present Value of the project. If the discounted income exceeds the discounted expenditure then the NPV is positive and represents a profit over and above the criterion rate; if, on the other hand, the discounted expenditure is larger than the discounted income then the NPV will be negative and the project would not service its loan or pay the minimum acceptable profit and is unlikely to be carried out.

It will be realised that this method can be used for two different purposes:

1 To assess whether a project is worth carrying out.
2 To see which of two or more alternative solutions is the best from the pointy of view of cost. (The examples on page 21 concerning the mechanical stoker used a criterion rate of return of 8%). When used in this way there may not be any income to set against expenditure, but unlike the 'rate of return' method the 'criterion rate' method will work in these conditions; it will of course always show a negative balance, and the scheme with the smallest negative balance will be the best choice.

Example of 'criterion rate' calculations

A simplified investment problem has been chosen as an example. A plot of land is being bought for £800,000 a block of flats is to be erected, and the whole development disposed of for £2,700,000.

In 'Case A' full drawings and bill of quantities are to be prepared and tender invited for a building contract of 15 months duration at a cost (including professional fees) of £1,320,000.

In 'Case B' the project is to be started before full documentation has been prepared and the building programme will be compressed into 10 months. However, the result of this rush will be a cost (including fees) of £1,480,000.

It is required to examine these two programmes to see which would be the more profitable. The cash flows are set out in Table 7.

Suppose that a criterion rate of 12% pa is chosen. This will be equivalent to 1% per monthly period and so the two alternatives can be discounted using the 1% DCF tables (Appendix D – Table 3).

Table 7. Undiscounted cash flows.

		Case A	Case B	
End of month:	1	–800,000	–800,000	(land purchase)
	3	—	–40,000	
	4	–40,000	—	(professional fees)
	5	—	–40,000	
	6	—	–60,000	
	7	—	–100,000	
	8	–40,000	–140,000	
	9	—	–180,000	
	10	–40,000	–220,000	
	11	–60,000	–180,000	
	12	–60,000	–160,000	
	13	–60,000	–140,000	
	14	–80,000	–120,000	(Payments to
	15	–80,000	–100,000	builder
	16	–100,000	—	and professional
	17	–100,000	—	fees)
	18	–120,000	—	
	19	–140,000	—	
	20	–120,000	—	
	21	–80,000	—	
	22	–80,000	—	
	23	–60,000	—	
	24	–60,000	—	
		–2,120,000	–2,280,000	
Sale		2,700,000	2,700,000	
		(month 24)	(month 15)	
Profit		£580,000	£420,000	

It will be seen from Table 8 that both schemes would be profitable assuming 12% per annum as a criterion rate of return, although the profit of Case A would have a present value of £108,040 while the profit of Case B would be slightly lower at £98,490. In these circumstances it would appear to be more profitable to go for the longer and less expensive contract arrangements.

If, however, the criterion rate of return was 15%, the NPV's would be reduced to £140,320 (Case A) and £147,540 (Case B). Both schemes are still profitable, but with higher interest rates the shorter and more expensive scheme now shows the better profit.

In Fig. 5.7 the NPV and rate of return for the two schemes is shown graphically. It will be seen that Case A is much the more profitable at low interest rates, that both schemes show an equal NPV of about £164,000 at 14%, and that Case B becomes relatively more and more attractive as the required rates of return rise above this point. Case A ceases to be profitable if a rate of more than 21% is required, and Case B at 25%.

This demonstrates the point that if a high rate of return is required it is usually worthwhile to compress the design/building period even if this results in a less efficient building process and higher cost.

Table 8. Cash flows (discounted).

	Case A			Case B		
End month:						
1	−800,000	× 99.0p =	−792,080	−800,000	× 99.0p =	−792,080
3				−40,000	× 98.0p =	−38,820
4	−40,000	× 96.1p =	−38,440			
5				−40,000	× 95.1p =	−38,060
6				−60,000	× 94.2p =	−56,520
7				−100,000	× 93.3p =	−93,260
8	−40,000	× 92.3p =	−36,940	−140,000	× 92.3p =	−129,280
9				−180,000	× 91.4p =	−164,580
10	−40,000	× 90.5p =	−36,220	−220,000	× 90.5p =	−199,160
11	−60,000	× 89.6p =	−53,780	−180,000	× 89.6p =	−161,340
12	−60,000	× 88.7p =	−53,240	−160,000	× 88.7p =	−142,000
13	−60,000	× 87.9p =	−52,720	−140,000	× 87.9p =	−123,020
14	−80,000	× 87.0p =	−69,600	−120,000	× 87.0p =	−104,400
15	−80,000	× 86.1p =	−68,900	+2,600,000	× 86.1p =	+2,239,500
16	−100,000	× 85.3p =	−85,280		NPV =	£196,980
17	−100,000	× 84.4p =	−84,440			
18	−120,000	× 83.6p =	−100,320			
19	−140,000	× 82.8p =	−115,880			
20	−120,000	× 82.0p =	−98,340			
21	−80,000	× 81.1p =	−64,920			
22	−80,000	× 80.3p =	−64,280			
23	−60,000	× 79.5p =	−47,720			
24	+2,640,000	× 78.8p =	+2,079,180			
		NPV	£216,080			

Payback

Payback is a third DCF technique which is sometimes used in assessing commercial installations, and which compares the time taken to repay the costs of the schemes in question. It is not a very meaningful way of validating building projects, and is rarely if ever used by cost planners.

Care required with DCF techniques

These discounting techniques are very useful in enabling a project to be looked at from the point of view of profitability. By themselves, however, they only tell part of the story and it is vital that the cash flow situation should be considered as a whole before a decision is made. For this purpose the undiscounted figures for each period should also be looked at, either in tabular form as Table 7 or, better still, in diagram form as in Figs 5.2, 5.3, etc.

In looking at Table 7, for instance, we can see that Case B not only requires a higher capital commitment, but also requires it over a much shorter period. By the end of the first year Case A requires an investment of £1,040,000 whereas Case B requires no less than £1,920,000, and the client may not be able to raise this amount of money in the time

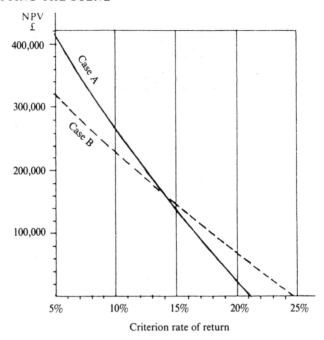

Fig. 5.7. Diagram showing NPV at different rates of return for Cases A and B.

available. From this point of view Case A might suit a client who was committed to a number of developments in progress simultaneously, while a client who liked to finish one development at a time and then go on to the next might prefer Case B. This question of keeping capital expenditure to a minimum, or of phasing payments, might be more important than a small difference in NPV.

A further factor which could offset any advantage which Case B might have over Case A would be the inflation of property prices. Since the price of houses and flats in many parts of the country has been increasing more rapidly than the general rate of inflation it is possible that the flats which would sell for £2,700,000 at the end of 15 months might sell for £3,000,000 at the end of 24 months, so destroying the previous calculations and giving an advantage to Case A.

It is worthwhile remembering that undiscounted cash flow statements and diagrams can be very useful in the social development field, where the funding of capital expenditure may be done on a quarterly or yearly basis, and it is useful to be able to plan which period the expenditure will fall into. This topic is developed in greater detail in Chapter 18.

While DCF appraisals are useful, they must be used with discretion. In any case, DCF assessments of investment can only really work if there are quantifiable receipts to set against expenditure; otherwise it merely proves that the less that is done, and the slower it is done, and the longer it can be put off, the better. Unlike straightforward cash flow statements, it therefore has little place in programming a purely social development, although it can be used for cost comparisons of alternative solutions within a given time scale.

Part 2
Cost and Design

Chapter 6
Building Costs – Product or Resources

The developer will normally need to involve himself with the building industry in order to get his building work undertaken.

The building industry and the civil engineering industry together are often referred to as the construction industry, and many firms (certainly most of the large ones) operate in both sectors, even though their staff and organisations may be largely separate. It is, however, very difficult to separate the two halves statistically, because of the overlap which occurs. For instance, the construction of the foundations of very large buildings could almost be classed as civil engineering (and you will find it so referenced in your library, if this operates the Dewey or UDC system), and there may be quite a lot of building work on some civil engineering projects.

A further subdivision of the building industry is into housing and other work, many main contractors having divisions which specialise in one or the other.

The building industry in the UK has changed very considerably over the last twenty years, in two ways.

1 Most firms of general contractors now undertake only a small proportion of their turnover using their own direct employees, the majority of their work being done by specialist or trade sub-contractors. Very often the supervisory staff and a handful of labourers will constitute the entire workforce of the general contractor on the site. The general contractor's role has changed from being primarily a provider of resources into being the provider of management and financial services.

2 Between the end of World War II and the late 1970s the building industry was geared largely to serving the public sector of the economy, and was considerably affected by periodic changes in government spending policies. In much of the country, and particularly in the south-east, the industry has had to accustom itself to the different norms of private enterprise clients, more interested in results than in procedures.

The UK construction industry is quite large, with a turnover in 1987 of 34 billion pounds. This represented nearly half of the country's fixed capital formation (i.e. plant, ships, aeroplanes, etc.) and about 8% of the gross national product.

It consists mainly of small firms, and even though the bulk of its turnover is carried out under the aegis of comparatively large and well-known national contractors they usually employ smaller firms to do much of the actual work.

Because it operates with a comparatively small investment in fixed assets in relation to its large turnover, the industry is very flexible and is able to accommodate itself to major changes in workload much more easily than industries which are plant-based. Its workforce can be reduced in hard times without the sort of political crisis caused by the closure of a car factory, and perhaps because of this it has often been used as an

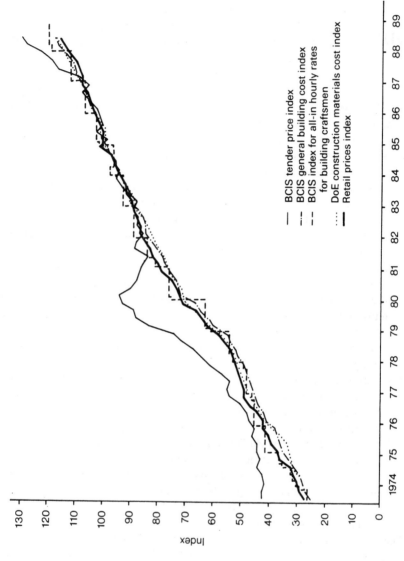

Fig. 6.1. Cost trends. 1st quarter 1974 = 100. *(Reproduced by kind permission of the Building Cost Information Service).*

economic regulator by government. Similarly it can expand its efforts rapidly in order to meet demand without requiring major investment – for instance, in the two years 1987–1988 the building industry was able to increase its output by one-fifth.

Nevertheless violent changes in demand do harm the industry. Labour which is laid off in difficult times does not always return to the industry afterwards, and the problem is compounded by difficulties of training skilled workers. Nobody can afford to do this when times are hard, and in boom times they are usually too busy. It is also difficult to keep a balanced management team of the right size when demand fluctuates wildly, and as stated above management expertise is now probably the general contractor's main stock-in-trade.

A further problem concerns the supply of materials, since material and component manufacturers are plant-based and find it more difficult to respond rapidly to changes in demand. If a boom is regional in character this problem can be mitigated by importing materials and components from elsewhere in the UK or Europe, but for bulk materials in particular (where transportation forms a large part of their cost) this can often be an expensive alternative.

It is therefore very important for the cost planner to look at the general economic situation when forecasting costs. For example, for a few years after the start of the last major slump in building in 1980 tender prices remained almost static although official costs of labour and material inputs continued to show an increase.

In the four-and-a half years between the 3rd quarter of 1980 and the 1st quarter of 1985 tender prices increased by less than 1%, but the official index of building costs rose by no less than 38% in the corresponding period

This imbalance was partly due, of course, to tenderers' willingness to take a cut in profits during difficult times but it was mostly due to the fact that real costs did not increase, whatever the official figures might say. Theoretical inflationary increases were totally counterbalanced because labour could be easily obtained and no longer had to be bribed to work, and with fear of the sack productivity tended to improve. Under the stress of competition materials arrived when they were supposed to, and merchants started to offer discounts instead of requiring extra payments.

The management of projects and the meeting of deadlines became much easier during a period of low demand, and it was no longer necessary to make a substantial allowance in tenders for risk.

Costs and prices

There are two ways in which building costs can be estimated or analysed, based in the first instance on the prices charged for the finished buildings or parts thereof, and in the second instance on the cost of the resources required to create them.

Most of the cost planning and cost control procedures traditionally used by quantity surveyors on behalf of the client or design team depend upon 'prices' for finished work-in-place, obtained from bills of quantities or elsewhere. These prices may be little more than notional breakdowns by the builder of his tender offer and may have more of a marketing than a cost basis.

Practical builders are often scornful of this approach, and it is often suggested by people from the construction side that the cost planner should be more concerned with 'costs' (or 'real costs' as the proponents of this argument like to call them). This is an important issue and one which we are now going to look at, but it is not the simple choice between fiction and reality that is implied in this argument.

In fact any figure in this context is simultaneously both a 'price' and a 'cost', it just depends where you are looking at it from. It can generally be said that 'the seller's price is the buyer's cost'. Thus the builder's price is the client's cost, the sub-contractor's price is the builder's cost, and the materials supplier's price is the sub-contractors cost of materials.

Today, with so much of the work on major projects being undertaken by sub-contractors on a price basis, and with the increasing use of management contracts, it is doubtful whether many builders now know much more about production costs than the quantity surveyor. But in a private-enterprise economy the whole notion that there is such a thing as 'real costs' is a mistaken one anyhow.

In fact both the finished-product price approach and the resource cost approach have their strengths and weaknesses, and the good cost planner should understand these and know when each should be used, rather than simply adopting the one normally used by his profession in the past.

The two approaches will be examined in more detail in Chapters 12 and 13, but meanwhile we need to understand how resource costs are in fact incurred.

The builder's own costs

The builder does like to think that he knows about production costs, and he thinks of them in resource categories, which are rarely thought of separately by the traditional quantity surveyor as he manipulates his all-in unit rates.

Builders' direct costs

1 Direct site labour

This category refers to costs relating to the tradesmen and labourers actually producing the work. At one time these costs would have been described as 'wages', being almost entirely composed of this single item, but today they will include substantial payments in respect of national insurance schemes, holiday schemes, etc. In the strictest sense of the word there may be almost no 'wages' paid to production employees at all, because for various reasons to do with taxation and perhaps the inherent British dislike of working for a master most such workers are 'self-employed' and are taken on as 'labour-only sub-contractors' for a fee.

Labour costs are of particular concern to the contractor because they have to be paid out weekly in ready cash as they are incurred and cannot be postponed or put on a credit basis as can most of his other commitments – we saw in the last chapter how important this is.

It is convenient to include the 'labour-only' sub-contractors under this heading, as they will usually require weekly cash.

2 Materials

The amounts comprising these costs will usually be paid by means of monthly credit accounts, payment being due at the end of the month following that in which the materials are delivered (so that materials delivered in January are paid for at the end of February – rather like Access or Visa). Such settlement by the builder usually entitles him to a 'cash discount', $2\frac{1}{2}\%$ being the most common figure although 5% may occasionally be allowed.

It is not unusual for a builder to delay payment beyond the 'cash' settlement date, sometimes for a further month or even two months. Since he may well lose his cash discount for the sake of one or two months credit this could be an expensive way of borrowing money compared to a bank overdraft, but it has the advantage of being ready and convenient and of not requiring collateral security. Within reason, and being careful not to let things get out of hand, builders' merchants and some sub-contractors are not averse to acting as financiers to the industry in this way (particularly in hard times). If the builder owes a substantial amount it is difficult for him to withdraw his custom or even reduce his level of buying substantially, as immediate settlement of accounts might be called for. The merchant has therefore a captive customer who cannot afford to be too fussy over prices or delivery dates.

In many cases the builder will have been paid by the client for the materials, or for the work in which they are incorporated, before settling with his merchants (as we saw in the last chapter).

Because of the increasing tendency to buy-in fabricated components rather than making things on site, a larger proportion of expenditure now falls into this category, and a smaller proportion into labour, than was previously the case.

3 Small plant

This item covers hand tools and small mechanical plant of a kind that can be directly associated with specific pieces of finished work.

Sub-contractors and major specialist suppliers

Because of the increasing tendency to employ specialist sub-contractors rather than using the builder's own labour and materials this category has expanded over the last twenty years and is now usually the largest head of expenditure. In terms of payment methods it very much resembles the 'materials' category, but sub-contractors tend to be less accommodating than materials suppliers with regard to extended credit.

Sub-contractors and suppliers 'nominated' or 'named' by the architect are also paid by credit account but as they are in a position to, complain to the architect if payment is delayed, and since such complaints cast doubts upon the builder's solvency, he will usually make efforts to pay them fairly promptly.

Again, the builder will have often have been paid by the client for the work before settling with his sub-contractors, and some large firms of builders attempt to formalise this by inserting a 'pay-when-paid' clause in their sub-contracts, a move usually resisted by its victims.

Site indirect costs ('preliminaries')

The term 'preliminaries' is sometimes used to describe these items because they usually appear in a bill of quantities under this heading. They comprise all the items of site expenditure which cannot be attributed to individual items of work but to the project as a whole, or to substantial sections of it.

Such costs include:

1 Salaries and wages of site staff

The salaries paid to management, supervisory and clerical staff employed on the site. In former times these payments would have been limited to a foreman and a few junior site staff, but today the total costs of site management will often be greater than those of the directly employed production workers.

2 Site offices, messrooms, and facilities

Again, at one time these were a comparatively minor item, but today the temporary site office buildings can form a major multi-storey complex.

3 Major plant

Large items of mechanical plant may be dealt with in one of two ways; either they may be charged to the job when brought on to the site and credited (less depreciation) when removed, or else an hourly or weekly hire charge will be made plus a charge for bringing the plant on site and removing it. Most large builders find it convenient to set up a subsidiary plant company, which will charge the plant out to the sites on a hire basis, while plant hired in from outside will be similarly dealt with and paid for by credit account. Lorry (truck) transport is usually arranged in a like manner.

It is unusual for a builder to own large items of plant except through a subsidiary company. There are substantial advantages in hire-purchase through a finance company, since the interest charges (less tax) will be lower than the return which the contractor expects to make on the working capital which he employs; it would therefore be uneconomic for him to invest any of this capital in plant purchase.

It is difficult to attribute the costs of major plant items to specific pieces of direct work, and they are usually treated as a site indirect cost. However they can sometimes be allocated to major cost centres, such as 'Excavation' or 'Concrete superstructure'.

Off-site costs

These are also called 'establishment charges' or 'overheads' and represent costs incurred in running the company as a whole, and which cannot be attributed to any one particular contract – head office expenses, builder's yard, salaries of central management and directors, insurances, and interest on loans. Off-site costs are usually allocated to projects as a percentage of the direct costs. This can operate unfairly against small simple projects which require little head-office input, and some firms have separate 'small works' departments run on more economical lines, to avoid loading their smaller jobs with a large overhead and therefore making it difficult for them to compete with smaller firms.

Profit

Strictly speaking, profit (also called 'mark-up') is not a cost, it is in fact the difference between the builder's cost and the client's price.

Two typical examples

The following cost breakdowns relating to two large office buildings in the Home Counties of England in the mid 1980s, undertaken under ordinary lump-sum contracts by a main contractor, may be of interest. The figures for Project B overstate the proportion of builder's direct work, since it was not possible to separate out some of the small sub-contractors undertaking traditional trade work.

	Project A	Project B
1 Builder's direct costs	21.0%	37.0%
2 Sub-contractors and major specialist suppliers	70.0%	49.5%
Total direct costs	91.0%	86.5%
3 Site indirect costs	6.5%	10.0%
Total site costs	97.5%	96.5%
4 Off-site costs	2.5%	3.5%
Total costs	100.0%	100.0%

Allocation of resource costs to building work

The builder would appear to be in a much better position than the independent cost planner as far as knowledge of actual building costs is concerned, but this is not necessarily the case. Keeping an accurate record of costs in the above four categories is not very difficult, but translating them into usable data for cost planning future projects is another matter.

He would in fact be ill-advised to use the actual costs from a particular project for cost planning a quite different one. There are two reasons for this:

1 Many of the factors which affect site costs on an individual project have nothing to do with the design of the building and will not repeat from job to job. These include the weather, industrial and personal relations, the skill with which the work was organised, accidents, late delivery of materials, defective work, and failure by sub-contractors.

2 In practice, costs are rarely kept in any greater detail than the 'activity' or 'operation'. An operation has been described by the Building Research Establishment as 'a piece of work which can be completed by one man, or a

gang of men, without interruption by others', such as the whole of the brickwork to one floor level, or the whole of the first fixings of joinery. Unfortunately, each operation is unique to the project, and the cost information cannot be re-used in this form.

System building, prefabrication, and cost

No discussion of building costs would be complete if it did not deal with the effect of system building and prefabrication on costs. This was seen in the 1960s as the answer to many of the industry's problems, but a number of unfortunate experiences caused the inevitable reaction against it.

System building involves varying degrees of discipline; in its most simple form it may only be a standardised form of more-or-less traditional building, saving time and money by the familiarity of the standard construction and detailing, the repetitive use of formwork units, standard components, and so on. It is difficult to obtain competitive tenders for this sort of work, because if prices are given by two or more firms each firm will quote for its own patent system and it will be almost impossible to compare the value of the tenders without a considerable amount of work. In these circumstances cost comparisons based on cost modelling techniques of the type described in this book may provide a truer picture than an attempt to check detailed quantities and prices, even where these are provided.

However, in its more highly developed forms system building may involve the delivery of the whole building in the form of a kit of parts ready to fit together in a few days (in the case of a house) or in a few weeks in the case of a larger building. The economics of this practice are not easy to evaluate, partly because of the advertising and hard selling which often go with it, and partly because the economics are in fact much more complicated than the more usual site building costs.

The cost advantages which the prefabricator possesses are, at first sight, considerable. He has the benefit of planned mass production under factory conditions, safe from the weather hold-ups which affect site output and free from the difficulties of supervision and quality control which occur when operatives are working all over a scattered site. He can employ expensive but money-saving plant which would be too cumbersome, valuable, or delicate for site use, and which in any case could not find full employment for its output on a single site. He can take advantage of modern methods of handling and transportation to bring his large fabricated units direct from his factory to the position on site where they are required. In fact we might tend to accept the commonly-held view that the lack of development of factory-based building techniques simply shows that the construction industry is hopelessly old-fashioned.

However, there are very sound reasons why prefabrication has had a much smaller impact upon the industry than was at one time expected. One reason has been the popular dislike of an environment composed of factory-produced buildings, although this might have been easier to overcome if there really had been a strong economic justification for them. In fact the economic gains have usually proved to be disappointing. What are the reasons for this?

First are the obvious reasons in favour of site production, although these were always known about and it was hoped that they would be more than offset by the above advantages. The site builder has none of the heavy overheads of factory production (he will not have to pay rent for his site workshops and will probably not even have to pay rates); many of the cheap techniques (such as bricks-and-mortar) which he uses are not suitable for off-site production, and the crude handling methods which he can use with the small and rough components used for site assembly are cheaper than the tackle required for the careful handling of large units. He is also able to provide a 'made-to-measure' building instead of one 'off-the-peg'.

With experience of pre-fabrication other economic problems came to light. A broken brick is simply a broken brick and there are plenty of unbroken ones to use, but if the corner of some special component is damaged another one will have to be ordered. The cost of replacement is the least part of the difficulty; the disruption to programme caused by the resultant delay can be far more serious.

It has also been found that the disadvantage caused by the fact that only the building superstructure can be prefabricated and the infrastructure has to be produced on site is much greater than was supposed. On low-rise projects such as housing, factories, health buildings, etc., the site preparation, levelling, foundations, drainage and site services, roads,paths,car parks, fencing, landscaping, etc., involve so much work on site and site organisation that the prefabrication of the basic superstructure is only dealing with part of the problem and not always the most significant part. This particularly applies if the finishings and services are not even included in the package.

This last problem is tending to increase rather than decline in importance, because today there are very few projects where the site is a level open field on which a factory-produced building can easily be placed – in fact the arrangement of the building or buildings is more usually dictated by the shape and configuration of the site. The building of the various in-situ connections between standard units (especially if these are at different levels on an undulating site), or the work necessary to accommodate them to irregular site boundaries, is piecemeal work which is difficult to organise efficiently, and it can more than swallow any cost savings generated by factory production of the main units. It such circumstances it might well have been better to build an in-situ building designed from the start to fit its location.

An attempt to design pre-fabricated bathrooms for houses and flats in the 1960s failed largely for this reason – when all the other partitions in the dwelling had to be built and finished on site, and all the other doors had to be hung there was not really any advantage in having had two partitions and one door fixed in the factory, and the work of handling this large unit, and fitting it into the in situ construction, was simply extra cost. The people putting forward this scheme just had not thought it through.

Prefabricated 'toilet pods' for commercial buildings have recently come back into vogue, but the situation is quite different. Where most of the floor area is simply open space for letting the toilets are quite different to the rest of the building, with their plumbing and generally more domestic human scale.

A more important and less obvious handicap which faces the prefabricating firm is the large capital investment which has to be made in developing and testing the system, establishing and equipping the factory, and most of all, stockpiling the components until they are bought by the customer. There is little opportunity to obtain the very high rate of return on capital which the site builder is able to obtain on his much more limited capital investment (as explained in the previous chapter), and builders have not found this to be a worthwhile field for deploying their financial resources.

Even where it is possible to produce a scheme for prefabrication which shows at any rate a reasonable return on capital, there are still two further difficulties to contend with. The first is the actuarial risk of some major snag developing in the system once it is tried out in the harsh conditions of the real world, involving both the costs of putting things right and the possible loss of custom resulting from bad publicity; it is not really practicable to make adequate allowance for this in a pricing structure.

The second difficulty is the one which has been most often quoted by firms who have been unsuccessful in this field, but it is nevertheless quite true. Factory produced buildings need a large and assured market in order to be competitive in cost, as do all factory products which involve a heavy initial expenditure on plant, tooling, and design, and preferably demand should not fluctuate too wildly year by year. These conditions are not met by the construction sector which is notoriously subject to boom and slump as a result either of economic forces or, more often, government fiscal policy. The flexible site-organised building industry has enough trouble as a result of 'stop-go', but it is absolutely fatal to the capital-intensive factory production of buildings. Those firms from outside the building industry who moved into this field were soon forced to abandon it.

It is interesting that the motor industry, often held up as an example to the construction industry, has wisely refused to get involved in factory production of buildings.

The only sector where prefabrication of complete buildings seems to have established the right conditions for its survival is in the field of temporary and light industrial buildings (sheds, garages, temporary classrooms and offices, church halls, single-storey factories and warehouses, etc.) where the prefabricated building dominates the market. However, in many ways the production, distribution, and erection of these light units does not involve the same difficulties that occur with larger and heavier components, but it should be noted that the firms in this sector are tending to use their experience to progress to larger and more complex structures.

A further justification for prefabrication is that the industry may be unable to cope with the volume of work required at busy times because of a shortage of labour, particularly qualified tradesmen, to work on site. This reason has in fact been used to try and justify some of the prefabrication disasters of the 1960s and early 1970s – 'we had no alternative'.

As we have just seen however, prefabrication cannot be the answer to short-term 'boom' conditions of this nature which are exactly the opposite of the settled market which it needs for survival. It was thought at one time that prefabrication might be forced on society simply because people were going to prefer working in the sheltered and well-organised conditions of a factory to working on a building site in all weathers; however this is turning out not to be the case and it would be difficult to reproduce the currently popular site self-employment pattern of work in a factory.

Prefabrication having gone through the stages of fashionable popularity and the ensuing disillusionment, it is now perhaps possible to look more dispassionately at its advantages and disadvantages, and to sort out those which are inherent in prefabrication from those which simply resulted from poor implementation.

Today prefabrication is in fact widely used, though not usually in the context of a complete building system except for the exceptions already mentioned. The structure of a modern building is usually produced on site, although it could be argued that the traditional steel frame is an example of prefabrication – and a particularly interesting one since the actual components are cut and finished to individual design requirements at the factory from standard sections.

However, it is in the fitting-out of the building that prefabrication is more usually encountered and the tendency is for its influence to increase. Today windows are not merely fabricated off-site complete with fittings, but are often delivered pre-glazed and decorated. Similarly doors often come already hung in their frames and fitted with hardware, and although wrapped in protective material these components generally require more careful treatment than is given to general construction materials. Also of course the hydraulic and engineering services largely comprise finished components joined together by piping or cabling on site.

As already mentioned, there has even been a return to prefabricated toilet modules, although in the context of high-grade office buildings rather than housing. The reasons for this include the ability to obtain a high standard of finish using factory processes, and especially the ability to speed up the work programme because the manufacture of these units can proceed in parallel with the structural construction of the building. These units are usually purpose-made and are probably more expensive than site-produced work – the saving comes from the telescoping of the construction period, and this may well be the sole advantage of prefabrication in many cases.

In any comparison between different forms of patent construction, or between factory and site production, it is also necessary to consider economic life and maintenance costs as well as first costs, and this will be even more difficult than usual because these factors will be pure guesswork as far as non-traditional materials or finishes are concerned. The prudent architect or cost-planner will need to be a good deal less optimistic than the salesman when assessing the probabilities. The very expensive maintenance, and sometimes premature demolition, which has proved necessary with some systems because of difficulties at the joints between components was quite unforeseen when the systems were evaluated.

Effect of job organisation on costs

It has already been pointed out that many major costs on building projects are not directly related to the quantity of work produced but are concerned with time and with occurrences (or non-occurrences) of various kinds. The main quantity-related cost is materials, so that given prudent buying and no unreasonable wastage there should be little variation in the cost of this part of the work whichever builder is appointed and however he organises the work. The real scope for saving (or wasting) money lies in the non-quantity-related items, and this depends on the way the job is organised and managed. This is where the real competition between contractors takes place, especially today when as we have seen the contractors main stock in trade is management skills.

A well-managed job is one where both plant and men have clear uninterrupted flows of work. Money is not spent on removing or returning men or plant from the site, or on unproductive time waiting on site because of gaps in the work flow, or on moving them unnecessarily around the site. Materials are channelled to the spot where they are needed at the right time and with a minimum of double handling.

It is not merely a question of the actual time spent hanging around – people work much more productively when they can see a clear task in front of them and where they feel they are participating in an efficient operation. It is a constant complaint by builders that they are usually unable to recover the cost of this aspect of disruption when delays or interruptions are caused by the client.

Here of course is scope for comparing the cost-planner's traditional price-orientated approach unfavourably with the resource-based method which can take account of method and work-flow. But the traditionalist's reply must be that he bases his estimates on successful tenders, and that few successful tenderers work out their prices on the basis that the job will be badly run, whatever may happen afterwards! In fact he might well claim that his method allows for good management without ever having to bother about the details of its implementation, and this facility is especially valuable in the early stages of estimating before the project has been fully designed.

Chapter 7
Methods of Procurement

Meeting the client's needs

'Procurement' is the term used to describe the total process of meeting the client's need for a building, starting at the point where this need is first expressed. It is necessary for us to remember that the easiest method of procurement is for him to buy or lease suitable accommodation which already exists, in which case he will not need the services of an architect, cost planner, or builder at all!

But if he decides that he wants a new building, or needs to refurbish an old one, he will have to enter into contractual relationships with one or more people or organisations inside the construction industry in order to get it designed and built. This applies even if he happens himself to be an employer of building labour – he is most unlikely to have the full range of needed skills and resources in-house. Since the nature and terms of these contracts will materially affect the cost control of the project we must now consider the options which are open.

Basic forms of building contract

Relationships between the client and the industry are almost invariably governed by contract. Although there are many different forms of contract they are basically of two different kinds:

1 Cost reimbursement

The contractor agrees that all his expenditure on labour, materials, etc., will be met by the client, on top of which he will charge a fee on an agreed basis (e.g., 'I will redecorate your drawing room for the cost of paint plus £10 per hour for my time').

2 Price in advance

The contractor agrees to carry out his obligations for a sum of money agreed in advance (e.g., 'I will redecorate your drawing room for two hundred pounds').

The two types of contract are illustrated with 'homely' examples because they are rarely used in their pure form on large projects. The lump sum 'price in advance' contract involves too great a risk to the contractor (the building may need deeper foundations than he bargained for, or inflation may send wages rocketing), while the cost reimbursement contract encourages waste and extravagant working by the contractor, who has not got to pay for anything that he uses. In fact, in its most primitive form where his profit is a percentage of the cost, he has a positive incentive to waste as much money as possible.

From the point of view of cost planning, however, there is rather an interesting paradox. The 'price in advance' contract gives the client almost no control over the details of methods, programming, or expenditure, but gives him a perfect forecast of his total cost. The 'cost reimbursement' contract, on the other hand, allows him to give orders about the tempo or methods or work, and he can obtain detailed allocations of actual site costs, but he can never forecast the total cost with the same degree of certainty.

We can summarise the situation as follows:

	Price in advance	Cost reimbursement
1 Requirements must be fully known and defined in advance of work	Yes	No
2 Binding cost commitment	Yes	No
3 Management involvement by client team possible	No	Yes
4 Production cost feedback possible	No	Yes

It should also be noted at this stage that the type of building contract which is chosen will have some effect on the type and scale of professional advice which is employed; for example, in the case of a large 'cost reimbursement' contract it would almost certainly be necessary to employ a quantity surveyor to check all the invoices and wage payments.

Building contracts in detail

We can now look at the various contractual methods of building, trying as far as possible to match them against the time and cost criteria which were set out in Chapter 2. While this list may be used to suggest the best approach to the individual problem it must be remembered, firstly, that no form of contract is proof against things going wrong and that, conversely, a keen combination of design team and building team will make a good job of things whatever the contractual arrangements. However, all other things being equal, a suitable form of contract will help. In the following list the time and cost requirements which are especially catered for by each of the contract types are shown in brackets as cross references to the numbered paragraphs on pages 13–15.

Cost reimbursement

1 Cost plus percentage (1, 2, 5, 6, 7, 8, 9, 10, 11, 20, 21)

In many ways the most convenient contractual basis of all; the contractor can be selected, the contract placed and work started before the scheme has been finalised and without any estimates or quantities needing to be prepared. Further, the contractor's management methods, in theory at any rate, can be used for the direct benefit of the client. The client also knows that the contractor will not make an exorbitant profit. The disadvantages of such a contract have already been mentioned; there are the usual cost reimbursement drawbacks of poor cost forecasting facilities, and low productivity arising from the fact that people are not 'working to a price' so can afford to do things properly (or slowly and extravagantly). In addition to these usual drawbacks, however, there is a positive incentive towards wastefulness because the contractor is being rewarded with a percentage of everything that is spent, so it is not surprising that this method of contracting has a bad cost reputation.

However, because of the virtues which have been mentioned it is a widely used method for such projects as:

(a) Emergency first aid and repair work
(b) Alterations and repairs to old buildings where the extent of the works cannot be foreseen until the contract has started
(c) Contracts where very high quality work is required
(d) Contracts where cost is not important but where the client wants control over the method of working.

2 Cost plus fixed fee (1, 2, 5, 6, 7, 8, 9, 10, 20, 21)

An attempt to mitigate the worst feature of the above contract by paying the contractor a fee based on the estimated cost of the project instead of a percentage of the actual cost, so reducing the incentive for wastefulness. Unfortunately, this means that a fairly detailed scheme and an estimate have to be prepared before work can start, so losing some of the advantage of the 'cost plus' system; it is also likely in practice that the so called fixed fee will have to be re-negotiated at the end of the job because of the major variations which are sure to arise in projects of the type for which such a contract would be used.

3 Target cost (1, 2, 4, 6, 7, 8, 9, 10, 18, 20)

An attempt to get the benefits of cost reimbursement without the disadvantages. A bill of quantities is prepared and priced by negotiation to arrive at a target cost; the contractor then carries out the work on an actual cost basis. There are various methods of arriving at the final price; if the actual cost is lower than the target cost the contractor and client usually split the difference between them, while if the actual cost is higher than target the contractor has to be content with his actual costs plus a small overhead percentage. The contractor therefore has an incentive to do the job as cheaply as possible and the clients gets a direct benefit if he does so; the main disadvantages are that the target cost is subject to revision in respect of the many likely variations (so that one might as well have an ordinary lump sum or remeasurement contract), and the quantity surveyor's fees are likely to be high because there is dual documentation (cost reimbursement accounts to be checked and bill of quantities to be prepared, priced, agreed, and updated).

This system however retains some of the benefits of cost reimbursement, especially the ability of the contractor and client to work closely together in the management of the project, and it has been used with particular success on a 'continuation' basis where the client has a succession of projects in which the same contractor can participate.

4 Management-based projects (1, 2, 6, 7, 8, 9, 10, 11, 12, 20, 21)

A method of procurement whereby a management team organises the project on the client's behalf, for a fee. This method was first used with success on major industrial projects where time and integration with complex engineering contracts were vital; it is now increasingly used for more general projects and particularly in refurbishment works where close relationships between the client's representatives and the construction team are very important.

There are three different approaches which may be used:

   (a) Management contracting
   (b) Construction management, and
   (c) Project management

and these are dealt with at greater length later in the chapter.

5 Direct labour (1, 2, 5, 6, 7, 8, 9, 10, 11, 12, 20, 21)

This is not, strictly speaking a 'contract' at all, because the client employs labour, buys materials, and engages subcontractors in a series of minor contracts, doing all the organising and bearing all the risk himself. It is not often used except by clients who normally maintain a building department, such as major industrial concerns and public authorities. As far as the client for a one-off building is concerned, his main difficulty would be in getting together an efficient management team on such a short term basis; for this reason it has been suggested that in the future his professional advisers (because they have a longer term interest in building management) might be able to organise this, but little has been done by such methods at the present date. Direct labour has the usual cost reimbursement advantages and disadvantages, although it is sometimes possible to set up a 'model system' of tendering and cost control. In the public sector it suffers from being the subject of strong party political controversy.

Price in advance

One of the difficulties about 'price in advance' contracts is, as we have seen, inflation. Because large building contracts will last for a year or more it is unreasonable to ask a builder to bear the total risk of inflation during this period, and most of the contract methods set out hereafter have two versions – one for use on small projects where the contractor bears the inflation risk, and another (for larger projects) where the risk is borne in whole ir in part by the client. It should be noted that while the former gives the client a firm cost forecast the latter may well be the cheaper, even on short term jobs; builders do not like this risk and tend to over-price it if they are given a choice of tendering on the two methods (we will have seen from the cash flows in Chapter 5 why the contractor is in a worse position that the client to take even a modest degree of extra risk).

1 Lump sum (1, 3, 4, 6, 7, 8, 10, 12, 13, 17, 19)

In a lump sum contract the builder commits himself to a price on the basis of the work shown on the drawings and in the specification. Although commonly used abroad, it is difficult to obtain prices in competition on this type of contract in the UK, except for very small schemes. A quantity surveyor is not usually employed and bills of quantities are not used; this means that preliminary cost forecasting, budgeting, and valuing of changes in the scheme may be unsatisfactory. But where the client's requirements are firm, and the drawings and specification are good, this is a very simple and effective contractual basis.

2 Contract based on bills of quantities

This is the normal mode for carrying out major building work in the UK and for this reason will have to be considered in most detail. Wide usage has developed a number of variations on the basic theme, as follows:

(a) *Competitive tender – JCT Standard Form of Contract* (1, 3, 4, 12, 13, 15, 18, 19, 20). Most major work in the UK is undertaken on this basis. The scheme is designed, and a number of contractors are asked to submit lump sum tenders based upon the pricing and totalling of a bill of quantities prepared on behalf of the client. The successful contractor's bill then becomes the instrument for financial administration of the project, so that the one document provides a simple means of contractor selection, price commitment, and contract management. This method of contract probably gives the lowest price of any, under UK conditions; although superior to any of the previous methods in regard to cost forecasting and budgeting it has a number of disadvantages in this respect which are set out in Chapter 14. However, it is this type of contract, with its competitive bill of quantities rates, which provides most of the raw data for elemental cost analysis, and indeed acts as a control against which the cost of buildings, erected under less competitive arrangements can be judged. It should be noted that, in practice, the provisions of the JCT Standard Form of Contract make the total price and time commitments rather less firm than they appear to be in theory.

(b) *As (a) but bill of approximate quantities* (1, 2, 5, 9, 12, 13, 15, 18, 19, 20). Used as an alternative where the scheme has not been fully designed, the Bill of Quantities being only a notional representation of the finished product. The work is measured as executed, and a final price agreed using the items and rates in the bill of quantities as far as possible. This method has the advantage of permitting an earlier start to be made, but control, both of time and money, is much weaker than in the previous case. A contractor who has under-estimated the cost of his commitments under such a nebulous arrangement as this is sure to be able to find some way of recouping his actual costs.

(c) *Negotiated tender* (1, 2, 3, 4, 5, 6, 7, 8, 9, 10, 11, 13, 14, 16, 17, 18, 19, 20, 21, 22, 23). A contractor is selected early in the design stage, either with a pin or else by means of a simple tender document in which he states his management costs, on-costs, and labour charges. The contract bill of quantities is then prepared and priced jointly with him as the design develops (he will usually be involved in the detailed design also). This assists cost forecasting and budgeting, as there is some contractual commitment to the cost assumptions made during design. An early start can also be made on site, and, because everything is negotiated, the contract terms and the arrangement of the bill of quantities can be tailormade to suit the actual requirements of the parties. This method of contracting is very suitable for use on a 'continuation' basis, where the same client, professional advisers, and contractor can establish a good working rapport; however, it probably does not produce rock-bottom prices and requires some degree of special expertise on both sides to obtain the best results. Partly for this last reason, the quantity surveyor may well be the most influential member of the professional team and may be appointed first.

(d) *Serial contracting* (1, 2, 3, 4, 5, 11, 13, 14, 16, 17, 18, 19, 20, 21). This method is only open to clients with a continuing work programme; contractors are asked to price, in competition, a typical bill of quantities, on which basis they agree to carry out any similar work which the client may require over a fixed period of two or three years. Usually the contractor is asked to take the rough with the smooth, and to quote average rates which will apply on both difficult and straightforward projects. Some local authorities use this system, which obviously relies rather heavily on the integrity and fair play of the client. It does not usually produce rock-bottom prices but gives superb cost forecasting control, as contractually binding detailed prices are available for the future schemes before they are designed.

3 Schedules of prices (1, 5, 9, 10, 11, 17)

Used almost entirely for repair and maintenance work, particularly by government departments. It has something in common with the serial contract, as the contractor binds himself to a 'price list' at which he will undertake work as required for a period of years. Sometimes used as a matter of convenience for small new works which arise during its currency of operation.

4 Design-and-Build (1, 2, 3, 4, 5, 6, 7, 8, 10, 17, 19, 21, 22, 23)

Under this system the contractor acts directly for the client, filling the roles of both the professional design team and builder. The big advantage is that the contractor is willing to commit himself firmly to a price and completion date at an early stage, because he has total responsibility for working out the scheme without interference from others. Disadvantages are the lack of competition and the drawbacks of production-oriented design; because of the first factor the client usually appoints a quantity surveyor to keep an eye on costs, and he may ask a consultant architect to advise on design – however, the latter role is so limited and frustrating that not every architect is keen to take it on. The contractor may have his own in-house design team or he may employ outside consultants; as these latter are responsible to him and not to the client this is not much more than a domestic detail.

Until 1989 design-and-build firms had an advantage compared to traditional procurement methods for new projects in that the design costs, being part of the building contract (unlike an independent architect's fees), were not liable to VAT. With recent changes in VAT legislation, under which most non-housing building work is subject to VAT this advantage has largely disappeared.

Although design-and-build has been dealt with here as a price-based method there does not seem to be any valid reason why such an organisation could not be employed on a cost-reimbursement basis where the project does not lend itself to a price-based contract – for example, refurbishment work.

Managed projects

1 Management Contracting

Of the three approaches to managed projects mentioned above management contracting is the closest to the traditional practices and structure of the contracting industry, and in its most basic form is little different to the old prime cost plus fixed fee contract, in spite of its high-sounding name. The main difference in theory is that the managing contractor does not undertake anything other than site management with his own resources, all direct works being sub-contracted by him, but this is now so much the general pattern of the industry that it is not easy to discern much real difference in practice.

2 Construction Management

This is a development of management contracting, where the construction management firm acts in a role more like that of a professional consultant than a building contractor, all the works contractors entering into direct contracts with the client rather than being traditional sub-contractors.

In many ways this type of project is rather less attractive to building contracting firms since it only uses one part of their organisation – and one which is greatly needed within their own business – and denies them the benefits of the project cash flow. Conversely it does enable consultancy firms, who could not undertake management contracting, to offer management services to clients. This practice is as yet more common in the USA than in Britain.

The advantages claimed for both the management contracting and construction management approaches, compared to the traditional JCT80 type contract, are considerable. Fast-tracking (making the design and construction processes concurrent instead of consecutive) becomes possible, construction-based expertise becomes available to the designer at an early stage, and the collaboration of client and designers with the builder, instead of the usual confrontation, is especially valuable on very large and complex time-sensitive projects or on refurbishment jobs.

It is fair to say that these advantages have not always been realised in practice, partly because of the difficulties faced by traditionally trained contractor staff (and contractor organisations) in adapting their methods to a new client-oriented role, and partly because the removal of adversarial relationships only occurs at the top management level – down where the action is the direct works contractors are normally employed on the usual lowest-price-for-the-work basis and the usual confrontations apply. Oddly enough it is sometimes said that these are worse than usual, because the trade contractors feel that the builder, who used to be a fellow-sufferer, is now insulated from cut-throat tactics whereas they are not.

However, it must be pointed out that the advantages of management contracting and construction management have undoubtedly led to their employment on large and difficult projects where any system would find itself in trouble, and where traditional methods would probably have got into much greater difficulties (even if it had been possible to use them at all).

But for all his new-found status the management contractor (and the construction manager) is still largely in the builder's traditional role, dependent for his success upon satisfactory performance by the architect, consulting engineers, and quantity surveyor, none of whom he controls, and his ability to guarantee the client an on-time and on-cost project is therefore limited. Sometimes therefore the management contractor may seek to control the designers, particularly as regards the timing of information flow, but this is an unhappy compromise and if the client is looking for a manager who will accept total responsibility for performance he needs to consider a full project management service.

3 Project management

This is a term which is often used very loosely but in its true sense involves a manager who comes in right at the top, interposing himself between the client and all consultants including not only the building and engineering professionals but those concerned with development advice and marketing. He obtains advice and services from all these, and in turn advises the client. He does not usually handle any money, other than his own fee, and all consultants and contractors have direct contracts with the client although they take their orders from the project manager.

It can be seen that the project manager is a true professional, in the sense that he has no entrepreneurial interest in the project, and he may come from almost any discipline. It is his ability to manage that is important – he can get all the technical input he needs from his consultants – though given a good management reputation there may well be a tendency for a client to pick a project manager whose background is related to those aspects which he considers to be the key ones on the development in question – a general practice surveyor if marketing is paramount, an architect where quality of design is the vital factor, or perhaps a quantity surveyor where overall cost is important.

The existence of a project manager does not necessarily imply that a non-traditional method of construction procurement will be used – it would be quite possible to use a general contractor appointed on bills of quantities for instance – but certainly it is more common for him to adopt a management contracting or construction management approach, if only because the type of project that demands a project manager usually lends itself to such methods.

One has to remember that at the end of the day total costs are likely to be greater with all these new methods than with traditional methods, and it is probably a mistake to employ project management methods on moderately-sized straightforward jobs which are in no particular hurry.

Finally, it is of the essence of professional project management, either at construction level or at total scheme level, that the management organisation must not carry out any of the other professional or construction tasks for which it is responsible to the client. A management contracting firm should not do any of the building work (and there are some very grey areas at present involving works contractors which are group subsidiaries of the management contractor) nor should a quantity surveying firm that is acting as project manager do the cost planning themselves. Similarly the project manager should not also attempt to act as construction manager – the roles are quite different – and it is perhaps for this reason that the role of scheme project manager has not often proved attractive to construction firms, in the UK at any rate.

The usual cost reimbursement advantages and disadvantages apply to all managed projects, although the management team will almost certainly let out most of the work to sub-traders and specialists on a 'price-in-advance' basis. As mentioned previously, where the construction management division of a building contracting firm is acting as the management team there is often not much difference between this and the 'cost plus fixed fee' contract (Type 2), although it is certainly more usual for a management contractor to include the cost of his site staff and facilities in the fixed fee than is the case with Type 2 contracts.

Summing-up

There are a variety of client requirements, and many contractual methods of satisfying them. The method of contracting for a particular project or group of projects should be chosen with the individual client's needs in mind, not just on the 'the usual basis'; it is, for instance, pointless to prepare a bill of quantities for a client who is concerned not with money but with convenience. We must also remember that the client may not really need a building at all – perhaps he should be changing his distribution methods instead of building a new warehouse. It may therefore be important that he should not go initially to somebody who has a vested interest in putting up buildings.

Chapter 8
Methods of Building Design

The building design process is a complex interaction of skills, judgement, knowledge, information, and time which has as its objective the satisfaction of the client's demands for shelter, within the overall needs of society. Ultimate satisfaction is obtained when the 'best' solution has been discovered within the constraints imposed by such factors as statutory obligations, technical feasibility, environmental standards, site conditions, and cost. (Optimisation is developed in more detail in Chapter 21.) Problems arise, however, in establishing which of the alternative options available is the 'best', as some factors such as personal comfort or visual delight are difficult to measure. Neither is it easy to translate all attributes into a common unit or value, for example excessive noise levels compared with higher initial cost. In practice, compromises in the client's demands are nearly always necessary to get within the constraints. The role of the design economist is to provide information with regard to initial and future costs so that the design team can make decisions knowing the cost implications of those decisions. It is not usually the economist's responsibility to actually provide 'value' as this must be the province of the team as a whole, of which, however, he should be a contributing member. In theory the team will pool their combined knowledge for the benefit of the client, whose representative should wherever possible be a member of the team and play his own part in the corporate decision-making process. As with any group activity the composition of the team should be carefully planned to avoid vociferous members unduly influencing the final decisions.

If the design economist is to be an effective member of the team he must have a clear idea of when the major decisions of cost significance are to be made so that he can provide information at the crucial time. In addition, he must have the techniques, knowledge, and experience to enable him to provide answers to questions of cost that will be posed as the design is refined.

It is, therefore, essential that he understands the manner in which the design team, and in particular the architect, thinks and operates – commonly referred to as the design method. Unfortunately each individual designer adopts a different approach, as there are many ways of solving the same problem. However, it is possible to identify some of the common techniques which are very often incorporated in a typical approach to the task of achieving good design. This knowledge can be used to select the form of cost advice most appropriate to a particular problem.

It is not possible to place all the responsibility for establishing a successful cost solution solely on the shoulders of the building economist, or quantity surveyor. The architect must of course cooperate in, and contribute to, the cost planning process. Cost being one of the several constraints that he will face in varying degrees, depending on the type of project and the client's financial resources. In this respect he should have a reasonable grasp of the factors affecting the cost of the project and the options that are available to him within that constraint. This knowledge will very often be gained by experience, e.g. especially if his firm specialises in one particular type of contract, e.g. housing, factories, hospitals, etc. Even without this experience he should be aware of the very basic design and cost relationships shown below.

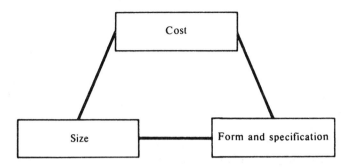

The advantage of illustrating this triangular set of relationships is that any two of the factors can be seen as function of the remaining one. For example, if the size of the building is fixed, together with the form and specification, then a certain cost will be generated. Conversely, if the cost of the building is established together with its size (as is the case with some government yardsticks), then this constrains the form and specification that can be chosen. On the other hand if the shape and quality standard of the specification is declared together with an established sum of money, then the amount of accommodation is the design variable which is limited. Since one factor must be the resultant, it is never possible to declare all three in an initial brief.

This is an over-simplistic view of the cost system but it is a starting point in the understanding of the complex relationships which exist between design and cost. It is the skill of the design team in achieving the right balance between these factors that makes a project a success or a failure.

Recognition for this team approach to design is included in what has come to be known as the RIBA Plan of Work. This is a model procedure dealing with some basic steps in decision making for a medium sized project, and is included in the RIBA Handbook of Architectural Practice and Management (Volume 2). The responsibilities of each member of the design team at each stage of the design process are identified. A design-tender-construct procedure is envisaged and from the quantity surveyor's point of view it anticipates and incorporates some of the cost planning techniques described in later chapters. The pre-tender procedure includes the following stages:

Stage A	Inception	Appointment of design team and general approach defined.
Stage B	Feasibility	Testing to see whether client's requirements can be met in terms of planning, accommodation, costs, etc.
Stage C	Outline proposals	General approach identified together with critical dimensions, main space locations and uses.
Stage D	Scheme design	Basic form determined and cost plan (budget) determined.

Stage E	Detail design	Design developed to the point where detailing is complete and the building 'works'.
		Cost checking against budget commences.
Stage F	Production information	Working drawings prepared for the contract documents.
Stage G	Bills of quantities	

It is of course recognised that a certain amount of overlap is bound to occur in practice. In recent years the RIBA Plan of Work has come under fire from a number of quarters because of the inherent inflexibility of its procedures, and because it tends to delay the letting of the contract which in times of rapid inflation can severely penalise some types of client. However, it was never intended as a rigid set of rules and the procedures adopted for any contract should be those which best suit the prevailing economic climate and the particular interests of the sponsor. The plan has been included at this point in order to give an over-view of the complete design process, but the emphasis in the rest of the chapter will be on the very early design period, prior to sketch design, when the critical decisions affecting cost are usually taken.

Comparison of design and scientific method

To understand the concept of design it is useful to compare design method with the traditional approach of scientific method as shown in Fig. 8.1. To establish a scientific law an observation is made in nature and an inference drawn, e.g. light passing through a prism breaks down into several colours. A hypothesis is set up (e.g. white light is a combination of several colours) and tests are applied to establish whether the hypothesis is true. If the results of the tests conclude that the hypothesis and original inference are correct then a scientific law can be established based on the hypothesis. If not, a new solution must be set up and tested until the tests corroborate the hypothesis. Design can be said to follow a similar pattern. The brief is observed and some inference obtained. A hypothesis is set up in the form of a model (e.g. a drawing) and this is tested by evaluation to see whether it 'works'. A check is made to see whether it complies with the interpretation of the brief, and if it does then the design is accepted. There are, however, some very important differences. The design team cannot 'loop' round the system producing new hypotheses (i.e. designs) ad infinitum. They work within the constraints of time and cost, and they must produce a final solution within the design period set by their client and within the fee structure by which they are paid. In addition, much of their creative work can only be evaluated in terms of social responses, such as aesthetic appeal, which at the present time do not have a satisfactory quantitative measure by which they can be tested.

Consequently, their objective tends to change from 'what is the best solution for our client' to 'how can we produce a design which satisfies as many of the demands of the brief in the time available'. To arrive at a satisfactory solution, some kind of strategy, very often incorporating 'rules of thumb' learned from previous experience, is used to narrow down on a particular range of alternatives. The responsibility of the design economist in this search for a satisfactory solution should be to indicate to the team where it should look in terms of form, quality, spatial standards, etc., to solve the cost

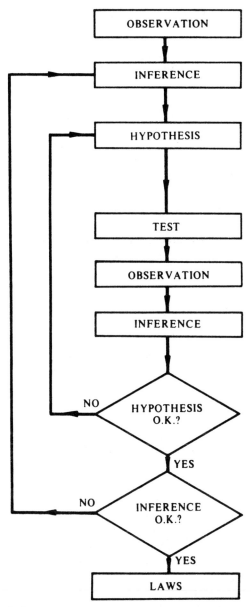

Fig. 8.1. Scientific method.

problem. This will avoid wasting time on abortive designs which will not meet the cost criteria, and will thereby increase the chance of arriving at the 'best' solution in the time available. In other words he should contribute to the overall design strategy.

A conceptual design model

A good deal of research work has been undertaken to establish a general pattern for design. The Building Performance Research Unit at Strathclyde University has put forward the following view, which is shared by several other writers.

They suggest that design consists of three stages as follows:

1 *Analysis.* Where the problem is researched in order to obtain an understanding of what is required.

2 *Synthesis.* Where the information obtained in analysis is used to converge on a solution at the level being investigated.

3 *Appraisal.* Where the solution is represented in some form which is then measured and evaluated.

Quantity surveyors have traditionally been involved almost exclusively in the latter part of the 'Appraisal' process, and yet the real decision-making role is in the *Analysis* and *Synthesis* stages. New techniques are therefore required to enable the design economist to contribute to understanding and solving the design problem. It is assumed by the Strathclyde team that decision-making becomes more detailed and refined as time goes on (from say, concepts of building form and spatial arrangement through to the eventual choice of ironmongery) and that at each stage different methods of problem-solving are adopted. Any cost technique must therefore follow a similar pattern, which may involve 'coarse' measurement and evaluation in the very early stages and a more reliable measurement and cost application when more information becomes available. It is with these thoughts in mind that cost models have been constructed which can input information at each level of refinement. These are still being developed and are dealt with in later chapters.

One of the criticisms that can be made of traditional cost planning methods is that they do very little to contribute to the pre-sketch design dialogue, where all the major decisions of form and quality tend to be taken. Current research suggests that there is a heavy commitment of cost prior to a sketch design being formalised. This may amount

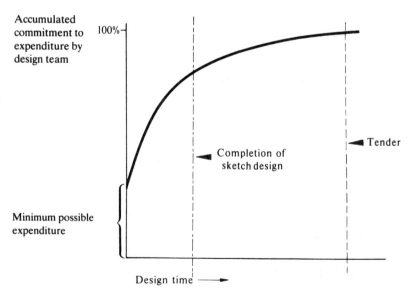

Fig. 8.2.

to over 80% of the final potential building cost, leaving perhaps only 20% to actually 'control'. Figure 8.2 illustrates the point in a hypothetical diagrammatic form. With an improved understanding of design method it may be possible to input information prior to sketch design which will reduce the number of abortive design solutions needing to be produced.

Design techniques

It is not possible to describe the vast range of techniques applicable to design in a chapter or book of this nature. Those interested in developing their understanding in this area of study should consult other publications, such as Design Methods by Christopher Jones or Design in Architecture by Geoffrey Broadbent. It is possible to identify some decisions which form the basis of most design approaches. In very simple terms these can be illustrated by some general questions relating to matters of principle affecting the design. These include the following:

1 'What are the constraints within which I have to work?'

The constraints on a project can be of three kinds, those imposed by physical factors, those imposed by external bodies and institutions, and lastly those imposed by the client and his advisers.

The physical factors very often relate to the site, and such things as the position of boundaries and easements, the method of access, the nearness to service supplies, any visual aspects and views and of course soil conditions and adjacent structures. In addition to the site, the physical performance of materials acts as a limitation and may exclude the use of a particular specification. These constraints are usually fixed and therefore provide very clear boundaries to the design problem.

Constraints imposed by external bodies, however, can often be the subject of negotiation. Planning requirements are usually the result of the policy of the local planning committee and are often open to interpretation – hence the facility for appeal when these interpretations conflict. Some building regulations can be waived if the appropriate authority can be convinced that an alternative construction form or arrangement is satisfactory. Resolving these problems can be a time-consuming business, and until they are settled the solution to the design problem has to remain flexible. However, once they are defined they again provide a clear boundary to the design problem.

Constraints imposed by the client and/or the design team tend to be far less stringent than the previous two categories, and can be compromised more easily. Even here some may be inflexible due to a specific demand which takes precedence over all other needs. An example may be the requirement to design a form which envelopes an expensive manufacturing process; because the plant is so expensive the form of the building must take second place. Another example is a cost limit which is subject to the financial standing of the client; this may however be imposed by an outside body such as a government department or finance company who will define the cost that it sees as being realistic. Most self-imposed constraints are able to be reconsidered in the light of experience and gradual evolution of the design solution.

Definition of the constraints is of enormous assistance in containing the design solution. They help in narrowing down the range of possible solutions, which are for practical purposes almost infinite without them. It is the responsibility of the design economist to account for these controlling factors in his budget and, therefore, set a realistic strategy of cost. Ignorance of these issues will possibly result in abortive effort and a less than satisfactory service to the client.

2 'What are the priorities of the scheme?'

In a sense this is the question that sets the self-imposed constraints and whose solution should provide value for money. If priorities can be ranked and given their due importance in solving the design problem then it should be possible to spend the client's money in accordance with these requirements. this would be the ideal design and would provide the optimum solution. Unfortunately 'ranking all the priorities' is an extremely difficult task. For example, should maintenance-free windows take priority over an improved reduction in noise levels between rooms in a commercial office block? It is difficult to compare the two, and it is even more difficult to award a satisfactory weighting. However, this kind of decision is at the root of good cost budgeting and, if possible, it should be made in conjunction with the designer. The theoretical aim should be to spend money in the same order and importance as the priorities.

Whether this can ever be achieved is a debatable point, as an item which is not given a high priority by the client may yet be essential, for example easy-to-clean windows on a multi-storey block. To provide this item may be more expensive than providing an item of greater ranked importance. This does not mean that the higher ranked item is of less value to the client but just that because of external economic forces and the nature of buildings he has to pay more to achieve his objective. This emphasises the difference between the two concepts of 'value' put forward by the economist Adam Smith; value in exchange and value in use. In most buildings of a public or social nature it is value in use which is being considered, whereas in a speculative housing development it is value in exchange (which is much more dependent on scarcity factors) that is the major concern. There is, of course, a link between the two because the greater the degree of user satisfaction, the more likely it is that the client will be prepared to pay more for his building. Value for money is achieved when the priorities of the client, e.g., profit, symbolism, welfare, worship, etc have been successfully balanced with the initial and future costs allowable for the development.

3 'How much space is required?'

The purpose of building is usually to provide space for a particular activity within which the climate is modified so that the activity may function more efficiently. Other considerations such as the environmental and social impact of the building and its activity arise out of this prime need. It is therefore usual for the client's brief to give an indication of the usable area required, but even where this is stated there is considerable flexibility allowed; for example the circulation area (corridors, waiting, lifts, public areas, etc.) is not usually included in the list. In addition, the multiple use of space (e.g. assembly halls doubling up as dining halls in schools) may allow a more efficient use of a certain area and reduce the overall need. Part of the design problem is to discover the most efficient use of the spaces required to satisfy the client's brief and to arrange the areas in such a way that circulation is kept to a minimum.

As there is a strong correlation between the area of a building and its cost, the design economist should be an active participant in the discussion of areas and their spatial arrangement. Statutory requirements with regard to means of escape in cases of fire, healthy, and hygiene, etc. in conjunction with the client's requirements for efficient movement in the building will provide an indication of the minimum areas allowable for circulation and ancillary purposes. A careful study of these needs will provide the constraints for area and thereby indicate the balance between quantity and quality of building that can be achieved within a given cost limit.

4 'What arrangement of space is required'

This question is heavily influenced by the amount of space needed. In answering it, a good designer can exercise his knowledge to enormous advantage. Indeed determining spatial organisation has been recognised as one of the most fundamental of the architect's range of skills.

Despite the advent of computer programming techniques which attempt to optimise the positioning of space, the ability of the architect to take into account a large number of factors and bring them into a suitable relationship has not yet been surpassed.

A number of techniques have been developed to assist him in this important task, and perhaps the most common is the association matrix. In this design aid the relationship between spaces is identified in a table rather like a 'mileage between towns' indicator in a road atlas. The figures in the table however will relate to the 'cost of communication or movement' related to a unit of distance between spaces rather than measured distance. For example, the salary cost of the managing director spending time in walking one metre may be five times that of his administrative clerk. It follows that his office should be given priority in being closer to the centre of communication. Problems arise in the technique when the subject has no direct wages cost, for example the casualty patient in a hospital who does not cost the hospital administration any salary. Without an artificial weighting factor, hospitals might be designed with the patient always taking the longest route!

The simplest form of the matrix is that in which a simple weighting system is used to define the ease of movement between spaces. To illustrate the use of such a table let us take the example of a new administration/accommodation wing added to the existing appliance garage of a fire station. The spaces required by the fire authority are as follows:

Blankstown: extention to existing fire station

Space	Area (m²)
(a) Administration area including watchroom	120
(b) Lecture room	80
(c) Firemen's dormitory	140
(d) Television room	25

(e) Recreation area	150
(f) Mess and officers' lounge	200
(g) Administration/visitors' toilet and wash room (male and female)	25
(h) Visitors' entrance	20
(i) Firemen's entrance	20
(j) Firemen's ablutions area	40
(k) Appliance garage (existing)	—
	820

The first step is to identify a suitable scale for the ease of movement required between spaces. The following may be considered reasonable:

Weighting	Degree of 'ease of movement'
5	Essential
4	Desirable
3	Tolerable
2	Undesirable
1	Intolerable

By considering the relationship of each space to the other spaces an association matrix can be established using the above key.

It can be seen from the developed matrix (Fig. 8.3) that the crucial space is the existing appliance garage. This is fairly obvious as the efficiency of the station is centred around the ability of each fireman to reach his equipment quickly when an emergency call is received. The next step is to arrange the spaces in a diagrammatic form to show the desirable 'clusters' of accommodation. This is usually shown by means of a 'bubble' diagram identifying the spaces and their required links. In arranging the groups of spaces, consideration needs to be given to important constraints, such as the site, which may not allow ideal groupings to be made. For example the parking and practice area required behind the fire station may restrict the usable site area and force a two-storey solution. The resultant bubble diagram incorporating the site constraint and association table may be as indicated in Fig. 8.4.

The bubble diagrams reinforce the relationship between spaces and emphasise the view that the appliance room is the controlling factor in any arrangement. In addition, the size of the spaces is illustrated, and a visual indication of the likely room clusters is given ready for incorporation into a suitable building form. The efficiency of spatial arrangements in building is not a field that has involved the cost adviser to any great extent in the past, apart from commenting on the ratio of circulation space to usable floor area. In buildings in which ease of movement between spaces is a priority, such as those housing industrial processes, this area of study may prove fruitful for cost research and investigation as the layout has a major effect on the overall economy of the scheme.

5 'What form should the building take?'

Answering this question involves another of the essential design skills of the architect. He should be able to translate the functional spatial arrangement of the bubble diagram into a building form that will reflect the relationships determined. In doing so he will

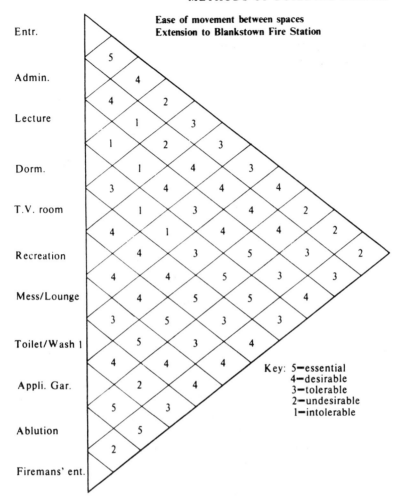

Fig. 8.3. Design method association matrix.

also be aware of the constraints which usually reduce his options considerably, the largest constraint in many cases being the site itself. In others, planning requirements or cost limits may be paramount.

In the case of the Blankstown Fire Station the available site area is considerably reduced by the need for practice yard at the rear. The orientation is also fairly limited because of the desirability of having a frontage on to the High Street. The resultant plan is an attempt to segregate administrative areas from recreation/work areas and at the same time provide speed of access to the appliances from any part of the building or surrounding areas.

Figure 8.5 illustrates the first sketch of a design that may be suitable. Note that from force of circumstances the architect has had to use open space in rooms for through access to the appliance garage. He has also split the visitors/administration toilet facilities between ground and first floor to obtain male and female facilities. In addition

Extensions to Blankstown Fire Station
Note: Numbers of interconnecting lines shows strength of association

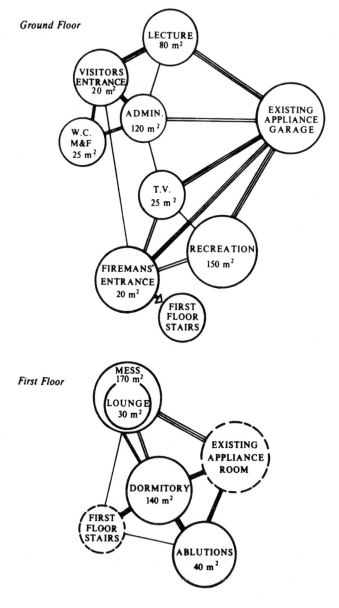

Fig. 8.4. Bubble diagram.

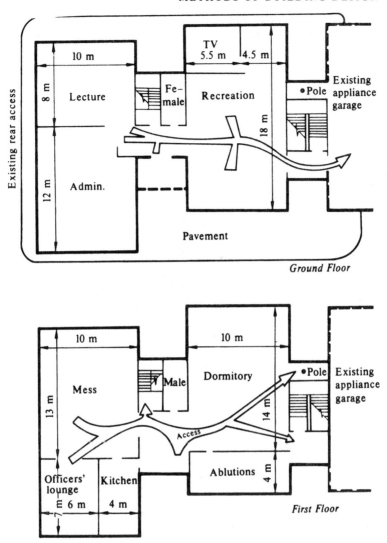

Fig. 8.5. Sketch plans.

he has attempted to isolate noisy areas from the quiet/rest rooms and by setting the recreation block back he has provided passing motorists with a better view of the garage doors – very useful in an emergency.

It is interesting to note that it is at this stage that the cost adviser is usually asked for an estimate, yet it is clear that there is already a heavy commitment to space, circulation, and form standards. In addition, the specification standard will probably have been indicated in the brief and therefore the degree of cost control that can be exercised from this point on is severely limited. This reinforces the view that cost advice is most effective if given prior to the sketch design. In larger, more complex buildings a request to design a more compact form once the architect has already produced a solution is likely to involve considerable effort and also resentment. One of the aims of good cost budgeting must be to avoid abortive effort and therefore to advise where cost-acceptable solutions are likely to be found.

6 'What is the level of specification?'

This is very often the decision that has the least external constraint. It is also the design decision that suffers most when cost reduction is required at a late stage of the design process.

At the present time there are very few numerical techniques that attempt to measure quality. It is therefore a decision that is based largely on intuitive judgement arising out of the previous experience of the design team. The client's brief will have attempted to give an indication of need, perhaps in the form of a requirement for soft or hard finish in certain rooms, maintenance-free exterior, and so on.

Standards of environmental comfort and the need for prestigious public areas, etc., will also have been conveyed by the client. In these circumstances the designer has to narrow his selection to those materials that will fulfil the function required and to make a selection based on comparative performance. One aspect of that performance will be the economic considerations of installation and durability that will generate capital, maintenance, and operational costs. One role of the cost adviser should be the production of alternative estimates for each major solution, in order that the cost implications of a decision can be known.

Generally

The above series of questions is a gross over-simplification of the design problem and the text has been orientated towards those factors which have cost implications. The object has been to give the student, in disciplines other than architecture, a grasp of the early stages of the design problem and the potential for economic advice, and in particular to dispel the common misapprehension that the architect starts work by sketching a building form neatly divided up into rectangular rooms and passages.

However, it should not be assumed that a linear chronological order has necessarily been implied by the sequence of the above questions. In some cases a linear sequence may be the method adopted for the solving of a particular problem, but in the vast majority of cases there will be a good deal of overlap. There may be simultaneous decisions covering all the points discussed and possibly reversal of the process. The above procedure has been largely a process of designing from the 'inside-out'; from internal space requirements to external form. In buildings where external form is important or the site constraints are very tight then it may be necessary to design from the 'outside-in'. Each individual designer will develop his own methods and techniques which will suit his approach – not every architect, for example, will use numerical methods to assist in deciding on circulation priorities.

There are, of course, a large number of design decisions still to be made and these include the working and arrangement of the building services, the sizing and choice of the structure, the fixing of components, and the detailing of the specification and so on. Some of these decisions will be made in conjunction with those already outlined and in fact may depend upon those decisions and vice versa. Design cannot be neatly contained in water-tight compartments of sequential decisions, and this does make the input of information by other consultants more difficult.

An understanding of design method will, however, help the cost adviser to know when his advice is likely to be most helpful in its contribution to solving the design problem. It will also assist in determining the degree of refinement in cost advice that will match the stage of detail in the design process.

Recognition of design methods in cost information systems

It is generally recognised that any cost information system must be compatible with design method. However, unless cost is of paramount importance it should not dictate the approach the designer takes – if the financial tail wags the project dog this will not normally be conducive to satisfaction of all the client's requirements. The cost adviser should also be aware that striving to achieve optimisation in cost is only a part of the total objective. While all consultants would like to achieve optimisation in their own subsystem of activity it is the performance of all systems acting together that will determine the degree of client satisfaction.

Two examples will suffice to demonstrate recent attempts that have been made to marry cost advice to design method in the pre-sketch design stage, when the major cost-significant decisions are made. The systems are not fully developed but they give an indication of current trends.

Since these are the first of many cost control systems at different levels which we shall be encountering it is necessary to make a point which will be repeated on appropriate occasions. Systems such as the following represent overall strategies within which a cost control process operates. It is very rare, however, for the cost adviser to have sufficient resources available to be able to carry out every single step of the strategy in respect of every aspect of the design, and it is doubtful that this would be a very economical thing to do even if he had. Just as a general commanding an army cannot afford to attack all along a front but must concentrate his efforts at the most sensitive points, so the cost adviser must use his resources where they will do most good. Making a wise decision on this question is probably one of the most important and difficult parts of his job, and is one of the things which makes it a truly professional task.

1 An evaluative system

In this system the cost adviser attempts to evaluate the cost implications of design knowledge and decisions immediately they are postulated. Design is considered to be the arrangement of spaces into a building form. The procedure could be as follows:

1 Designer receives brief with room sizes given	Cost adviser evaluates cost of alternative finishes and room contents including service outlets.
2 Designer organises rooms into groups and location in a building shape and form	The form generates a type of production method and site transportation which the cost adviser evaluates.
3 Designer looks at the structure needed to support the chosen form	Cost adviser evaluates cost of alternative forms of support.

4 Designer clads the structure	Cost adviser evaluates alternative specifications of external walls and roof.
5 Designer organises circulation and service runs etc.	Cost adviser evaluates alternative arrangements of lifts, service runs, and corridor space.

There are two problems with this method. First, it assumes a 'linear' design method, with each event occurring after the other, although in practice these decisions are very often taken simultaneously'; second, it depends on the generation of information by the designer before evaluation can take place, and therefore gives no indication as to where to look for a good cost solution prior to a committed decision being made on building form. It is, however, a significant step forward in providing an evaluation of the decision-making process prior to sketch design.

2 A strategic cost information system

By use of this method the cost adviser attempts to explore a range of possible solutions available to the designer using a structured 'search' process in which the designer is involved. The object is to identify a cost 'strategy' which the designer can employ in his synthesis of building form and which will avoid abortive re-design at a later date. The procedure may take the following form:

1 Design team receives brief with accommodation area and quality standard given	Cost adviser explores the cost of different components of the building according to changes in the major design variables (e.g. the area and shape of a bay in a structural frame).
2 Design team selects the specification and parameters for each component from the explorations undertaken	Cost adviser explores the use of these components and parameters in buildings of different shape and height according to a predetermined series of building descriptions, e.g. plan shape, number of storeys, density of partitions, etc. A cost table is produced of the feasible solutions available within the indentifiable constraints.
3 Design team identifies the lowest total cost from the table and uses it as the point of reference for selecting any other alternative.	Cost adviser contributes to the discussion of an alternative solution.
4 A solution is chosen from the table by the design team	Cost adviser provides a breakdown of the major component costs which is then adopted as the point of reference for the initial cost plan for use in traditional budgeting. The design variables incorporated in the selected solution are communicated to the designer.

5 Designer uses the parameters and description as guidelines in the preparation of his design solution

Cost adviser uses traditional budgetary procedures to maintain control.

The descriptions conveyed to the architect for his design strategy should not be so rigid that a 'strait-jacket' is imposed resulting in only one possible solution. Rather they are 'coarse' measures which still allow the designer reasonable scope for using his skills of modelling and spatial organisation within the strategy adopted by the team as a whole. The choice of descriptors is important in making sure that the creativity of the designer is not stifled.

The advantage of this method is that it attempts to undertake a systematic search through possible solutions in order that the design team may know where the better cost propositions are likely to be. It identifies a 'least cost' solution which can then form the point of reference for any alternative selection. It is, therefore, possible to obtain a gauge of value by comparing a specific choice with the lowest cost. The approach is developed further in Chapter 21.

There are however quite a number of problems. The setting up of 'models' that will cope with the large number of components and design variables is an onerous task, even though the models need not be very refined at this stage of the design process. It is also extremely difficult to produce a satisfactory range of descriptors which will be simple enough for the design team to incorporate into their thinking and also comprehensive enough to allow a reasonable evaluation of the building as a whole.

Both the evaluative system and the strategic approach depend heavily on the use of computers to provide the information sufficiently quickly for decision-making purposes. The advent of micro-computer technology has meant that machines are now available, at less than the cost of one man-year, that will cope with these types of problems. The real expense is in writing the program to suit a particular type of building and it may mean that quantity surveyors of the future will need greater familiarity with computer techniques if they are to provide this improved service to the building client. Without the ability to re-use computer data and systems most cost advice is likely to be too expensive to be provided on any but the largest projects.

Summary

An understanding of design method enables the cost adviser to know:
1 The time when his advice will be most effective.
2 The type of advice that needs to be given.
3 The objectives of the design team additional to minimisation of cost.

It would appear that the earlier the advice is given the greater the chance of the advice being incorporated into the final design solution. The need is for techniques which contribute to the analytical understanding of the problem and assist in the convergence on the best solution in the shortest possible time. The development of cost models and computer techniques seems to show a possible way forward in achieving these objectives.

Chapter 9
Introduction to Cost Modelling

In the majority of cases in which an industrialist is about to manufacture a new product he first of all builds a prototype. He does this for a number of reasons, such as:

1 To identify and solve three-dimensional problems that were not apparent in the drawings.
2 To identify the tools required for production.
3 To help estimate the cost of production.
4 To evaluate its functional performance (this may involve testing to destruction).
5 To test the product's marketability.
6 To provide a sample for a customer as evidence of the quality standard to be achieved.

By building, manipulating, and testing the prototype he can iron out and avoid future problems when the product is actually manufactured, supplied, or sold. However, where the final product is large or costly it is not always feasible to build a prototype, especially with a product such as a building where only one production model is to be built and the prototype costs cannot be written-off over a large production run. Most new structures are 'one-off' designs of considerable capital and operational cost and are intended to be very durable, so that they would in any case need to be tested over a very long period of time to establish their efficiency. It would, therefore, be impracticable as well as uneconomic to build a prototype to test its functional and cost performance as a whole. It may be possible to build samples of major components (and this possibility may need to be taken into account in cost planning a building which is to incorporate some measure of innovation), but because of the time needed for construction, the great expense and the individual nature of buildings, it is just not realistic to construct the whole project a number of times to see whether or not it will work satisfactorily.

It might be interesting at this point to mention the Sydney Opera House, a building of great prestige and enormous innovation, where the construction and testing of prototypes and mock-ups was undertaken on an exceptional scale. The result of this was that very few problems were encountered in the use and maintenance of the building, considering the potential for trouble in a project of this nature, but of course a considerable penalty was paid in terms of construction time and cost.

If physical prototypes are not normally possible the design of a building must still be assessed in some way to try and ensure that the demands of the client are satisfied.

For the sake of expediency and cost it is, therefore, necessary to construct models that represent the real situation in another form, or to a smaller scale, so that a realistic appraisal of performance can be made. These models can be, for example, physical (as in a three-dimensional architect's balsa model); mathematical (as in a heat loss equation); or statistical (where some collected information indicates a certain trend). Cost is one of the measures of function and performance of a building and should therefore be capable of being 'modelled' in order that a design can be evaluated. In recent years a considerable effort has been made to construct models that will help in the understanding and prediction of the cost effect of changing design variables.

Cost modelling may be defined as the symbolic representation of a system, expressing the content of that system in terms of the factors which influence its cost. In other words the model attempts to represent the significant cost items of a cash flow, building or component in a form which will allow analysis and prediction of cost to be undertaken according to changes in such factors as the design variables, construction methods, timing of events, etc. The idea is to simulate a current or future situation in such a way that the solutions posed in the simulation will generate results which may be analysed and used in the decision-making process of design. In terms of quantity surveying practice this usually means estimating the cost of a building design at an early stage to establish its feasibility.

Traditional cost models

If the above definition is understood it becomes clear that quantity surveyors have been using a form of modelling technique for a number of years. In their measurement for bills of quantities they have been representing the building in a form suitable for the contractor's estimator; and when prices are applied to the measured quantities the bill becomes a representation (or model) of the cost of the building. By altering the quantity of the measured items or changing the price according to variations in specification it would be possible to evaluate the effect on cost of manipulating certain design variables.

The bill is prepared at a very late stage of the design process however, and any information obtained from changing the quantities of price rates would come too late to avoid abortive design effort. An estimate obtained from a bill would also be too late to give the client an indication of his likely cost commitment at the outset of the project or allow any cost control to be exercised. There is, therefore, an obvious need to use much simpler models at an earlier stage of design to overcome these problems. The complexity of the model will depend very largely on the amount of information the designer can give the cost adviser. It is not much use trying to take into account shape and layout of the building if all that has been determined is an idea of the approximate area of accommodation. Conversely, there is little point using an over-simplified costing technique when the building form and specification is known and sufficient time is available to do a more thorough job.

As in many instances in this world the choice of technique is a case of picking the right horse for the course. Figure 9.1 shows some of the more traditional models that have been developed over the years to suit various stages of the design process. The pyramid is an attempt to show that more detail is required in the structure of the model as we descend the list.

The first two model types are basically a 'single price-rate' methods of estimating cost as introduced in Chapter 1; the size of the building is measured in one form or other and the result is multiplied by a single price-rate to give the estimated cost. However, it will be remembered that such an estimate suffered from the defect that it was not

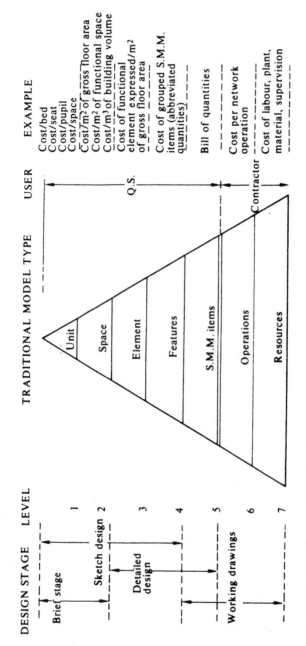

Fig. 9.1. Traditional cost modelling.

possible to relate the cost of the work shown on the drawings to the estimate during the subsequent progress of the design. Attempts to deal with this problem at a later stage by a model based on 'approximate quantities' (at level 4 of the pyramid) had other disadvantages: the quantity surveyor required extra time and a substantial additional fee for the work involved, the drawings had to be fairly complete and the whole of the building work had to be dealt with; even if the approximate quantities when priced by the surveyor showed a far higher total than the original estimate the discrepancy could not be traced, and rational measures to remedy the situation were, therefore, difficult. This particular model is therefore not structured in a suitable way to be compatible with the design process nor is it sufficiently sensitive to allow for ease of adjustment.

What is obviously required is a technique whereby the original estimate can be prepared on the basis of a known and economic standard of construction and finish, and whereby this estimate can be constantly checked against the working drawings and any necessary adjustments made to one or the other. By the time the job goes out to tender the probable cost will, therefore, be known within close limits, and during the progress of the work this monitoring of cost can continue. Such a technique is known as design cost planning or sometimes more simply as cost planning. These terms are somewhat misleading because the word 'planning' suggests that it is the preliminary forecasting which is all important, whereas the really vital part of the whole operation is the control of cost as the design develops and the building goes up. A better term therefore might be design-cost-planing-and-control, to remind us of this; it is too cumbersome a term to use in practice but we must remember that this is what we mean when we talk about cost planning. A cost plan without control is simply an 'approximate estimate'.

The model by which design cost planning has been achieved is that of dividing the building into functional 'elements' (level 3 of our pyramid). Elements are defined as major parts of the building which always perform the same function irrespective of their location or specification. For example, internal partitions always vertically divide two internal spaces; a roof encloses the top of a building and keeps the weather out and the heat in; the sub-structure transmits the building load to the subsoil; and so on. these elements have some relation to the design process and are easily understood by all parties including the client, aiding communication. Therefore, the model of the cost of the building based on elements has the advantage that each major part of the design can be estimated separately and used as a basis for checking when that part of the building has just been designed. By setting a budget based on elements, budgetary control procedures can be adopted that will allow monitoring and correction of the design should it start to go outside the cost constraints. A balance in terms of cost can also be achieved between the various parts of the building and this is a first step to providing value for the client.

Level 5 of the model pyramid shows the position of the bill of quantities in terms of the detailed modelling used in traditional techniques. There has been an attempt in recent revisions of the rules for measuring bills of quantities (the Standard Method of Measurement of Building Works) to orientate the measurement of the building to the way costs are incurred on site. However, the vast majority of items are still largely measured 'in-place'; that is to say they are measured as fixed in the building with no allowance for waste and no identification of the plant and tools required to install them.

This is to avoid the possible situation arising where the quantity surveyor tells the contractor 'how to do the job' or makes assumptions as to his efficiency. Consequently, the bill of quantities is still primarily a document for obtaining a contractual estimate in a short space of time, rather than a document for management and cost control of construction on site. It is not possible, for example, to ascertain the lengths of timber required for roof joists (although this must have been known by the measurer), or the hoisting requirements for materials and components, just by looking at a bill, unless it has been annotated for the purpose. However, it is probable that the rules will gradually be changed to more closely resemble the production process.

The last two levels of 6 and 7 are indeed more closely related to construction. If a contractor prepares a 'network', he outlines the activities or operations in the order that they will be undertaken on site. For example, external cavity wall brickwork, which may be one item in a bill, will be broken down into floor levels and possibly into zones marked on the floor plan. This aids site management and the organisation of labour, plant, and material. These basic resources form the most detailed level of modelling as they are the ingredients of the production process. In fact all the other models used for forecasting cost entail the assumption that the resource requirements of the building have been considered, although in most cases the resource assessment is so distant that it is hardly recognisable. In the case of levels 1 to 4 on the pyramid the data used is actually based upon an analysis of the bill of quantities, which is somebody else's view (the estimator's) of what he intends to charge for the resources plus profit and overheads. Because of the large number of assumptions upon which a bill price is based it is most unlikely that the real mix of resources for the project has been determined. The factor that allows each technique to work is the knowledge that the overall cost of the job, which has been analysed to provide data for the model, is the 'going rate' for that building. If adjustment is made for current market conditions affecting price levels it should be possible to build an identical building on the same site for a similar price. This is the theory behind all estimating, but of course there are usually vast numbers of differences between one job and another and this is what makes forecasting very much a matter of professional judgement. The model is there to provide the structure or bare bones on which the cost adviser hangs his intuitive assessment. However, the major part of intuition is based on experience which in turn is based upon the amount of data passing through the advisers's hands. The sources and reliability of this data are the subject of Chapter 10 but it is sufficient to note here that its use requires considerable care. Forecasting is not just a 'number crunching' exercise no matter how refined the model, and although the calculations based on the techniques outlined below are easy, their success depends on the interpretation of analysed price rates. It is in this latter area that the professional adviser earns his money.

Single price estimating models

1 The unit method

This technique is based on the fact that there is usually a close relationship between the cost of a building and the number of functional units it accommodates. By functional units we mean those factors which express the intended use of the building better than any other. For example it may be the number of pupils in a school, number of vehicle spaces in a car park, or number of theatre seats. Several cost yardsticks operated by government departments were based upon this relationship and provided a test of what was reasonable for a given scheme. Cost per person or bed space for housing development and cost per bed for hospital wards are further examples of this test of reasonableness. Assuming that the standard of the building compatible with the unit cost is 'socially acceptable' (see Chapter 10) then this method of making sure that extravagant schemes are not foisted upon the tax or rate-payer can provide a very good yardstick for the purposes of public accountability. After all, by costing a single functional unit instead of, say, floor area, the designer has greater flexibility in choosing between the quantity of building and its quality, and the funding organisation has the assurance that it has not paid beyond what is reasonable.

To take an example of the more general use of the technique, suppose a multi-storey car park for 500 cars had recently been constructed for the sum of £3,000,000. The cost per car space would obviously be:

$$\frac{£3,000,000}{500} = £6,000$$

If a similar car park is to be built but for 400 cars then it might reasonably be assumed that the cost would be:

$$400 \times £6,000 = £2,400,000$$

all other things being equal.

Unfortunately not all things are equal and consequently a number of judgements must be made about prevailing price levels (due to inflation and market conditions) at the proposed date of tender, site difficulties, specification changes, different circulation and access arrangements, etc. Once again as with all single price estimates these adjustments to the analysed figure are based on the cost adviser's judgement and are critical to success.

2 The cube method

This is a traditional method. All quantity surveyor's offices used to keep a 'cube book'. When a contract was signed the amount of the accepted tender was divided by the cubic content of the building and the resultant cost per cubic foot entered in the book. When an estimate was required for a new project the cubic footage would be calculated and an appropriate rate per foot taken from a previous job in the cube book. The method has largely died out as a result of a number of factors, the main one being that the cost of a building is usually more closely related to its floor area than its cubic capacity.

Even with a 'blunt instrument' such as a cube, it was necessary to have common standards of measurement so that the rate obtained from one job might be fairly compared with another. The RIBA pre-metric set of rules was the one commonly accepted; these involved multiplying the plan area measured over the external walls by the height from the top of the concrete foundation to halfway up the roof if pitched, or to 2′0″ (0.60m) above the roof if flat. (See Fig. 9.2.) Extra allowances were made for

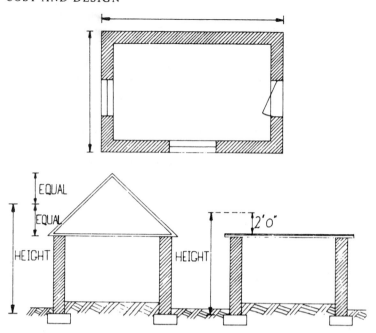

Fig. 9.2. RIBA pre-metric rules for cubing of buildings.

projections such as porches, bays, oriels, turrets and fleches, dormers, chimney stacks, and lantern lights.

This formula has little to recommend it except uniformity; the additional allowance for a pitched roof (which may well be cheaper than a flat roof) and the allowance for fluctuations in foundation depth are both very arbitrary.

The Standard Form of Cost Analysis (published by the Building Cost Information Service of the Royal Institution of Chartered Surveyors) still makes provision for cubing but uses a different set of rules.

3 The superficial area method

This is an alternative single-price-rate method in which the total floor area is calculated. The rule usually accepted is that the area of the building inside the external walls is taken at each floor level and is measured over partitions, stair wells, etc. In cases where columns to frames are situated inside the external walls these columns are ignored and the dimensions taken from the internal face of perimeter walls. Where there is no external wall at floor level (as in the case of freight sheds) the dimensions are taken from the external faces of columns.

More recent than the cube method, it first came into use for such projects as schools and local authority housing where the storey heights were reasonably constant. It has the virtue of being closer to the terms in which the client's requirements are expressed, as the accommodation is more likely to be related to floor area than to cubic displacement. Sometimes when a very early estimate is required the only data available may be the approximate floor area.

4 The storey enclosure method

This was a single-price-rate system which never achieved wide acceptability because it was quickly superseded by the elemental estimate. It is of interest for historical reasons as it was the first attempt to compensate for such factors as the height and shape of buildings.

In this method the areas of the various floors, roof, and containing walls are measured, each then being weighted by a different percentage and the resultant figures totalled to give the number of storey enclosure units. The rules are summarised below:

1 Twice the area of the lowest floor, unless it is below ground level in which case it is multiplied by three. The area is measured between the internal faces of the external walls.
2 The area of the roof measured flat on plan to the extremities of the eaves.
3 Twice the area of the upper floors, plus the addition of a factor of 0.15 for the first floor, 0.30 for the second floor, 0.45 for the third floor and so on.
4 The area of external walls. Any part of an enclosing wall below ground level is multiplied by two.

Note: It was recommended that lifts and other engineering services should be excluded from the calculation and worked out separately. This is common practice in single-price-rate estimating and applies equally to the cubic and superficial methods.

The paper presenting the method (in the early 1950s) claimed that it had performed far better than the other single price approaches but lack of use has meant that it has not been possible to verify this claim in practice. The system depends heavily on the 'weightings' built into the rules, and these are unlikely to apply equally to every building. In addition the units tend to be abstract, not relating to the physical form of the building or the client's accommodation requirements, and therefore the technique suffers from deficiencies similar to those of the cube method.

Comparison of single price methods

Table 9 shows three of the above single price methods applied to a multi-storey office block. The actual cost of each building is known and the analysis of building A is used in each case to forecast building B. The buildings have exactly the same dimensions, except that A has seven storeys and B has eight storeys within the same envelope.

It can be seen that the cubic method has not taken into account the extra floor and there is, therefore, a shortfall of approximately 12% on the estimate. The superficial method on the other hand has over compensated for the extra floor by not taking into account the fact that the envelope remains the same but it has still produced a very reasonable figure. The storey-enclosure method by considering both envelope and floor has produced an even closer estimate well within acceptable limits for this kind of early exercise.

In each case a change in specification, site, or location would result in additional problems, and call for the skill of the cost adviser in using his judgement to compensate for the changes. In addition changes in plan shape and storey height would affect the cubic and superficial methods to a greater or lesser degree. These alterations, or variables, are enormously complex and difficult to assess. It is for this reason that single-price-rate methods tend to be rejected except for the very earliest estimates.

Table 9. Comparison of single price estimating models using a cost analysis of Office Block A to forecast Office Block B.

Dimensions

Office A

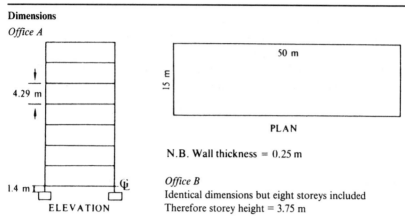

PLAN

N.B. Wall thickness = 0.25 m

ELEVATION

Office B
Identical dimensions but eight storeys included
Therefore storey height = 3.75 m

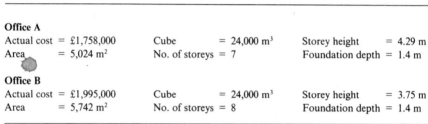

Office A

Actual cost = £1,758,000	Cube = 24,000 m³	Storey height = 4.29 m
Area = 5,024 m²	No. of storeys = 7	Foundation depth = 1.4 m

Office B

Actual cost = £1,995,000	Cube = 24,000 m³	Storey height = 3.75 m
Area = 5,742 m²	No. of storeys = 8	Foundation depth = 1.4 m

Cubic metre

Office A analysis

$$\frac{£1,758,000}{24,000} = £73.25/m^3$$

Office B forecast

£73.25 × 24,000 = £1,758,000

Error

Underestimate of approximately 12%

Square metre

Office A analysis

$$\frac{£1,758,000}{5,024} = £349.92/m^2$$

Office B forecast

£349.92 × 5,742 = £2,009,000

Error

Well within acceptable range

Storey enclosure

Office A analysis

Lowest floor	717.75 × 2 =	1,435.50
First floor	717.75 × 2.15 =	1,543.16
Second floor	717.75 × 2.30 =	1,650.83
Third floor	717.75 × 2.45 =	1,758.49
Fourth floor	717.75 × 2.60 =	1,866.15
Fifth floor	717.75 × 2.75 =	1,973.81
Sixth floor	717.75 × 2.90 =	2,081.48
Roof	750.00 × 1 =	750.00
External Walls	3,900.00 × 1 =	3,900.00
		16,959.42

$$\frac{£1,758,000}{16,960} = £103,65 \text{ SEU}$$

Office B forecast

As above	16,959.42
add	
Extra floor (7th) 717.75 × 3.05	2,189.14
Total	19,148.56

19,149 × £103.65 = £1,984,800

Error

Well within acceptable range

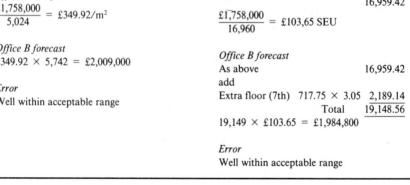

Elemental estimating

Although the storey-enclosure method could have brought increased accuracy it was still no more possible to relate the working drawings (when received) to the estimate than it had been with the earlier methods. The omnibus 'rate' bore no definite relationship to the actual cost of any of the measured areas but was only a theoretical average of them all. In the discussion which followed the first presentation of the paper, James Nisbet (who subsequently became Chief Quantity Surveyor to the War Department) put his finger on the real weakness in the storey enclosure approach to the problem. He asked whether, if the walls, floors, and roofs were all going to be measured separately, it would not be simpler and more accurate to price each separately according to an assumed specification.

This basically is the 'elemental' approach. The building is split up into a number of major components, or elements, which can be easily measured from sketch drawings and which are priced separately; these elemental costs are then added together to give the total of the estimate. In its simplest form where no change of specification or location occurs the method merely involves adjusting for the revised 'quantity' of the element. For example, supposing an external wall element costs £40,000 and the area of the wall is 1,000 m². The cost per square metre of wall (known as the elemental unit rate) is £40.00. If there are 1,500 m² of the same type of wall on the proposed project then the cost will be:

$$\frac{£40,000}{1,000} \times 1,500 = £60,000$$

A cost index is then used to update the prevailing price levels to the proposed tender date and the new estimate for that part of the building has been established.

Having set out a series of estimates, one for each element (known as cost targets- it is then possible to consider each element in turn instead of trying to cope with the whole building. Costs can therefore be watched as design develops.

The initiative was immediately taken by several official departments in pressing for elemental estimating. The Ministry of Education had used elements in checking tenders for schools and in Hertfordshire the County Architect's Department was using a similar system.

In 1955 – largely through the influence of Mr C. M. Nott, Chief Quantity Surveyor of Hertfordshire County Council – the Bedfordshire and Hertfordshire branch of the RICS moved a resolution at the triennial conference of quantity surveyor members of the RICS suggesting that the profession should devise methods of analytically examining the cost of building, and should prepare methods of applying such analyses to the control of costs at design stage.

The resolution was supported by senior surveyors in public service, but, amazing as it may seem today, was only carried by a very narrow majority. Nevertheless, once the lead had been given, the RICS responded quite enthusiastically, setting up a Cost Research Panel and the Building Cost Information Service and encouraging cost awareness among its members.

Among the earliest users of cost planning were the County Architect's Department of Hertfordshire County Council and the Works Directorate of the then War

Department. The pioneer work of these, and others, was so successful that design cost planning became the normal practice in a profession which once was reluctant even to make a start with it.

Spatial costing

In the late 1960s and early 1970s a variation on the superficial and elemental methods was very much in vogue under the title of spatial costing. In this method the basic concept was the cost of a room of a certain type expressed in terms of its floor area. There is a high correlation between the cost of room finishes, fittings, ceilings and walls, and the activity for which the room is designed. If, at the brief stage of design, all that is available is the accommodation schedule (i.e. the list of required rooms) then it should be possible to forecast the cost of these items quite reliably as a function of the areas and, if possible, shapes of each room or space. The other major items, of structural frame, foundations, external envelope, and main services would then be considered separately. The wall costs attaching to a room would be the cost of the finishings plus 50% of the structural partition, similarly with the floor and ceiling. The extra cost of external walls over and above the 50% partition costs, and the extra cost of roof over the ceiling costs, are added on a measured basis as are the other exclusions which have been mentioned.

A major difficulty is dividing up the normal bill of quantities measurement of the building into the room categories for cost analysis purposes. The cost of preparing detailed analysis has been a contributory cause of the lack of data available to the profession for elemental cost planning. The room concept requires more complex analysis and is therefore even more expensive to use. It also lacks the simplicity of the elemental form and the clear communication of what is meant and included. Computer models which symbolise a particular building and can build up rates from the basic resource costs are more likely to provide a better solution in the future. They may by-pass the need for analysing old or 'historic' costs and will be more flexible in use.

Objectives of modelling

Models for estimating and planning costs have evolved gradually. Their adoption by the profession at large has led to the establishment of what might be called 'traditional' techniques. These have been discussed and referred to in this chapter, together with those that have been suggested but have not yet been widely used.

At this stage it is perhaps useful to consider what a good cost model should be attempting to achieve. Traditional and future models can then be tested to see to what degree they comply with this set of requirements.

The broad objectives can be listed as follows:

1 To give confidence to the client with regard to the expected cost of his project, ie economic assurance.

2 To allow the quick development of a representation of building in such a way that its cost can be tested and analysed.

3 To establish a suitable system for advising the designer on cost that is compatible with his own build-up of the design. It should be usable as soon as the designer makes the first decision that can be quantified, and should be

capable of refinement, to deal with the more detailed decisions that follow thereafter.

4 To establish a link between the cost control of design and the manner in which costs are generated and controlled on site. This involves dealing with the cost of resources at as early a stage as possible to aid communication between the design team and the contract manager.

Linked to these objectives will be some guiding principles which can be applied to the way we approach and verify the model. These may be summarised as follows:

1 The degree of refinement of the model should be tailored to suit the stage of design refinement. In other words it should call on as much design information as is available at the point in time when the model will be used. The cost data applied to this information should represent the degree of reliability that can reasonably be expected from an estimate at that stage.

2 The cost data in the model should be capable of ascertainment, and continuous updating and evolution in the light of changes in external market and environmental conditions, without too much time being involved.

3 The representation of the building or component in the model should bear an understandable relationship to design method (e.g. the arrangement and use of space) and if possible the manner in which the costs are incurred (e.g. the production method). Ideally the model should show the relationship between the client's objectives expressed in the brief and the subsequent cost of resources used to achieve these objectives. This is, however, an extremely difficult task as many of these objectives will be of an intangible nature and at the present time it can only remain a guiding principle.

4 The model should cope with constraints imposed on design and be able to test the feasibility of a proposed solution within these constraints in order that definite decisions can be made.

5 The results of the model should allow the designer to incorporate this knowledge into his drawings, specification and quantities, so that they form part of the strategic decision-making process.

Computer developments

Micro-computer techniques now enable the economic modelling of proposed building designs in the manner already discussed to be carried out rapidly and exhaustively. However, the recent development of computer-aided architectural design systems has added a new twist to the situation. The more sophisticated of these systems model the project in 3-D as the architect designs it, producing not only elemental cost models but thermal performance models, day-lighting models, and so on. It is a very simple matter to write a computer program which will produce an outline elemental analysis from three-dimensional coordinates of the various corners of the building mass (most quantity surveying students could do this in a few hours); the difficulty lies in associating this with cost data sophisticated and reliable enough for the architect to be able to use by himself as he designs, or alternatively, to perfect a means by which the quantity surveyor or cost planner can directly interact with him while he works (writing letters or even telephone will not be good enough). The danger is that a model prepared from over-simplified or incorrect cost data will look just as convincing on the screen, and be just as easy (or even easier) to call up by somebody who does not understand the complexities involved in cost forecasting. Architects, cost planners, and computer software houses must come to terms with this.

Chapter 10
Cost Data

Introduction

In the section on traditional cost models (Chapter 9) it was postulated that the actual techniques used were merely the structure around which the professional cost adviser applied his judgement to make the techniques work. The model itself was ill-equipped to produce a reliable answer, and indeed reliability at any stage of refinement was entirely dependent upon the costs applied to the measured quantities in their various forms. If the right cost figure was applied to any of the single-price methods then they would be more reliable than if the wrong figure(s) were applied to the more detailed techniques. The factor that made detailed cost planning techniques more satisfactory was the control that was exercised as design developed. Even this control, however, was based on unit rates applied to abbreviated quantities (another model) and was, therefore, dependent on the reliability of cost information. At the end of the day any estimate was dependent on the prevailing market conditions at the time of tender, and this again required information in order that market trends could be detected.

So at the root of all this forecasting and control activity we find the need for cost data to supplement the numbers, areas, volumes, etc. which have been used to describe the building. It is this data that is critical in determining whether an estimate is reliable or not. One of the major problems in more recent models, in which computer techniques are used, is the need to ensure that the information on cost held within the memory of the computer (i.e. the data base) is reliable and relevant for all conditions of use of that model. Very often human intervention is not envisaged in the computer operation and professional judgement is left until the results have been tabulated. This is very often too late because the adviser has little chance of knowing why a particular figure has been generated without considerable investigation on his part. Also if he is going to adjust a result to what he feels is reasonable, he might as well have used his own judgement in the first place using a traditional model. There is a well-known computer maxim 'Garbage in, Garbage out', which reinforces this view that wrong data used in any technique will produce incorrect results. The study of 'what information should we use' and 'where does it come from' is essential for the correct understanding of the use and development of costing techniques. First we need to understand what it is used for.

The use of cost data

If we were to look at the work being undertaken in a typical quantity surveying office we would find cost information being used for four main purposes:

    1 The control and monitoring of a contract, for which the contractor has already been selected, through interim and final account procedures.

2 The estimation of the future cost of a project and the control of design to ensure that this figure is close to the tender figure.

3 The 'balancing' of costs in a cost plan to ensure that money has been spent in accordance with the client's priorities.

4 The negotiation of rates with a contractor for the purposes of letting a contract quickly.

Arising out of these activities are other uses which might well be the concern of a special section in a quantity surveying practice or research department of a polytechnic or university. These uses relate to the deeper understanding of the external political and economic trends, and the relationship between design decisions and the degree to which they affect cost. These studies usually require a more detailed level of investigation, involving the analysis of large quantities of data or the classification and structuring of the data in a particular way to assist in the development of evaluation models.

To incorporate all uses of data we can probably classify them under four main headings:

1 *Forecasting of cost*

Under this heading would come such information as cost per square metre for various types of building, elemental unit rates, bill rates, and all-in unit rates applied to abbreviated quantities. In the examples listed the data would almost certainly arise from an analysis of past projects, or 'historic costs'. These would be updated by the use of a building cost index, itself a form of cost data, and be projected forward to the proposed tender date by an intuitive or calculated prediction technique. In recent years the problems associated with historic costs (see later in the chapter) have spurred researchers and practitioners to develop techniques which rely on current resource costs (labour, plant, and material) to forecast the cost of a project. This certainly helps in the negotiation of contracts and also in the reliable modelling of the building process. Resources are where the cost is generated, and the adviser is therefore dealing with the origins of cost. However, the problem lies in the acquisition of information on resource costs by anyone who is not a member of a large building organisation, and in the time-consuming nature of the task of synthesising costs from detailed resource inputs. In fact techniques employing resources to forecast cost have various ways of simplifying the problem to avoid the user being overwhelmed by too much data. These usually involve concentrating on the major items and quantities at the expense of those with less cost significance. It should be noted that even these techniques rely heavily on 'historic' information regarding the *normal* time and quantity requirements for the resources needed for a particular item or building.

2 *Comparison of cost*

In this use of data the need is not so much to discover what the building or component will actually cost at the time of tender but to make a comparison between items with similar function, or buildings of different design, to decide which is the better choice. The problems of the tender market are not so critical here unless there is evidence that there will be a change in the cost relationship between the alternatives by the time the decision takes effect. The criterion in choosing data for this task should be that it is structured in such a form that if the design or specification of an item is changed the use of the cost data will reflect the true change in the cost of the commodity.

3 *Balancing of cost*

In determining a budget for the cost control of a building it is necessary to break down the overall cost into smaller units. These smaller units are used not only for cost checking purposes but also to allow a cost strategy to be developed for design. This strategy will attempt to spend money in accordance with the client's requirements, by allocating sums of money to the various major components of the building. Data for this use is usually obtained from past projects, and it may be in the form of actual costs or as a proportion of the total cost of a similar project. At the stage that this information is used the design will not be developed (and may not even have commenced) and therefore the categories used for this breakdown will tend to be broad and probably not exceed forty items for any one project. The data can therefore be described as 'coarse' as opposed to refined.

4 *Analysis of cost trends*

Of paramount importance in any prediction technique which seeks to project costs into the future is the information that tells us what is happening to costs in the industry over a period of time. By looking at the way in which costs for different items are changing in relation to one another, or changing between one point in time and another, it is possible to have a better chance of selecting the specification which will suit the client's financial requirements over the short and long term. It will also allow us to obtain a more reliable prediction of what the market price to the client will be when he eventually goes out to tender. These trends may be shown as the change in the cost of materials and labour, or they may be a detected change in the total cost of a particular type of building or component. When data is used for the detection of cost trends the cost is very often related to a base year cost, in which case the presentation of the information is in the form of an index. It should be clearly noted that the detection of a cost trend does not necessarily imply that it will continue into the future. Indeed it is very unwise to extrapolate a cost movement without taking into account all the political, social, and economic factors that contribute to a change in cost levels. The number of variables involved in such an evaluation is enormous, and consequently the establishment of what future costs will be is nearly always left to the professional experience of the cost adviser. He will probably take account of economic reports, but even these differ widely in their predictions. It has been said that where you get two economists discussing a particular economic problem you are likely to get three different opinions!

If these are the major uses of cost data, we can now look at the problems in retrieving and storing the information for these particular applications.

How reliable is cost data?

Nearly all data used in cost planning techniques has been processed in some way, very often with a corresponding loss in accuracy and certainly with some loss of context. The published information received by the quantity surveyor appears to relate to a 'typical'

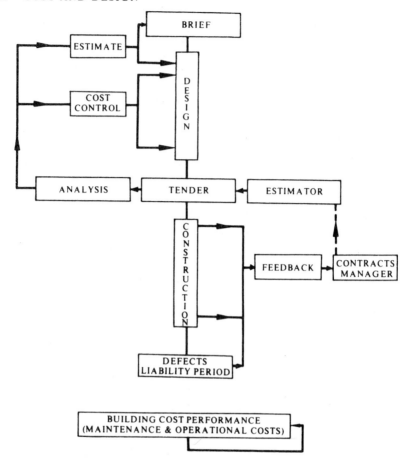

Fig. 10.1. Traditional cost retrieval.

building in a 'typical' location with only a brief summary of the contributing factors to such a cost (i.e. the explanatory variables) included. The feedback of such cost information centres in the main around the bill of quantities.

Figure 10.1 shows a diagrammatic representation of traditional cost retrieval and planning practice.

In theory, information on wages, materials, and plant is collected on site and fed to the contracts manager; he then passes this raw data on to the estimator who uses it together with data from other projects to forecast his next tender price. The consultant quantity surveyor then analyses the contract bill of quantities to provide him with data which he then uses to forecast the cost of the next similar building and to control the cost during the development of design. When this new design has been built, the feedback cycle starts all over again. Notice that in the diagram no feedback into the general system is shown from the area of building performance cost (i.e. maintenance and operational costs). Although some public authorities in recent years have attempted to obtain data on these costs and pass this information to the design team for consideration in their choice of detailing and specification this is by no means common practice. In the private sector, the time involved, the lack of clients involved in large programmes of building, and the tax skeletons in the cupboard, have all contributed to a reluctance to collect and divulge occupation costs.

The above procedures, although they represent the usual practice, are fraught with problems. Let us consider the contractor's estimating feedback cycle first of all. The difficulties are as follows, and are further considered in Chapter 13.

1 Although records of labour and material costs are kept on site, the information is kept primarily for the payment of wages and checking of invoices and secondly for monitoring progress on site as an aid to contract management. These purposes dictate a format of recording which is not compatible with the way in which the estimator prepares his estimate. The grouping of items into 'operations' for reference to the contractor's programme creates difficulties in rearrangement for the pricing of a bill of quantities based upon the Standard Method of Measurement requirement to measure items as fixed 'in place'. Consequently feedback to the estimator tends to be at the very general level of, for example, '. . . we were a bit low on plaster rates in this contract Jack . . .' rather than more specific data.

2 Even if site data could be related back to the estimator it is unlikely to be of very great benefit. Performance of the labour force varies considerably from day to day according to such factors as weather, supervision, obstruction by other trades, and psychological pressures. It is unlikely that individual performance on one contract is likely to be repeated on another. The estimator can only hope to get close to the total labour content of the project and hope that good site management will avoid wastage of this resource.

3 For work where 'labour only' sub-contractors are used no record of the performance of labour will be available, and therefore an increasingly large area of work is no longer subject to scrutiny and analysis.

4 Material use and costs will also be a variable between sites, and here again no firm data can be expected for the estimator to use. The amount of materials used will depend on such factors as ordering, site control, degree of vandalism, and of course the competence of the work force.

5 Plant cost will depend on good site management, the amount and type of plant available from stores, or on hire, and the amount of disruption affecting the contract period. Once again it may be difficult for the estimator to relate data obtained from one site to another.

The problems with feedback mean that very little data is kept in the contractor's office. The data base for the majority of estimators consists of a 'small black book' of labour and material constants, a list of addresses and telephone numbers of suppliers and sub-contractors, and a good deal of experienced judgement!

No wonder that it has been suggested that a tender represents 'the socially acceptable price' rather than a scientific appraisal of the resource needs of a project. Indeed there is some evidence to show that estimators bid randomly within plus or minus 10% of the actual contract cost excluding profit. This is considered to be the limit of their ability to reliably forecast what the job will cost and is discussed later in this chapter.

The problems with the retrieval of site data set the scene for the problems the cost planner will find when he analyses the priced tender document. The vast majority of his cost information arises from the bill of quantities and yet as we are beginning to see, this document may be based on incorrect assumptions which just happen to work in most cases and are convenient for the estimator to use. It has been said that the only reliable figure in the bill from the client's point of view is the tender figure. If we start to break this composite sum down into smaller units for estimating purposes then we are liable to find error in each individual sub-section when considered in isolation from the other sections. The greater the number of categories the greater the degree of error for the following reasons:

1 We know that the rates in the bill of quantities are not true costs, but they are not even true prices in the sense of a price on a supermarket shelf. The contractor is not offering to 'sell' brickwork at so many pounds per square metre, his accepted offer is for constructing a whole building and the rates are simply a notional breakdown of his total price for commercial and administrative purposes. There is thus no reason why any individual rate should be justifiable in relation to either cost or competition.

2 The way in which the contractor prices his preliminary items of plant, scaffolding, etc. will vary from one firm to another. Some will include these items in the work rates whilst others will place them in the preliminaries section.

3 The same applies to the contractor's profit and overheads. Consequently the 'cost' of any individual trade, element, or bill item will vary according to these two factors.

4 It may be in the contractor's best interest to 'load' those parts of the building which are executed first, such as excavation and earthworks, to improve his cash flow (i.e. obtain his money quicker on interim valuations to help finance the rest of the work). Although if this is overdone it is likely to be challenged by the quantity surveyor before accepting the tender, these sort of discrepancies can occur on a sufficient scale to distort any analysis of the bill.

5 In a similar fashion, the contractor may anticipate variations to the contract and reduce the price of likely omissions whilst increasing the unit rate of any items which are likely to increase in quantity.

6 In addition to the deliberate pricing method of each contractor there are all the variations in unit rates caused by different assumptions being made by each estimator with regard to the resource requirements. These variations tend to be highest in high-risk trades such as excavator and carpentry, and lowest in those most easily controlled such as glazing and concrete work. The assumptions made will relate to the estimator's view of his firm's expertise and economic structure.

7 In an estimate for a complex product such as a building it is inevitable that some mistakes will be made. These will tend to cancel each other out but sometimes there is an error of cost significance in a single item which may not be spotted by the quantity surveyor.

If other factors such as the contractor's previous experience of the particular design team (e.g. pricing the architect or quantity surveyor!), knowledge of the locality, and experience of the type of building proposed are taken into account it can be seen that data obtained from bills is likely to be highly variable even for contractors who are tendering for the same job. If we now consider the bill rates of contractors tendering for different projects then we can expect variability due to all the above factors plus a good few more. These additional items will relate to the site and contract conditions for each particular job and will include:

1 *The site conditions*

Under this heading we can include the problems of access; boundary conditions especially the problems of adjacent buildings; soil bearing capacity and consistency; topography and the orientation of the building (which may well affect manoeuvrability on the site and the type of resources, particularly plant, that can be used).

2 *Design variations*

The way the building is designed has an enormous effect on the efficient use of resources, and on production method and time. For example if there is a good deal of repetition of formwork then it is reasonable to expect lower unit rates if this fact has been communicated to the estimator. Similarly wet trades requiring drying and curing time will possibly create more delay than if a dry form of construction is used. The extra costs, if any, will be represented in the rates or will result in a redistribution of prices, for example, more money in preliminaries for the longer supervision required and the plant not fully employed, and less money in the work sections of the bill.

3 *Contract conditions*

The impact of these is very difficult to anticipate particularly with regard to their effect on unit rates. In general it can be assumed that the more onerous the terms as far as the contractor is concerned the more likely it is that a higher estimate will be given. However, when a contractor is short of work or particularly wants a job, say for prestige purposes, then the effect of 'tough' clauses may be less dramatic than would normally be expected. Any contract which requires more working capital for the project, or additional risk, is almost certain to incur a cost penalty which will be passed on to the client. In addition, conditions relating to the length of the construction period, particularly a shortening of the time, and also the phasing of the works, both of which may result in uneconomic working, will influence the estimator's rates and particularly the preliminary items.

4 *Size of contract*

For each contractor there is an optimum size of contract that will suit his particular structure and resources. Large firms very often create 'small works' sections of the main company to deal with those contracts which are small or of a specialist nature, such as restoration work, and which cannot carry the overheads of a giant corporation. Small firms on the other hand find it difficult to gear themselves up to a multi-million pound project with its specialist plant, complex labour relations, and sophisticated supervision and control requirements. The size of the firm in relation to the contract for which it is bidding will therefore affect its approach to the estimate. The unit rates for a large project should include an allowance for the economies of scale that could be expected. However, the problems of site working, the lack of advanced mechanical production plant, and the nature of the industry do not always allow these economies to be made.

5 *Location*

This factor will obviously affect the problems of accessibility to the production resources. The transport of plant, men, and materials to site, with perhaps accommodation of the workforce on remote sites, makes this an important consideration. On top of these problems are those of local climate which may affect the starting on the site, the degree of protection required and the interruption of the works programme. Despite these problems it should be noted that only the Housing Cost Yardstick, of the old principal government cost limits, identified location as a cost variable and this particular yardstick divided the country into regions. The other yardsticks made no allowance for regional variations but did take the problems of a particular site into consideration by use of an 'abnormal' allowance. A contractor's estimator is likely to pass on the problems of a particular area as an extra cost to the client if he has had experience of that location. This again will create variation in the rates due to different allowances being made.

All the problems listed contribute to the wide variability in bill rates. Derek Beeston of the Property Services Agency found that the variability tended to be different for each trade. He looked at the major items in each work section (or trade) and compared the unit rate for each identical item in a large sample of bills of quantities from different Department of the Environment projects. He expressed the degree of variability in the form of the 'coefficient of variation', which enables a comparison to be made between the variability of items of different value and is given by the formula:

$$\text{Coefficient of variation} = \frac{\text{standard deviation*}}{\text{arithmetic mean}} \times 100\%$$

So in concrete work for example the cost per cubic metre may be £30.00 and the standard deviation £5.00:

$$\text{Coefficient of variation} = (5/30) \times 100\%$$
$$= 16.67\%$$

*This and other statistical terms are defined in Appendix C.

This can now be compared with plaster work with a cost of £3.00 per square metre and a standard deviation of 30 pence:

$$\text{Coefficient of variation} = (30/300) \times 100\%$$
$$= 10\%$$

The standard deviation on its own would not allow the degree of variability in each trade to be compared.

The results of his investigations showed the following:

Excavator	45%	Joiner	28%
Drainlayer	29%	Roofer	24%
Concretor	15%	Plumber	23%
Steelworker	19%	Painter	22%
Bricklayer	26%	Glazier	13%
Carpenter	31%	All trades	22%

It is no surprise that excavator, with its high risk problems due to weather, soil conditions, accessibility etc., should have the highest variability. The ranking of some of the others is perhaps unexpected and a detailed investigation would need to be undertaken to discover the reason for their particular performance. However, the table does illustrate very clearly the problems of using bill data. Once again success depends on the skill of the user in determining what rate to use against a particular measured quantity.

The contractor's bid

If we accept the problems associated with obtaining reliable feedback from site and the rather arbitrary value given to unit rates in the bill of quantities then we might wonder how the contractor arrives at his 'bid' for a particular contract when he is in competition with others. The theory is that the estimator *knows* how long it will take a team of craftsmen and labourers to execute a job; that he *knows* the precise amount of material required including wastage; and that he *can define* the exact output of each item of plant to be used and the time it will be on site. He then brings together all these factors for each trade and adds on the *measured* cost of supervision and overheads to arrive at the net cost of the job (excluding profit). The process is essentially *deterministic*; the method assumes that the estimator can determine the exact amount of resources and their cost. However, every contractor knows that this is not possible.

Research in the field of estimating has suggested that while the above process is theoretically adopted it is in fact very often used to justify a target that the contractor has in mind from the very beginning. Brian Fine, who used to work for one of the largest UK building companies and is now a management consultant has suggested that contractors may bid at what they consider to be the 'socially acceptable price'. This is not well defined but is considered to be the price which society is known to be prepared to pay for a particular type of building. For commercial projects this price would probably be related to the income of the office block or output of the industrial plant, whilst for projects of a social nature it may be related to the known funds available or in the case of government funding it may be a defined cost yardstick. It is considered that this value of the building is relatively easily calculated and that generally the building is buildable at around the socially acceptable price. By bidding at this value the estimator overcomes the problem of being unable to predict resource requirements for the project. As the client is probably basing his anticipated cost on what has been successfully built before it is therefore likely that the estimator will have put forward a reasonable estimate. He then uses his 'constants' to justify his bid.

It is unlikely that many contractors consciously go through the above process, and indeed they very often react strongly to the suggestion that they do. However they will occasionally admit that they can write down the cost of the job before they start pricing it, or that they work out the yardstick before pricing a government job! Many clients assist the contractor in this respect by naming the value of the contract when enquiring if the contractor is willing to tender. In addition the building grapevine of sub-contractors, who probably quote several times for the same job but for different contractors, will ensure that the anticipated contract sum is well circulated. It is strange that despite the wide variation in bill rates the final tender figures are considerably closer together then one would be led to expect. A coefficient of variation for the spread of tenders on a project tends to be in the range of 5–8% and this can be applied to a wide range of projects.

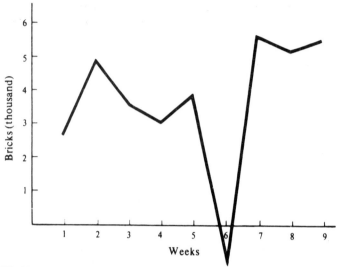

Fig. 10.2. Weekly output of bricklaying gang.

Support for this non-deterministic argument as to how contractors estimate is strengthened by the following observations:

1 The name of the project appears to influence the cost of the job. A good example of this phenomenon is the decision by a number of hospital boards to let nurses' homes as a separate contract to the main hospital development. The reason for this decision was that the nurses' homes were costing appreciably more than comparable students' accommodation in the education sector. Contractors had apparently applied the same criteria of complexity, difficulty, and high cost that is associated with hospital projects to the far simpler problem of the nurses' home. Hence the high cost and of course the suggestion that the *real cost* was not being estimated.

2 As stated previously in the chapter, it is not possible to establish accurate constants of labour, plant and material that can be applied to the job that is to be estimated. A typical output of a gang of bricklayers on site may be as shown in Fig. 10.2.

    Data of this nature may be of very little use in estimating the next job. On the graph in Fig. 10.2, week 6 results in a negative output, i.e. the gang have had to pull down something they constructed the week before because it was unsatisfactory or the subject of a variation order. This is not an uncommon experience, but will it happen on the next job? If confident predictions cannot be made then an intuitive assessment in the form of the socially acceptable price is perhaps the answer.

3 In a competitive tendering market it appears that contractors take their place in the bidding order by chance. As would be expected with chance, in the long term they appear to win contracts in direct proportion to the number of contractors they are bidding against. J. C. Pimm of Management Procedures (Building) Ltd looked at four companies, the number of contracts they had bid for, and the number of competitors identified by the total number of bids. From this information and knowing the number of contracts actually won, he produced the following table:

Company	No. of contracts won	No. of bids	Bids per contracts	Per cent if won by chance	Per cent actually won
A	41	249	6.1	16.4	17.0
B	36	183	7.0	14.2	15.4
C	19	88	4.6	21.6	21.1
D	35	202	5.8	17.3	17.1

It can be seen that there is a very high correlation between those actually won and the number expected to be won by chance, assuming that each contractor took his place randomly in the bidding list. This is supported by Brian Fine who suggests that contractors can only estimate within plus or minus 10% of the actual cost. If this is so then each contractor needs to add on a 'mark-up', to the estimate, to allow for the fact that if he is awarded the contract he is likely to have made the largest error. If this is not allowed for, then the mark-up will come out of the expected profit, or end up as a loss. The actual amount to be added should depend upon the number of contractors bidding. The probability of the lowest tenderer having made a large error increases with the number of competitors. In practice, very few firms consciously allow for this factor, but they probably note its eventual impact on the profitability of a project.

4 If the estimation of cost was really a deterministic exercise then the constants used in the estimate would be absolute and unchanging unless, of course, they were amended as the result of ascertained cost feedback. However, what happens when an estimator or his firm loses a number of bids in succession? They reduce their rates! This may be done on a lump sum basis, or the amount may be spread through the rates in the bill. For a firm (or its estimator) to survive they must get work. This practice suggests that the estimate does not mean very much in terms of an accurate prediction. If the firm does get the job and they think it will be 'tight' they put their best management on the project and very often they get a better return than on a job where they foresaw a large profit and did not maintain such tight control! The lower price that they are bidding could be the socially acceptable price.

It would not be fair to suggest that the concept of socially acceptable price has been thoroughly substantiated on the basis of the few arguments considered. However, doubt has been cast on the deterministic estimating process and this in turn will influence the way in which we look at and use cost data. Obviously the factors of supply and demand related to the capacity of the industry, and its workload, will influence the level of the contractor's bid, and the relationship between these factors and the socially acceptable price has not yet been well defined. However, it is important to realise that market conditions do reflect the economic standing of the country and this aspect may well have more impact on 'historic' cost data, and the price society is prepared to pay for its buildings, than any other.

The structuring of cost data

In addition to the difficulty of acquiring reliable cost data there is the problem of what to do with it once acquired. The need for a system of cost planning to have access to cost data at a number of different levels of complexity postulates a hierarchy of cost groupings or even a series of interlocking hierarchies, because some of the criteria to which cost data will be applied are of a different type and not merely a different scale.

One such hierarchy, which we will encounter in Chapter 12, is based upon the concept of design cost elements. This is a highly pragmatic breakdown of a project for design cost planning purposes which, apart from its inherent faults, suffers from the disadvantage that it does not meet the requirements of any of the other functions in the design/build process.

Another hierarchy is the Standard Method of Measurement format used for bills of quantities, which is only of very limited use for other purposes. However, because the information provided in the bill of quantities is generally of a high quality, is tied back into the contract, and is available to the builder without further trouble, its format does tend to pervade process documentation rather more than might be expected.

Operational cost centres which are useful for actual control on site are difficult to relate, except at the highest levels of generality, to the estimating and commercial systems of the industry. Because of the obstruction to communication between builder and designer which is caused by the price mechanism there has been little attempt to link site operation costs to design criteria.

A number of efforts have been made during the last twenty years to devise an integrated system of groupings of building work which would meet the needs of all participants in the project, from the point of view of design process, costing, communication, organisation, and feedback. It is argued that such a system would reduce duplication of effort (it is alleged that the same piece of building work may be physically measured up to fifteen times for various purposes), improve communication between the parties, and permit a far higher quality of costing and cost feedback. The CI/SfB system whose use is encouraged by the RIBA is one such effort, but is somewhat too design-orientated to have found much favour elsewhere. A much more complex version of the same basic system, CBC, was developed in Scandinavia and relied heavily on computer use. In Britain the Department of the Environment set up a Data Co-ordination study with substantial resources and a large measure of contribution from industry and the professions, it being their very reasonable view that this was the only way of obtaining results that would be useful, and acceptable, to all. Unfortunately this work has now ceased, without producing much more than a number of very interesting reports and the *Construction Industry Thesaurus*, which is used mainly for library and information purposes.

Little real progress has therefore been made in this field, but there is plenty of background reading available.

Sources of data

So far in this chapter we have dealt with some of the more general problems of cost information associated with its use, retrieval, and structure. Although these problems should be identified it has to be recognised that the quantity surveying profession has been using data obtained from bills of quantities for scores of years with reasonable success. Contractors, too, have been able to prepare realistic estimates from the very general feedback that they get from site. The problems associated with data should not therefore be allowed to obscure the fact that the present data sources available to the industry have allowed a simple and reasonably effective system of cost forecasting and control to develop.

From the point of view of the client's cost adviser, information can be obtained from two main sources. Either he can use the data passing through his own hands in his office or he can obtain the information from published material found in the technical press, price books, and information systems. Of the two there is no doubt that a quantity surveyor would usually prefer to use his own data, for the following reasons:

1   The data is from a project of which he knows the background. He is therefore aware of all the problems associated with the project, its location, the market conditions, complexity, etc., which influenced its price. We have already said that traditional cost modelling and the successful use of data requires sound professional judgement. This judgement is better served by using information which is known to the quantity surveyor by his involvement in the project rather than that received second hand, in an abbreviated form, through a published source.

2   The further detailed breakdown of any structured information is available should he require it. To take the example of an elemental cost analysis, the firm will have available the priced bill of quantities, the dimensions, the working drawings, and the contract should they wish to find further information on why an element cost a certain amount or what were the prevailing market conditions at the time. It is very unlikely that this level of detail will be available in the published data, however well prepared. To take another example, several quantity surveyors prepare their own cost indices because again they know the weightings and costs included and these can be manipulated to suit the needs of a particular project.

3   There is a shorter time lag between receiving raw data, processing it, and being able to use the structured information in the office. This is particularly relevant to cost indices where the published information may be anything up to six months behind the times.

4   The choice of classification and structure of the processed data is under the control of the firm. It is therefore possible to ensure that the details which are most relevant to that particular practice are emphasised and identified rather than having to accept a standardised published format. If errors and ambiguities occur it is also easier to spot them in your own system than in somebody else's.

Having established the preference of most cost planners for their own information it would be unrealistic to suppose that any but the largest firms, or organisations which deal mainly with one type of building programme, could exist upon the data passing through their offices. The range of projects undertaken by most practices is considerable, varying from government and local authority buildings to commercial and leisure buildings for individual and corporate clients. The variety of specification, location, size, and shape of this range is also large and it is therefore most unlikely that the office will have up-to-date information which it can apply to all its new projects. The common unit rates which each quantity surveyor will gather from his handling of project information will help him in preparing his approximate quantities estimate, but a wide variety of data at the 'coarser' level of elements and single-price rates is unlikely to be readily available. While he can probably make intuitive assessments, particularly of cost per square metre, it is helpful to have a check available in the form of published information to reassure him of his own judgement. Indeed it is probably the primary role of published information to provide data which allows the practitioner to check on his own knowledge and to provide a context for his own decision making. The following are the major published sources of this type of data. It will be seen that they apply almost entirely to conditions in the UK and the absence of a similar range of information on costs in other parts of the world makes the cost planner's task overseas much more difficult. Such information as exists is often of a more condensed statistical nature.

1 The technical press

There has been a steady improvement in the quantity and quality of cost information appearing in the journals and magazines concerned with design and building management. This is largely due to the increased demand from clients for better cost control, the tighter yardsticks and cost limits, and the recognition that there was previously very little disseminated information in the cost area. With this stimulation of interest, editors have found it profitable to give more space to cost forecasts and analysis than that provided in the past.

The type of information that can now be found in these publications includes elemental analyses; resource costs and measured rates; cost indices with future cost projections; regional economic reports; analysis of the economic performance of the industry; and cost studies of different types of building. Indeed with the rapid increase in the level of resource costs and measured rates over very short time-spans the technical press has, in many instances, replaced the standard price books as a source for quick reference material. Most journals are published weekly or monthly and this allows a comparatively short time lag between the reception and publication of current information. Typical examples of these journals are *The Architects' Journal, Building, Building Trades Journal*, and *Building Specification*.

2 Builders' price books

A traditional source of useful information has been the price books which are normally published annually by several well-established organisations. These too have expanded the scope of their contents to keep pace with current demand. However, in periods of high inflation they have suffered the big drawback of having to prepare their information well in advance of the year of publication. This time lag of several months means that the rates included are forecasts of what is likely to happen at a sometimes unknown point of time in the year ahead. Despite these problems it is not unusual to find an assortment of these books on the shelves of all the building professions. Some of the most familiar are *Spon's Architects and Builders Price Book, Laxtons Price Book, Wessex Price Book* and *Hutchins Priced Schedules*.

Their popularity stems from the comprehensive nature of the data included and indeed the list of sub-sections is extensive:

(a) Professional fees for building work.
(b) Wage rates in the industry.
(c) Market prices of materials.
(d) Constants of labour and material for unit rates.

(e) Prices for measured work (in accordance with the SMM).

(f) Building cost index.

(g) Approximate estimate rates and comparisons.

(h) Cost limits and allowances.

(i) European prices and information.

As with data from any source, considerable care should be exercised in the use of the information. For example, in the 'Prices for measured work' section of the 1984 'Spons' the contract envisaged is situated in the outer London area, about a million pounds in value, and it is assumed that a reasonable quantity of work for all items is required. The prices include 10% for overhead charges and profit and if preliminaries are required to be included then at least 15% should be added on top of the published rates. The user is expected to make all allowances for changes in the location, size of contract, and market conditions.

3 Information services

With the advent of cost planning techniques and the need for a wide range of information on different types of project, it was realised that most offices would require information additional to that normally found in any one particular practice. The suggestion was made that if firms could be persuaded to supply their own elemental cost analyses to a central body then a large pool of information could be gathered which could then be disseminated to the contributing members. If these analyses could be supplemented with other more general information on trends and economic indicators then a useful service would be provided for the profession. Therefore, in April 1962, the RICS set up the Building Cost Information Service (BCIS), under the direction of Mr Douglas Robertson, to undertake this work. The service has since become well established and a valuable source of knowledge in the industry. Since 1972 the service has been available to organisations outside the quantity surveying profession. Although the original intention was to establish a data bank of elemental cost analyses this was never keenly supported by members, the reason for their lack of contributions being probably the expense and time involved in the preparation of an analysis which they themselves might never use again. The process of preparing an elemental analysis from a traditional bill of quantities is explained in Chapter 12, and its cost can easily run into hundreds of pounds. You would have to be pretty sure that this expenditure would be money well spent before committing your firm to this time and effort. The problems of confidentiality of cost information obtained from contract bills of quantities have been largely overcome by obtaining permission from the contractor before publishing the cost information. The number of published detailed analyses is now down to less than fifty per annum, and with the decline of theoretical elemental estimating (involving the adjustment of elemental unit rates) it is unlikely to grow much beyond this figure. In recent years, therefore, the information service has concentrated on other areas of cost study which could not easily be undertaken in the professional office, and it is in these other areas that it now provides the major contribution to the practitioner. A summary of the major sections of the service is given below:

Section A *General and background information*

This is a very broad and comprehensive section ranging from the statistics relating to the performance of the construction industry to current resource costs. It includes cost indices, building activity indicators, labour hours and wages, regional trends, and a briefing on the current economic and political climate.

Section B   *Publications digest*

This section contains a summary of important papers, articles, and books related to the economics of buildings.

Section F   *Cost studies and research and development papers*

These papers are analytical studies of particular aspects of building cost or of cost planning systems.

Section G   *Detailed cost analyses*

This section contains the analyses submitted to the service and presented in accordance with the standard rules for elemental analysis developed by the BCIS. Each analysis contains a general description of the project including market conditions, contractual arrangements, number of tenderers, etc., a list of elemental costs, a brief specification, and a schematic drawing.

Section H   *Concise cost analyses*

These are much briefer forms of analysis with costs broken down into only seven major elements. The job description and specification is also reduced to give only a review of the major design variables. The analyses include summaries of the detailed forms found in Section G, plus summaries of published analyses found in the technical press.

Section Z   *Reference indexes*

This is a comprehensive cumulative reference index to sections A–H.

Some time after the BCIS became established it was realised that there was a similar need relating to the cost of maintenance. Therefore, in April 1971, after a pilot service had been operating for a year, the Building Maintenance Cost Information Service was launched. It was sponsored in collaboration with the School of Architecture and Building Technology at the University of Bath by the former Ministry of Public Building and Works (now part of the Department of the Environment) and had similar aims with regard to cost information for maintenance as did the BCIS with regard to capital cost. It is now an independent organisation to which anyone can become a subscriber. The service collects information from many relevant sources and disseminates it by regular mailings, but its unique data are derived from its own subscribers who, like those of the BCIS, are expected to contribute to the service. The arrangement of the service is also similar to that of the BCIS but in lieu of the detailed analyses there are three new sections. The full list is as follows:

Section A   *General background information*

Section B   *Publications digest*

Section C   *Case studies*

These studies relate to maintenance management and control as exercised by various private and government organisations.

Section D   *Occupancy cost analyses*

The maintenance equivalent of the BCIS elemental cost analyses, these sheets break down the annual cost of maintaining and operating a building into major categories of expenditure such as cleaning, decoration, etc. Details of specification and a general description of the building are given to assist in using the data.

Section E   *Design/performance data*

There is a strong relationship between design and building maintenance, particularly in the early years of occupation, and this section identifies some of the major problems and suggests remedies which can be applied. The section is supplemented with sketches for ease of comprehension.

Section F   *Research and development papers*

Section Z   *Reference indexes*

Other cost information services are available but these two are definite leaders in their respective fields.

Since April of 1984 the BCIS has been operating a computer-based service accessible from a member office through a micro-computer, modem and telephone line. Initially this just provides data available in the manual system, but eventually a more comprehensive system with the possibility of members sending information up the line may be established.

4 Government literature

As the government gradually became the major client of the building industry it steadily increased its involvement in the research and development aspects of cost forecasting and control. Local authorities have very often led the way in developing new contractual arrangements, cost planning, and computer methods. In addition, central government has promoted a number of working parties in the industry and has sponsored research in several areas relating to cost. Most of the major government departments which undertake large building programmes have standardised their procedures for budgeting and have made recommendations as to normal cost planning and control practice. These guidelines are sometimes related to cost yardsticks which provide a good deal of cost information (especially at the brief stage) in their own right. A very useful publication that appears twice a year is the *Construction Forecasts* for the following two years published by the National Economic Development Office (NEDO) through HMSO. Although publicly financed, NEDO is not a government department or agency and it is not involved in implementing government decisions. The information it supplies is extremely useful in assessing the likely demand for building in each sector of the industry in the short term period, which information can then be related to likely cost predictions. At one time attempts were made to prepare a forecast over longer periods and on a regional basis but, although this was a useful government strategic planning tool, circumstances changed too fast and frequently for it to be of real benefit to the practitioner. NEDO is also of course responsible for providing the indices for the adjustment of fluctuations using the Formula Price Adjustment method. Although government departments have had to curtail their activity in recent years due to cutbacks in expenditure they still provide a ready source of information on a wide variety of topics.

5 University or polytechnic research

As quantity surveying has developed into a degree discipline it has been recognised that there is a need to establish a foundation of research which will be of benefit to the profession. The educational institutions are therefore beginning to establish programmes of work which will assist in developing a general theory for cost prediction and control, and to investigate the reasons for, and the degree to which, costs change related to design and economic variables. Papers on this research are beginning to be published and it is envisaged that, providing funds are forthcoming, this will provide a larger source of cost information in the future.

6 Technical information systems

For a large number of years there have been office library systems available to the building professions which contain a collection of the current trade literature relating to building products. Their major purpose is to provide a ready reference for the practitioner on specification and performance of a wide variety of building products. At one time it was not uncommon to find current price lists included, but with the problems of updating this information so frequently such cost information is very seldom published and circulated except on request. The use of such services for costing purposes, admittedly never their major role, has now largely died out.

Whilst the firm's own data and published data represent the two main sources for the cost planner, there is a further source available for some types of work, which in some ways is a combination of the two. This is the obtaining of information from specialist sub-contractors, specialist consultants, or even sometimes a building firm. Like published information this suffers from remoteness, but because there is usually an element of personal contact involved it is then possible to explain and discuss exactly what is involved. Such information can be very useful in connection with various forms of roofing, flooring, windows, doors, cladding, finishes, framing, landscaping, and in particular engineering services. It is helpful to know the people with whom one is dealing, since there is clearly little other guarantee of the soundness of such advice.

If prices are obtained from specialist sub-contractors there are several additional points to watch. Not only must the cost planner remember to add the builder's profit and perhaps discount allowance to the figures quoted, but if the field is one with which he is not familiar he should obtain an indication of the extra facilities which the builder may have to provide. In some sectors of industry a price for site installation simply covers the cost of sending a specialist supervisor, the builder having to do most of the hard work. Other people may be almost entirely self-sufficient.

Generally

An attempt has been made in this chapter to paint, with broad strokes of the brush, a picture of the use, problems and sources of cost data. It is a vital primary need of the cost adviser and one which will receive far more attention in years to come. As modelling techniques develop so will the need for different data and information structures that will suit the new methods. The direction for these ideas will almost certainly be along the lines of representing the way costs are incurred on site. It therefore seems likely that more attention will be given to the feedback of information from site and subsequently the presentation of resource costs in such a way that they can be related to design development.

One last point, do not forget that the data included in bills of quantities is confidential and the contractor's permission must be sought if any analysis etc. is to be published.

Chapter 11
Cost Indices

One of the most important items of cost data, particularly with regard to forecasting techniques which rely on historic (i.e. past) information, is a cost index. The object of the cost index is to measure changes in the cost of an item or group of items from one point in time to another. A base date is chosen and is usually given the value of 100, all future increases or decreases being related to this figure. For example suppose the base date cost of employing a craftsman on site is £80.00 per week and the current figure is £120.00 per week. £80.00 would be given the value 100 and £120.00 would be represented by the figure 150, derived in the following manner:

$$\frac{(120 - 80) \times 100}{80} + 100 = \text{Cost index in relation to base } 100 = 150$$

As the base is 100 the number of points of any subsequent index above 100 (in this case 50) also represents the percentage increase since the base date. However, if we are comparing costs over time in which the data we wish to update is not at base year cost, but has an index of say 120, then to arrive at the percentage increase in cost where the current index is 150 the following calculation is adopted:

$$\frac{150 - 120}{120} \times 100 = 25\%$$

Notice that the answer is not the difference in the number of points on the index scale, i.e. 30.

Use of index numbers

There are a number of uses to which index numbers can be put in the construction industry and these can be summarised as follows:

1 Updating elemental cost analyses

This is perhaps the most common use for the quantity surveyor and is an essential part of the theoretical elemental cost planning process. Tender information of past projects can be brought up to current costs for budgeting purposes, although it should be realised that great care is required when updating information beyond a period of, say, two years.

2 Updating for research

It is extremely difficult to obtain large quantities of cost data relating to the same point in time in order to analyse trends and patterns for cost research. By bringing cost information, obtained at a number of different points in time, to a common date by the use of an index, a much large sample of data can be examined.

3 Extrapolation of existing trends

By plotting the pattern of costs measured by an index it may be possible to extrapolate a trend into the future. However, there are very great dangers in attempting this. During the 1960s it was possible to identify an increase in cost of about 5% per year. Extrapolations were undertaken using this figure for projects one or two years ahead. This worked quite well for a number of years until 1970 when the cost of building to the client suddenly rose by about 10%. Many quantity surveyors (and other professions) had not foreseen this rise and were heavily under-valued in their budgets. Since the 1970s the erratic behaviour of inflation rates has meant that extrapolation purely on trends is not possible.

4 Calculation of price fluctuations

The National Economic Development Organisation has produced a series of indices for building trades based upon the resource costs of each trade. By applying the indices to work undertaken during a specified period, it is possible to evaluate, to an acceptable degree of reliability, the increase in costs of resources to the contractor since the tender date. This enables the financial control of contracts which contain a fluctuations clause to be exercised more speedily and with less ambiguity. Formula Price Adjustment, as it is known, is now well established and will probably be used more widely in years to come.

5 Identification of changes in cost relationships

If a cost index is prepared for the different components of a building, or for alternative possible solutions to a design problem (e.g. steel *v* reinforced concrete frame), then it is possible to see the changes in the relationship between one component and another over time. It may then be possible to identify when one solution appears to be a better proposition than another.

6 Assessment of market conditions

Quantity surveyors are particularly interested in the price their clients have to pay for a building. If the index will measure the market price, as opposed to the change in the cost of resources, then a measure of current market conditions can be obtained which is of enormous benefit in updating and forecasting cost.

Methods of constructing an index

Over the last two or three decades there have been a number of attempts at providing a reliable cost index. They include the following:

1 The use of a notional bill

In this method a typical (or synthesised) bill of quantities is chosen and repriced at regular intervals by a competent quantity surveyor based on his experience of current rates. Unfortunately rates vary so much that it is extremely difficult to assess the current tender situation in such a way that a reliable index can be obtained from the resultant totals. This method is now to all intents and purposes obsolete.

2 Semi-intuitive assessment

This method is based upon trends in resource costs combined with tender reports by quantity surveyors. By receiving reports on current tenders from each of the project quantity surveyors in his organisation an experienced practitioner can adjust his knowledge of the changes in material and labour costs to prepare an index for the current market situation. This works quite well where the market is stable or steadily changing, but these judgements are extremely difficult to make at a time when the economy is subject to 'stop-go' conditions. This method has been superseded by the tender-based index described later.

3 Analysis of unit price for buildings of similar function

If data could be found giving, say, the cost per square metre of all schools of a certain type being built in the country at one point in time, then the average of these costs could be used to provide the basis of an index. If the exercise were repeated at regular intervals for that type of building then a regular index could be established. Problems arise, however, due to such large design variables as specification and shape. Such an index would also be only applicable to buildings which are homogeneous in function and standard and for which there is a regular building programme to provide the data. This method is used by the Nationwide Building Society to establish trends in house prices, and as such provides a fairly reliable indication of price movements in that particular market.

4 Factor cost index

If a typical building is analysed into its constituent resources and the cost of each resource is monitored over time, then a combined average index can be prepared which measures the change in the total cost of the building over the same period. Each resource (labour, plant, and material) would need to be given due importance in the index according to its value in the total building. The construction of, and problems associated with, this type of index are described in more detail later on in this chapter.

5 Tender based index

There is a great need for quantity surveyors to have a measure of the market price their client has to pay for his building. The accepted tender figure based on the pricing of a bill of quantities is a record of the market price for a particular building at a specific point in time. If the measured items in the bill are repriced using a standard schedule of base year prices (to give a base year 'tender'), then an index can be constructed by comparing the current tender figure with the new derived base year total. Short cut methods are available to avoid repricing the whole bill. Details of this method are provided later in the chapter.

The factor cost index

It has already been explained that this uses changes in the cost of resources to build up a composite index. To illustrate the method let us look at the construction of a simple index for the cost of a brick wall. We will assume that the following table represents the cost of the constituents at base year level in the first column, and then at today's date in the second column. The final column represents the index resulting from these figures (calculated as described previously).

Cost per square metre of wall

	Base year	Current	Index
	£	£	
Bricks	20	28	140
Mortar	2	3	150
Labour	8	10	125
		÷ 3	415

Average index = 138.34

However in the above example no account has been taken of the fact that bricks as a proportion of total costs are ten times more important than mortar. All the resources having been given equal status. Consequently a very rapid rise in the cost of mortar would have a disproportionate effect on the composite index, and this would not be representative of the increase in the cost of a brick wall as experienced by the contractor. To overcome this problem the resources need to be 'weighted' in accordance with their importance as follows:

	Index	Base year weighting	Extension
Bricks	140	20	2,800
Mortar	150	2	300
Labour	125	8	1,000
		30 ÷ 30	4,100

Average weighted index = 136.67

The effect of the weighting is to reduce the index in this example because mortar (which has increased most in price) does not now share equal status with bricks and labour.

A further problem arises at this point. Suppose for some reason the labour force becomes more productive over the time period, by say 25%. Then at base year costs the value of the labour would have been £6.00 instead of £8.00. This will affect the amount of labour involved in building a brick wall and consequently its cost. This change, although affecting the cost to the contractor and possibly client, will not be represented in the index unless the weightings are changed.

	Index	Current year weighting		Extension
Bricks	140	20		2,800
Mortar	150	2		300
Labour	125	6		750
		28	÷ 28	3,850

Average weighted index = 137.5

This is a very real difficulty with the factor cost index because it is not easy to judge productivity over time, particularly with regard to the construction of whole buildings. In fact in the majority of cases, weightings in building factor cost indices are assumed to remain constant until a revision of the index is undertaken and a further analysis of the importance of each resource is made.

Where an index uses the base year weightings for each calculation then it is known as a Laspeyres index. Where the index uses weightings obtained from the current year or point in time that is under consideration then it is known as a Paasche index. Both names derive from their original authors.

Another factor not considered in the above examples is the contractor's profit and overheads. Profit, particularly, can be a function of market conditions and it would be difficult to devise a reliable quantitative measure for this variable. This is why a number of factor cost indices have incorporated an additional component, called a 'market conditions' allowance, very often based on the professional judgement of its author.

The brick wall was chosen to illustrate the essential ingredients and problems of an index of this type. When constructing an index for a complete building the task becomes more complex although the same principles apply. A typical procedure can be summarised as follows:

1 A typical building (or group of buildings) is selected for analysis into its constituent proportions of labour, plant, and material.

2 Analysis takes place and the building resources are allocated under various headings to suit the representative cost factors for which information is available. For example if use was being made of published information for material indices (such as those prepared by the Department of the Environ-

ment) then the different materials in the building would be analysed according to the structure of that published data.

3 The different types of labour (corresponding to different wage-fixing bodies) would be identified and the basis for evaluation of a unit of labour cost identified. Usually this would include the following:

(a) Changes in the hourly or weekly wage rate as determined by agreement of the parties to the wage-fixing body.

(b) Changes in employer's 'on costs' and contributions such as holidays with pay, national insurance contributions, etc.

(c) Changes in the agreed standard working week.

(d) Changes in the average hours worked per week, as recorded in government statistics. This item and the preceding one will affect the degree to which overtime rates are paid for each unit of labour output.

In addition two other factors may be considered:

(e) Changes in productivity. One method of obtaining a gauge of productivity is to take the total quantity of materials used by the industry, priced at 'standard rates' (i.e. constant figures) and divide by the total building labour force. If the output per man increases over a particular time period then it is assumed that this is the result of productivity and the labour weighting should be reduced accordingly.

(f) Change of location. It may be necessary to assume a particular geographical position for the typical building in order that regional differentials can be ignored.

The labour unit cost for each type of craftsman and labourer would be computed and then weighted according to the importance of each in the total labour force. The resultant weighted average for labour would be carried forward for further weighting with other resources.

4 Material rates for comparison with base year prices are selected. Although it is possible to identify a long list of cost significant materials and calculate a separate index for each it is more usual to rely on published statistics for this type of information. The Department of Trade and Industry regularly publishes indices of materials for a range of material classifications. It is then merely a question of analysing the typical building according to the published material types to obtain the weighting to be applied to each material index.

5 Plant types, together with their weighting, are identified and a standard method of evaluation adopted. This may be based on an average of hire rates for the particular item or it may be calculated by considering purchase price, depreciation, maintenance, starting time, etc to establish a standard resource unit cost which can form the basis of the index. In many published forms the change in the cost of plant is ignored.

6 When the weighted index for each of the three basic resource types is computed they are then brought together and weighted again, this time in accordance with the importance of each within the total building cost.

Typical values for weighting are:

Labour   35–45%

Material 50–60%
Plant 5–10%

The final weighted index represents the change in cost between one point in time and another for the typical building chosen. Any subjective judgement to take into account market conditions, productivity, etc can then be made if it is considered necessary.

It is important to realise what the cost factor index is actually measuring. It measures the change in the cost of resources to a contractor for a 'typical' building. It does not directly measure the change in the price the client must pay, and although the 'market conditions' allowance attempts to rectify this situation the source may not necessarily be reliable. Neither does it measure the change in cost of a specific building under construction by a quantity surveyor. It is unlikely that the 'typical building' will have the same mix of resources as the client's project; however on many occasions the difference will not be cost significant.

To try and minimise this difference the 'typical' building may be more of an economic model than a likely real-life building, having, say, a mixture of light concrete block and clay block partitions so that a change in the cost of only one of the alternatives will not unduly distort the index on the one hand nor be ignored on the other. In spite of these drawbacks the method is very suitable for identifying trends in resource costs and relationships. It is particularly useful for evaluating cost fluctuations in contracts which allow for reimbursement of any changes in cost to the contractor that occur during the contract period.

The tender-based index

The main shortcoming of the factor cost index from the private quantity surveyor's point of view is that it takes little or no account of the tendering market. A much more attractive proposition appears to be an index which takes as its source of information the tender document itself. This should record what is happening in the market place and therefore be much more useful in updating prices for a design budget.

In essence the tender figure is compared with a figure produced by pricing the same tender document using standard rates at the base year prices. From the two resultant figures an index can be calculated showing the increase or decrease in cost to the client within the current tendering market. There are drawbacks with the method, such as the questionable validity of bill of quantity rates, and the remote chance of obtaining priced bills for jobs which are comparable except for date of tender.

If, however, a large enough sample of projects is taken then the difficulties can be dealt with by 'trampling the problem to death'. If an appreciable quantity of projects is analysed it is hoped that all other differences except current levels of cost will cancel themselves out. An obvious difficulty here is the work involved in analysing all the rates in scores of bills. Fortunately it has been found that much time can be saved by taking only the few largest items in each work section and this results in a negligible degree of error.

The procedure for preparing and using a typical tender-based index can be summarised as follows:

1 Prepare a priced list of typical BQ items at the prevailing base year pricing levels

(usually with an allowance for preliminaries included). This is of course a time-consuming task and the current published forms of this index tend to use the Department of the Environment's Schedule of Rates, which is readily available. This is a very comprehensive schedule and is more than adequate for the majority of buildings.

2   Take the priced document for the lowest tender received and note the format (e.g. work-section, elemental).

3   For each section of the format (e.g. excavator, concrete work) pick up the largest value item in the section, then the second, and so on until a total of 25% of the section value is achieved. (It is of course the value of the item as extended in the cash column, and not its unit price, with which we are concerned). This procedure is repeated for all sections with the exception of preliminary items, PC sums, profit and attendance on PC items, and day-work percentage additions.

4   For each of the items selected find the corresponding unit price in the base schedule of prices. The difference as an addition or subtraction from the base year rate is then obtained by comparing the bill unit price of each sample item with its equivalent base year price.

5   The extended base year value is calculated for each item (i.e. weighted by its quantity). The total of all the extended items at base year prices is then compared with the total of all the extended items for each section to obtain an index.

6   The index obtained for each section is then weighted according to the value of that section as a proportion of the whole tender bill for the project being considered (less preliminaries, PC sums, contingencies, etc) and a combined index for all sections obtained.

7   The preliminaries are usually dealt with by considering them as a percentage addition to the other items in the bill, excluding dayworks and contingencies, provided that the rates in the base schedule include an allowance for the preliminary items. In that case the index figure arrived at so far is not a true one, since net bill rates have been compared with gross base rates. So, if the 'false' index of bill items is 130 and the value of the preliminaries in the current tender is 10% (as a percentage of the remainder of the bill) then the final index would be:

$$130 + \left(130 \times \frac{10}{100}\right) = 143$$

8   To obtain an average index for publication rather than for one particular job, the index numbers for all the tenders which have been sampled in a specified time period are averaged by taking either the geometric or arithmetic mean. The difference between the mean of the current and proceeding quarters is used to gauge the movement in tender prices. The geometric mean is usually taken when relative changes in some variable quantity are averaged, because the arithmetic mean has a tendency to exaggerate the 'average' annual rate of increase.

A number of simplifications have been made in the above procedure for the purpose of clarity. For example, in practice, problems arise with items that cannot be matched (these are generally overlooked in the item selection process), and also where a sample amounting to 25% of a section's total cannot be obtained. The rules governing the index being used should always be carefully consulted in these situations to discover the method to adopt.

The resultant index has several advantages over a simple factor cost index from the point of view of the client's quantity surveyor:

1 It measures the change in the cost to the client of a particular project over time, taking full account of market conditions in addition to the change in cost to the contractor.
2 It is relatively simple to operate once a base schedule of prices has been obtained.
3 It allows comparison of the price obtained by tender for a specific project with the national or regional building price trend. This could assist in deciding whether to call for fresh tenders for another group of contractors, or not.
4 It allows the relationship between the market for buildings of different function and locality to be plotted.
5 It is not based on other indices and therefore any inherent inaccuracies are not compounded.

There are however one or two problems which have to be acknowledged:

1 To overcome the problems due to the variability of price rates in bills of quantities a large number of projects are required for each index. It is suggested that at least eighty are required for a suitable sample, but this condition is seldom met. Unless access to this quantity of bills is easily available the trend being plotted may be erratic and not typical. Very few organisations have access to this number of projects and therefore most firms cannot prepare their own index by this method.
2 The index relies heavily on the base year schedule which will have to be regularly revised to take into account new products, new methods of measurement etc. This is a time-consuming and costly task.
3 Because of the lack of projects at any one point in time the average index may have to rely on an unbalanced sample containing more jobs of one particular function and location than is considered desirable. This may lead to errors in the trend plotted.

However, despite these problems there is no doubt that the tender-based index represents a major breakthrough in the measurement of trends in market prices to the client.

Published forms

There are a large number of published indices which are available to the construction industry and the major sources are listed below, classified under the two major index types.

Factor cost indices

There are a large number of these indices available to the industry in published form,

Table 10. Weightings for BCIS cost indices – series II revision.

Work category	Per cent weighting			
	Steel framed	Concrete framed	Brickwork construction	General construction
1. Demolitions	—	—	—	—
2. Excavation	1.89	1.98	1.43	1.68
3. Hardcore	1.18	0.70	0.94	0.94
4. General piling	—	0.35	—	0.09
5. Sheet steel piling	—	—	—	—
6. Concrete	8.08	10.03	4.63	6.84
7. Reinforcement	2.15	7.71	1.85	3.39
8. Structural precast units	2.60	3.69	1.22	2.18
9. Non-struct, precast components	2.67	5.80	1.17	2.70
10. Formwork	3.53	6.20	1.13	3.00
11. Brickwork & blockwork	8.22	10.20	15.62	12.42
12. Stone	—	—	—	—
13. Asphalt work	1.01	1.82	0.49	0.95
14. Slate & tile roofing	0.90	0.20	3.15	1.85
15. Asbestos Cmt sheeting	0.18	—	0.05	0.07
16. Plastic Ctd steel roofing	3.65	0.16	—	0.95
17. Aluminium sheet roofing	0.59	0.15	0.01	0.19
18. Felt roofing	0.52	0.88	0.50	0.60
19. Felt roofing on metal	3.03	—	0.59	1.05
20. Carpentry, board, flooring	3.17	2.37	7.45	5.11
21. Hardwood flooring	—	0.05	0.06	0.04
22. Tile & sheet flooring	0.95	1.01	1.57	1.28
23. Jointless flooring	—	—	—	—
24. Softwood joinery	3.22	1.60	8.35	5.38
25. Hardwood joinery	0.37	0.44	1.24	0.82
26. Ironmongery	0.77	0.89	1.89	1.36
27. Steelwork	0.04	0.23	0.24	0.19
28. Steel windows/doors	3.11	0.88	0.25	1.12
29. Aluminium windows/doors	0.63	3.82	1.35	1.79
30. Metalwork	1.09	1.50	1.56	1.43
31. Cast iron pipes	0.24	0.30	0.12	0.20
32. Plastic pipes	0.39	0.26	0.96	0.64
33. Copper pipes	0.24	0.51	0.88	0.63
34. Steel pipes	0.30	0.06	0.06	0.12
35. Boilers, pumps, etc.	0.08	0.07	0.67	0.37
36. Sanitary fittings	0.63	0.43	1.87	1.20
37. Insulation	0.23	0.11	0.80	0.48
38. Plastering	0.83	1.16	2.99	1.99
39. Beds & screeds	1.93	2.57	1.58	1.92
40. Dry partitions & linings	0.73	1.54	2.47	1.80
41. Tiling & terrazzo	1.39	1.91	1.28	1.47
42. Suspended ceilings	1.13	2.43	0.55	1.16
43. Glass	2.34	1.00	1.52	1.60

Table 10. Continued.

Work category	Per cent weighting			
	Steel framed	Concrete framed	Brickwork construction	General construction
44. Decorations	2.45	1.29	3.40	2.64
Electrical	8.42	8.10	9.52	8.89
Heating ventilation and air conditioning	11.31	11.31	13.70	12.50
Lift	0.82	3.85	0.45	1.39
Structural steel	12.99	0.44	0.44	3.58

By kind permission of the Building Cost Information Service.

and in addition there are several operated by private firms. Of those that are published perhaps the following are the most familiar:

1 The BCIS building cost index

This is a long established index available from the Building Cost Information Service, which was originally based on a breakdown of four types of project into their constituent resources. With the introduction of the NEDO Price Adjustment Formula, which produces a factor cost index for each trade or work-section, the BCIS decided in 1977 to construct six indices using NEDO information together with weightings obtained from a study of forty bills of quantities, The bills were divided into three main building types in order that an index could be prepared for each. These were:

BCIS Index No 6   Steel framed
BCIS Index No 7   Concrete framed
BCIS Index No 8   Brickwork construction
In addition indices were prepared for the following:
BCIS Index No 5   General building cost index comprising 25% steel frame, 25% concrete frame, and 50% brickwork construction buildings.
BCIS Index No 4   General building cost index (excluding M & E) which is similar to the previous index but excludes electrical, heating, ventilating, and air conditioning and lift installations.
BCIS Index No 9   The mechanical, electrical and lift cost index.

All indices exclude any element of external works. The weightings obtained from the analysis for indices 6, 7 and 8 are shown in Table 10. To construct the index for any point in time it is necessary to multiply the NEDO index by the weighting and then calculate the arithmetic mean of the extended totals. The base date would be that of the NEDO index. Note that following the publication of the NEDO series II work categories the index has been extended.

2 'Building' housing cost index

This index is based on fixed weightings of specified labour, materials, and overheads found in typical house building and is published in 'Building' magazine.

3 NEDO formula

This index, or more correctly this series of indices, is prepared by the Property Services Agency of the Department of the Environment who have based it upon a large number of bills of quantities in order to arrive at typical proportions of labour, materials, and plant in each of 34 (series 1) or 48 (series 2) work categories. The indices are based on market prices, nationally agreed wages, etc and therefore do not take into account the problems of a contractor on a particular site. Although primarily provided for the assessment of price fluctuations on contracts where there is a fluctuations clause they can be adapted for use in a composite building cost index as already demonstrated with the BCIS format. The indices are published by HMSO.

4 DOE cost of new construction index

This index is prepared by the Department of the Environment and published in Housing and Construction Statistics (HMSO). It relates to the cost of construction completed in a given quarter and attempts to take into account the movement in material prices and labour including an allowance for productivity based upon the output per operative.

There are a number of other cost factor indices which appear in the technical press and builders' price books. The derivation of the indices is not always known and therefore they have not been included here, and should be used with caution.

Tender-based indices

As tender-based indices are a comparatively recent innovation there is not the long history of results that could be found with factor cost indices. However the number of institutions and practices producing an index of this type is steadily growing.

1 DOE (DQSD) index of building tender prices

This index gives the level of pricing contained in tenders for new work accepted by the Department of the Environment in the UK compared with a fixed price base. However it should be realised that since 1974 many of the contracts used in the index have contained the NEDO formula method of adjusting fluctuations instead of being tendered for on a fixed-price basis, and this tended to downturn the index for some years. The construction is similar to that already described in this chapter, where a sample of items is taken for each work-section up to 25% of the value of the particular trade. This index was the original tender price index and has formed the basis for the others which have followed. It is published in Housing and Construction Statistics (HMSO).

2 BCIS tender-based index

The method of construction is similar to the DOE form just described. Two indices are however prepared, one relating to fluctuating contracts and the other to fixed price. This goes some way towards overcoming the problem of the different allowances made in the tender figure for future variations in price. The sample of jobs from which the index is derived is not restricted to government projects, and therefore it can be expected to cover a wider range of building type than the preceding index. For a small fee the BCIS will undertake the preparation of an index for a priced bill from any of its members in order that a tender can be compared with the national average.

3 Davis Langdon and Everest tender price index

Again the method of construction is similar to (1) above, the only difference being that the index is based only upon schemes in the London area, includes non-government jobs, and is probably based upon a smaller sample. The index is published in *The Architects' Journal*.

Other indices are available which are in general use but do not strictly fall under the previous two headings. One example is the Nationwide Building Society's Index of House Prices which measures the change in cost to purchasers of different house types in different regions of the UK based on mortgage information. It should be used for guidance purposes only as house prices vary widely depending on such factors as size, location, specification, and fashion; it is of course substantially affected by variations (usually upward) in land values which may be considerably greater than building cost fluctuations.

Another example is the DOE price index of local authority house building, where a number of priced items are extracted from bills of quantities for traditional one- and two-storey houses in each quarter, and the index prepared from the average change in price of these items only. The items are considered to be of major cost significance and factors are applied to each item according to base year weightings, thus following the approach of a Laspeyres index. To obtain a national price index it is necessary to employ fixed regional weights, not only because the amount of building varies greatly from one region to another, but because price levels, and the rates of change in prices, vary too.

Comparison of index performance

The degree to which an index performs or moves depends on two factors:
  1 The market factors influencing the components represented within the index.
  2 The weightings applied to the components within the index.

This is true whether we are considering a tender-based or a factor cost index. In the latter case it is comparatively easy to see that if you measure an increase of 20% in the price of one material then if all other factors remain unchanged the index will go up by an amount equivalent to the weighting of that material in the total index, multiplied by 20%. With the tender-based index this process is obscured by the contractor's pricing method, and the fact that material cost is not shown as a separate item. However, the increase in the cost of this material will be included in the unit rates by the contractor

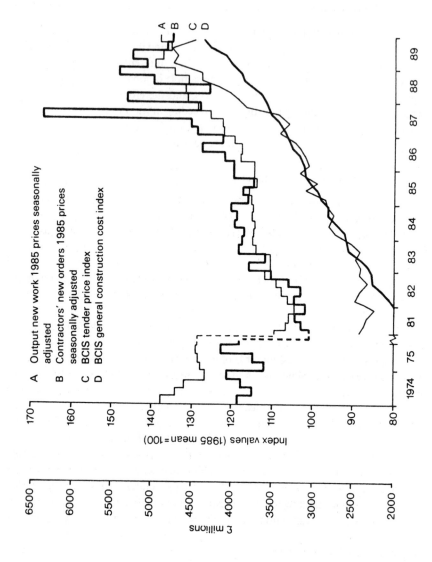

Fig. 11.1. Relationship of cost indices to trends in new orders and output of contractors' work. (*Reproduced by kind permission of the Building Cost Information Service*).

and will be subsequently weighted by the quantity of the items containing that material. It will also be weighted a second time when the work sections themselves are weighted. The tender-based index will, in addition, allow for any extra charge in the rates for increased overheads, profit, or above-normal resource costs because of the current market situation. It is these features that produce the distinctive performance of a tender-based index compared with the factor cost type.

By way of comparison let us look at the performance of the BCIS General Construction Cost Index (factor cost type) with the BCIS Tender Price Index (tender-based) for the years 1981 to 1989. The movement of each index is shown in Fig. 11.1 superimposed upon the industry activity indicators of contractors' new orders and contractors' output of new work expressed at constant 1985 prices.

The industry activity indicators are also shown for the years 1974 and 1975, to illustrate the cyclical nature of the industry's workload and the fact that the late 1980s boom was no larger in real terms than that which existed 15 years previously.

In the intervening period the British government took steps to avoid overheating the economy and cut public sector expenditure in a drastic way, mainly by reducing the public sector building programme. The industry then went into recession, and although it subsequently stabilised, the return of the Conservatives in 1979 with their emphasis on monetary controls to defeat inflation, led to government expenditure on building being cut yet again. However, the private sector came to the rescue and output gradually rose through the 1980s culminating in the sudden surge in the latter years of the decade, fuelled by commercial development.

It will be seen that, characteristically, tender prices were falling or constant during the end of the recession in the early 1980s (even though costs were rising) as contractors slashed their profit margins to obtain work. As the economy picked up in the middle of the decade the two indices moved together quite closely (though the tender price index, again characteristically, was the more erratic of the two). But when the volume of orders escalated from 1987 onwards the tender price index did the same, as contractors, finding orders easy to obtain but fearing strong competition for scarce labour and material resources, increased their prices. At the end of 1989 the start of a further recession was signalled, and in fact this arrived and hit the industry very hard the following year.

You will also notice the typical situation that, because of the long time occupied by construction projects, in a declining market the value of output is higher than the value of new orders but that when the market is buoyant the reverse is the case.

In any comparison of index performance the important fact that must be realised is that each index is measuring a different movement of cost or price. Even where the form of the index is the same, the difference in the weightings attributed to each component by individual indices will result in a different value for each at the end of all the calculations. In using an index it is essential that the user is quite clear what trend in cost he wishes to measure, and then picks the form and weighted structure that best suits his objective

Generally

It has already been stated that there are a number of problems associated with the construction of cost indices. These problems can also create difficulties in the use of the index and these together with some more general points are summarised below:

1 The index will usually be measuring a trend for a typical or model building type, and may not necessarily be measuring the change over time for the particular project that you are developing.

2 Where the base date is several years old then the question should be asked as to whether the index is being based on outdated criteria. This particularly applies to a factor cost index using base year weightings. It may be that the balance of resources has changed considerably in the intervening period and this is not represented in the index. Over quite a short period the 'mix' of resources on a typical building may alter as a result of the very changes in cost that the index is trying to record; for instance a rise in metal prices may cause a substitution of plastic for lead and copper in plumbing and roofing work. In the longer term changes in technology may not only make the base 'mix' untypical but may also distort its prices; as new techniques and materials come into greater use their price tends to decrease proportionately, whilst the cost of obsolescent technology tends to rise faster than the general rate of increase. However, every index must have some history in order that trends can be identified.

3 Regular publication of the index is required, otherwise the user is put in the position of having to forecast the immediate past (i.e. the time between the last publication and the present) as well as the future. Three months is probably the maximum delay that should be expected.

4 The index should be published at regular fairly short intervals in order that trends can be determined. General trends over long periods of time can be ascertained from a yearly index. However a quarterly or monthly index is essential if an indication is required of a sudden change in the building market, or of course if seasonal variations are to be measured.

5 The basis of construction of the index must be known in order for it to be used intelligently. The choice of a tender-based or factor cost index will be dependent on what the user wishes to measure, but in addition the weightings of any index are important in judging whether the results can be applied to the project under consideration.

6 Short term changes in an index are unreliable since the inherent errors in the system of compiling the index may well be the equivalent of several points on the scale.

7 In the tender-based indices, a good sample of bills must be used if bias due to regional variation and building function is not to distort the results.

8 When plotting a trend in an index it is sometimes advisable to use a logarithmic scale for the index against a natural scale for time. The advantage of this method is that if costs are rising at a regular percentage every year then the index values will be shown as a straight line. Any deviation above or below this line will show immediately as a change in this regular pattern and will assist in detecting a new trend. Figure 11.2 shows the principles involved using a 20% proportionate increase as an example. If a natural scale is used for the index then an exponential curve results for the 20% increase and comparison is more difficult.

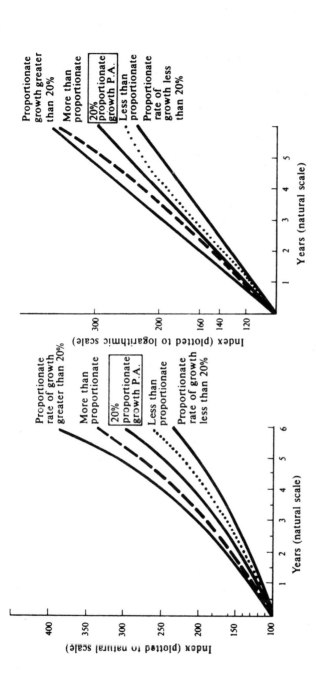

Fig. 11.2. Showing typical cost increases plotted against natural (LH) and logarithmic (RH) vertical scales. A proportionate rate of growth is one which increases annually by a constant percentage of its immediately previous value (e.g. an increase of 20% on an index value of 100 gives 120. A subsequent increase of 20% would give 144, not 140). Note the characteristic shape – straight, concave, convex – of the index graphs in the second diagram, making them much easier to identify visually.

The cost planner obviously has a duty to keep himself aware of past, current, and future economic trends, and he will use this awareness in coming to a conclusion on the extrapolated index value which he will use in estimating for a future project. Of course no one has yet proved that he or she can accurately forecast the future, an this particularly applies to future index numbers. Providing reasonable care has been exercised the quantity surveyor should not be blamed if his forecast turns out to be wrong due to changed economic or political factors that were unforeseen at the time of making the estimate. In recent years it has become common practice and far more sensible to give a range of anticipated future costs based on the possibility of certain events occurring. In any case it is most important that everything should be made explicit when the original estimate is given. The details should include the index which has been used, the point or range that has been chosen together with the reasons for this choice, and of course the time scale for the project which has been assumed when making the forecast.

Which type of index to use

By their very existence it is obvious that all cost indices have a role to play in assisting the cost adviser to determine the market forces affecting his project. The choice of index will however be determined by the use to which the index is put.

1 For short term forecasting, allowances for fluctuations, etc, then the factor cost index can be used.
2 For general estimating and cost planning purposes the tender-based method will more reliably measure the change in the cost to the client. It also has the advantage of being able to measure the performance of a tender, region, or building type against the national average.
3 For very long term forecasting and budgeting the use of a unit price index based on cost per square metre or unit of accommodation may be more applicable.

Part 3
Cost Planning Techniques

Chapter 12
Product-based Cost Models

A product-based cost model is one that models the finished project rather than the process of its construction. It therefore has to be based on data relating to finished work.

As stated in Chapter 9 the simplest form of such a model takes no account of the configuration, or details of design, of the building but is based simply upon the floor area of the proposed project (gross or net), or upon its volume, or upon some user parameter such as number of pupil places for a school or number of beds for a hospital.

It is customary to exclude site works from the single price-rate calculation and to estimate them separately, since their cost obviously has no relationship to the size of the building, and the engineering services are also sometimes treated in the same way (with rather less justification).

After World War II, and before cost planning systems became established, single price-rate estimates increasingly acquired a reputation for inaccuracy, but this was because there was no follow-up process for keeping the estimate and the design in tune as the design developed.

Such an estimate merely attempts to forecast that a building of a certain size can be built for a certain sum of money, it cannot analyse whether a particular design is going to meet that cost. Of course, it is possible to weight the estimate subjectively on the grounds that the proposed solution looks to be at the expensive end or the low-cost end of the market, but here we are well into the realms of guesswork.

However, although it cannot say that a particular design will achieve the required result, the 'single price-rate' approach has an important role as the first stage of a system which in the end will do that. It has the great advantage that it can be used before any design has taken place, whereas even the most simple process-based models require a tangible design as their basis.

The bill of quantities

The term 'cost model' is a relatively new one, but the traditional bill of quantities is in fact a very good example of a product-based cost model. Because of its insistence on measuring 'finished work in place' the bill has often been criticised by production-oriented people (although it never purported to be a production-control document), but it does have the virtue of providing a total cost model within a single document.

There are two important points in favour of the product-oriented approach to this definition of the project:

1 What the client will be contracting for is finished work in place, not a production process. The model is therefore couched in contractual terms.
2 The design team can define and categorize the finished work, according to an industry-agreed convention ('The Standard Method of Measurement of

Building Works'). This is not the case with the production processes, which depend on the methods of the particular constructor as much as on the details of the design, and which cannot easily be analysed or categorised outside the context of the particular project – there is no agreed basis for doing this.

Although the bill of quantities is a useful cost model, it is not usually available until the design of the project is completed and it is therefore of little use for cost control of the design. Its great importance however is as a source of cost data for subsequent projects.

Since the bill of quantities is a major source of cost data it is important to be quite clear about its role. It is essentially a marketing document, not a production control document, and the rates in it are prices not production costs. Although there clearly has to be a relationship between the builder's prices and his costs this really only applies at the level of the total project – he can't afford to undertake the project for less than his costs, and competition will usually ensure that he cannot charge an unreasonable profit on top of them.

However, this does not apply to individual pieces of work, because the rates in the bill of quantities are not separately-offered prices in the sense of the supermarket shelf. You can only buy the brickwork at so much a square metre if you also buy the carpentry at whatever price is charged, so these rates are nothing more than notional breakdowns of the total price, and are made with commercial rather than cost control objectives in mind.

Very often the client's quantity surveyor will object if the rate for a particular item is very different from that usually charged. But he has very little sanction to enforce his objection, except the rather impracticable one of advising the client not to accept the tender.

Elemental Cost Analysis

Elemental cost analysis is perhaps the best-known product-based cost model, and provides the data upon which elemental cost planning is based. This technique is currently used by the quantity surveying profession at large and, in spite of a number of failings which have become apparent over the years, has made possible a degree of control over costs which was previously unknown. It is the experience gained with this system, and the attitude of mind which it has engendered, which has led the way to more fundamental approaches to cost control.

Because elemental cost analysis was developed in order to provide data for the preparation of a design cost plan it is first necessary to look at the requirements of the latter. This is done in the context of the traditional competitive-tendered type of contract, which is still popular but which at the time the system was developed was almost universally used.

What are the essential features of a cost plan? The obvious requirement (common to any form of estimate prior to tendering) is that it should anticipate the tender amount as closely as possible, but there are two particular requirements that must be fulfilled before the estimate becomes a cost plan.

Firstly it must be prepared and set out in such a way that as each drawing is produced it can be checked against the estimate without waiting until the whole design is complete. This enables any necessary adjustments to be made before the drawing is used for the preparation of quantities. The second particular requirement of a cost plan is that it should be capable of comparison with other known schemes in order to see whether the amount of money allocated to each part of the building is reasonable in itself and also is a reasonable proportion of the whole. From the point of view of economical building this second requirement is possibly even more important than the first.

It can be seen that this philosophy is especially suited to work in the public sector where, as we have seen, the only real cost criterion is comparison with what has been done before. At the time it was developed such work dominated the building programme – it is possible that a totally different approach might have been used had profit development then been dominant. However, once a system has become established it tends to form the basis for future development as circumstances change, and this is what has happened with elemental cost planning, still widely used today.

How can the costs of one building project be compared with another? The first suggestion which would occur to a quantity surveyor would be to look at the priced bills of quantities for the two contracts. Suppose we look at the summary pages of the bills for two different schools.

	Clay Green School £	Woodley Road School £
Preliminaries	57,300	129,000
Excavation	24,240	28,980
Concrete work	129,915	272,925
Brickwork and blockwork	154,305	101,145
Masonry	10,950	—
Asphalt work	21,150	—
Roofing	74,655	39,045
Woodwork	228,675	290,895
Steelwork and metalwork	189,660	197,850
Plumbing, engineering and electrical work	319,500	399,825
Floor, wall, and ceiling finishes	136,725	133,275
Glazing	14,640	25,530
Painting and decoration	39,135	37,170
Drainage	43,260	33,495
Site works	109,890	79,470
	1,554,000	1,768,605
Insurances and summary items	5,250	37,950
Contingencies	30,000	30,000
	£1,589,250	£1,836,555

These trade totals tell us very little. We can see that the second school is dearer than the first (it is in fact larger), but it is difficult enough to try and compare the rather varied trade breakdowns without having to make repeated mental adjustments for the difference in size between the two buildings. So a first step is to divide each trade total by the floor area of the respective schools in order to obtain comparative prices per square metre for each trade.

	Clay Green (2,000 m²) Cost per m² £	Woodley Road (2,500 m²) Cost per m² £
Preliminaries	£28.65	£51.60
Excavation	12.12	11.59
Concrete work		
Brickwork and blockwork	77.15	40.46
Masonry	5.48	—
Asphalt work	10.57	—
Roofing	37.33	15.62
Woodwork	114.34	116.36
Steelwork and metalwork	94.83	79.14
Plumbing, engineering, and		
electrical work	159.75	159.93
Floor, wall, and ceiling		
finishes	68.36	53.31
Glazing	7.32	10.21
Painting and decoration	19.57	14.87
Drainage	21.63	13.40
Site works	54.95	31.79
	777.00	707.44
Insurances and summary items	2.63	15.18
Contingencies	15.00	12.00
	£794.63	£734.62

We now have the costs on a more truly comparable basis. Clay Green has a steel frame with timber pitched and tiled roofs and mainly brick faced external walls with 'window holes', while Woodley Road has a reinforced concrete frame with felt flat roofs on composite decking and large wall areas of metal windows with concrete panel infilling.

As we would expect, Woodley Road shows a far higher total for concrete work, and a lower figure for brickwork, roofing, and steel-work. At Clay Green there are some asphalt covered concrete flat roofs and the crawlway floor duct is tanked in asphalt, whereas at Woodley Road the floor ducts are waterproof rendered.

The glazing figures allow for the larger window area at Woodley Road, while the drainage and site works figures are consistent with Clay Green being on a larger and more rural site with a consequently greater extent of works. The differences in floor, wall, and ceiling finishes and decoration are simply due to a higher standard of specification at Clay Green.

However, while the effect of these differences in specification can be traced to some extent in the trade costs per square metre, the inclusion of parts of several elements of the building in one trade section makes it impossible to carry comparisons very far. For instance, the figures for Woodley Road for concrete work are affected by:

1 The reinforced concrete frame in lieu of structural steelwork.
2 Some concrete walls in lieu of brickwork.
3 The concrete facing panels in lieu of faced brickwork.

We do not know how much of the difference is due to any one of these causes. Similarly the steelwork figures are affected by the omission of the steel frame and the increased area of higher quality metal windows at Woodley Road. These are compensating differences and in fact the two figures do not reflect the scale of the variations between one school and the other in this section. Other sections are similarly affected – the rates for woodwork are almost meaningless without detailed breakdowns.

Although the figures which we have obtained are interesting, they do not enable us to make any really valid cost comparisons. We do not know whether the steel and the concrete frames are competitive in cost, and we cannot tell how much is saved by a felt roof instead of a tiled one.

In order to be able to make such comparisons we shall have to split up the bill of quantities in a different way. We shall have to divide the work into 'elements'.

An element has been defined as 'That part of the building which always performs the same functions irrespective of building type', and, we might add, irrespective of specification. What list of elements shall we use?

The possibilities are nearly endless. We could use as few as six, for example:

1 Sub-structure and ground floor, complete with finishes.
2 External and internal walls, complete with finishes.
3 Upper floors including staircases, complete with finishes, and proportion of frame.
4 Roof, complete with finishes and proportion of frame.
5 Services.
6 Site works.

Alternatively we could use any greater number within reason – some authorities use over 40. Points to bear in mind in arriving at a decision are:

1 The definition of an element as stated previously. Any element chosen must be capable of bring defined exactly, so as to ensure uniformity between the elemental breakdowns of any number of contracts even if the breakdowns are done by different people.
2 The element must be of cost importance.
3 The element must be easily separated, both in measuring from sketch drawings and in analysing bills of quantities.
4 The list of elements chosen should be capable of bring reconciled with those used by others, for comparison purposes. Unless the cost planner belongs to a very large firm or other organisation he will frequently need to make comparisons with analyses published in *The Architects' Journal* and elsewhere, or with analyses obtained from the Building Cost Information Service of the RICS.

To take an example of the last point, in most published forms of cost analysis the cost of parapet walls and copings is included under 'roof'. The reader might prefer to include it under 'external walls' (it would make the analysis of traditional bills easier) but he would then be unable to compare his own cost information on these two elements with the published information, as the basis of measurement would be quite different.

The Standard Form of Cost Analysis

As the use of elemental cost planning has increased, some differing forms of cost analysis have been developed, both by those independent authorities and firms who are building up their own cost records and by journals which publish cost information for the benefit of their readers. The weekly magazine *The Architects' Journal*, was one of the pioneers in this field, and produced a detailed set of rules for their published analyses, while the RICS Building Cost Information Service and others used somewhat different rules.

It was obviously desirable that a uniform set of rules should be established, so that users could benefit fully from cost data prepared outside their own organisations. The RICS therefore set up a working party to standardise cost analyses; this proved to be difficult because, in practice, it is not possible to define a set of independent functional elements which can be related to a bill of quantities, and so any standard cost analysis becomes a compromise between independent functions on the one hand and ease of producing the data from traditional documentation on the other.

In December 1969 the first Standard Form of Cost Analysis was published by the Building Cost Information Service of RICS; in addition to its sponsorship by the RICS the Form was also supported by the chief quantity surveyors of all the main government departments which were concerned with building (and this was at a time when these bodies directly controlled a major part of the non-housing building programme). It was this wide measure of support which gave the Form such importance, as anybody using a different format would soon have become isolated from the cost experience of the rest of the quantity surveying profession.

Twenty years later the SFCA has been little altered. As an example of a practical elemental breakdown, the headings in the current Standard Form of Cost Analysis are as follows: (Please note that the third level of detail is no longer used by the Building Cost Information Service in its published analyses – a tacit admission of the unreliability of sub-elemental analysis referred to later in this chapter).

1. Sub-structure

2. Superstructure
    2.A. Frame
    2.B. Upper floors
    2.C. Roof
        2.C.1. Roof structure
        2.C.2. Roof finishes
        2.C.3. Roof drainage
        2.C.4. Roof lights
    2.D. Stairs
        2.D.1. Stair structure
        2.D.2. etc.

2.E. External walls
2.F. Windows and external doors
  2.F.1. Windows
  2.F.2. External doors
2.G. Internal walls and partitions
2.H. Internal doors.

3. Internal finishes
3.A. Wall finishes
3.B. Floor finishes
3.C. Ceiling finishes
  3.C.1. Finishes to ceilings
  3.C.2. Suspended ceilings

4. Fittings and furnishings
4.A. Fittings and furnishings
  4.A.1. etc.

5. Services
5.A. etc.

6. External works
6.A. etc.

This summary is sufficient to enable the principle to be understood, and full details of all the sub-divisions are given in Appendix A. Most other forms of elemental analysis are basically similar, but in order to be successful they must incorporate the same sort of hierarchical principle. It will be seen that the Form can be used at different levels of generality – the six element groupings *Sub-structure, Superstructure, Internal Finishes*, etc.; the elements themselves 2.A., 2.B., 2.C., etc.; or the sub-elements 2.C.1., 2.C.2., 2.C.3., etc., but because the detail is grouped in this way an analysis into six items only will be quite compatible with a fully detailed analysis of another project. The construction of the list therefore required a careful selection of elements each of which was significant on its own but which could form part of a larger significant group.

One or two forms of analysis which have been used differ by including finishes, windows, and external doors in the external walling element, and including finishes, internal doors, and partitions in the internal walling element. This is more logical but means splitting finishes between external and internal walls, which is difficult to do when analysing a traditional bill of quantities.

We can see that if the costs per square metre of Clay Green and Woodley Road schools could be expressed in terms of these elements it would enable us to find the answers to the questions which we were asking (which is the cheaper frame, how do the two roofs compare in cost) and these figures would enable us to cost plan a third school basically similar to Woodley Road but with a steel frame.

The Standard Form and CI/SfB

It was a source of disappointment to many people that the elements themselves, their grouping, and their coding were not the same as in Table 1 of the CI/SfB system used by the architectural profession for coding and classifying design information. It would have been very convenient for the architect to have a record of typical costs filed with his design information.

However, the incompatibility of the systems is not quite such a disadvantage as might appear. We have seen that the so-called 'elemental costs' which we obtain at present are not true costs at all, but are only a breakdown of a bill of quantities in which money may have been allocated to the various parts of the work in a quite capricious manner, quite apart from the overall level of pricing of the bill itself. It would therefore be dangerous to detach these 'costs' from the analysis and index them for the use of an architect as though they were scientific data like thermal insulation values.

One day it may well be possible to do this; but at present it is vital that this cost data should only be used by a qualified cost planner who has the experience and knowledge necessary to assess and manipulate it. The cost analyses themselves however are referenced to CI/SfB Table 0 ('Built Environment') and so can be filed within the CI/SfB system.

Preambles to the Standard Form

In the Standard Form there is rather more information to be given about the size and nature of the building than was previously customary.

The gross floor area is measured in the normal way, that is the overall area at each floor level within the containing walls; but the basement floors (grouped together), the ground floor and the upper floors (grouped together) are each required to be shown separately.

The definition of 'enclosed spaces' means that open entrance areas, etc., are excluded from the gross floor areas, although even a light enclosing member such as a balustrade will suffice to include the area. It is possible that doubt might arise when a wall becomes so pierced by blank openings that it no longer acts as an 'enclosing wall'; obviously a mere series of columns does not meet the definition.

Although 'lift, plant, tank rooms, and the like above main roof slab' are to be included in the gross floor area, we have the option of excluding these if we are prepared to allocate their costs to 'Builder's Work in Connection'. If the plant room is only required because of the existence of a particular service it seems quite logical to adopt the second course.

Roof and wall areas in the Preambles to the Standard Form

These are both measured gross over all openings, etc., the roof being measured on plan area. As it is 'walls of enclosed spaces' which are required, parapets, gable ends of unused roof spaces, etc. would not be included in the wall area. The roof on the other hand is measured across overhang and would presumably include roofs over open entrance porches and other areas which do not count as 'enclosed spaces'. If there are substantial areas of open covered way it would probably be better to exclude them from the elemental analysis altogether, and deal with them under 'Site Works'.

The wall area is not shown on the Form, except in calculating the wall-to-floor elemental ratio. This is:

$$\frac{\text{wall area}}{\text{floor area}}$$

The lower the ratio the more economical the design.

Interdependence of elements

Although it is quite easy to define a cost-planning element it is more difficult to divide a bill of quantities up into elements which comply with this strict definition. A major difficulty is that an indivisible building element may combine several functions (some of which it shares with other elements).

For example 'external walls' may have all or some of the following functions:

1 Keeping out the weather.
2 Thermal insulation.
3 Sound insulation.
4 Supporting themselves (dead loads, wind loads).
5 Supporting floors and roofs.
6 Transmitting light and ventilation (curtain walls).

If an external wall only performs a few of these functions then it is obviously unreasonable to compare its cost with that of a wall which performs a greater number. It is therefore usually necessary to refer to the 'frame' and 'window' elements in order to arrive at a true indication of the wall's cost performance, and similar cross-references may be required when costing other elements.

Preliminaries, insurances, and contingencies

These may be shown as separate costs per square metre (treating them, in fact, almost as extra elements) or they may be allocated proportionately among other elements. This is an important point and deserves some consideration.

First, preliminaries and insurances. The cost of any or all of the following items (mostly at the contractor's discretion) may be included in the preliminaries or insurances sections of a priced bill of quantities for a new project.

Huts, temporary buildings, latrines.
Canteen, mess-rooms, tea-boy, welfare.
Huts and attendance on clerk of works, site architect and surveyor.
Mechanical plant, including tower cranes, excavating and concreting plant, lifts, dumpers, etc.
Scaffolding.
Non-mechanical plant and small tools.
Water for works.
Temporary electricity supply.
Consumable stores.

Temporary fencing and hoardings, temporary roads and standings, car parking space.

Agent, foreman, and other site supervisory staff.

Timekeeper and site clerks.

Watchmen and security lighting.

Heating of building for drying out.

Temporary weather protection.

Attendance on sub-contractors and artists.

National Insurance payments.

Superannuation.

Guaranteed week (wet time), travelling time and expenses, subsistence and lodging.

Redundancy payments.

Training levies.

Anticipated increases in costs of labour or material.

Bonus or other supplementary payments.

Making good damage and defects.

Fire insurances, third party insurance, and any other insurances required by the client.

Public liability or contractor's all-risk insurance.

Head office expenses (overheads).

Profit.

However, almost any of these items (and certainly any of the major ones) may be included in the rates for the building work instead of being shown separately; some contractors do not price preliminaries at all while others do price preliminaries but in such a way that it is impossible for the quantity surveyor to find out what is or is not supposed to be included. Sometimes at the last minute the contractor may have second thoughts about a tender and adjust it by adding a lump sum to preliminaries, or taking one off.

It must be realised that any large differences between the amounts of money inserted against the preliminaries items on one job and on another are less likely to be caused by genuine contractual differences (site conditions, access, etc.) than by the different pricing habits of the two contractors – remember that the allocation of costs within a contract bill of quantities is done largely for commercial purposes. Some of the principal items such as profit, supervision, scaffolding, plant, and overheads could affect the level of pricing by 15% to 20% according to whether or not they are included in preliminaries.

As these pricing habits are to some extent regional it may be possible for the quantity surveyor's department of a local authority, or for a firm of cost planners whose work is confined geographically, to consider preliminaries and insurances as a separate element with some degree of consistency. But it would be safer on the whole to add preliminaries and insurances to each element as a percentage in order to give a common basis of comparison. If we refer back to the summaries of the two schools, Clay Green and Woodley Road, we can see that the work-section prices for the latter appear low by comparison because preliminaries and insurances have been priced more fully.

This advice, of course, relates only to the analysis of bills of quantities and to the early stage estimates for a new project based upon such data. When preparing a more detailed estimate for a major new project preliminaries and insurances have to be considered on their merits. If the new project has some abnormal feature such as a difficult site or an uneconomically short contract period it will be necessary to give special consideration to these matters from the start.

We have also to consider contingencies. Unlike preliminaries the contingency sum is an arbitrary amount fixed at the architect's discretion. It is not really part of the contractor's tender but is an amount he is instructed to add to his tender in order that there may be a cushion to absorb unforeseen extras. It normally has no effect on the level of pricing of the bill and is better treated as a separate element rather than as a percentage on the remainder of the work.

Analysis of final accounts

Upon first consideration it might seem a good idea to analyse the final account instead of the tender, since this will give a more accurate picture of the actual cost of the building. The objections are twofold; firstly it would be much more difficult to analyse both bill and variation account than to analyse the bill alone; and secondly the analysis would not be available until perhaps three or four years after a tender analysis and so would only be of historic interest. The differences between the tender and final account are not usually great enough to invalidate an analysis obtained from the bill of quantities.

Cost analysis of management contracts

Management contracts and the like pose a difficult problem. There is little point in analysing the master estimate, since there is no contractual commitment to this – it is itself part of a cost planning system and may well have been prepared on an elemental basis. On the other hand, by the time all the costs are known we would be effectively analysing a final account, with all the problems just mentioned. But management contracts imply a production orientation to the project, and it is perhaps the resource-based cost information obtainable from these jobs which is more useful than the product-based costs.

System

Whatever methods of cost breakdown and whatever methods of cost planning are chosen they must be adhered to rigidly. Otherwise not only are the figures useless for reference purposes but the whole idea of working to a system is lost. The whole technique of cost analysis depends upon working in accordance with a fixed method; this is why standard forms should be used. There is plenty of scope for rough working but the forms must be used at all vital points, so that there is no possibility of preliminaries (for instance) being left out because the cost planner who did the previous estimate believes in showing them separately whereas the person using this estimate for reference supposes that they are included in the rates.

In most cases the forms and instructions issued by the Building Cost Information Service can be used, thus ensuring compatibility with information obtained from other sources.

Preparation of an elemental cost analysis

Cost analyses are most usually carried out today using a micro-computer. Special software can be bought, but many cost planners find that commonly-available spreadsheet packages enable them to construct their own cost-analysis system very easily.

The aim of cost analysis is to provide data for use in elemental cost planning; as little time as possible should be spent on it consistent with a fair degree of accuracy. Meticulous allocation of trivial sums of money, or the identification of insignificant changes in specification, should be avoided. If there is no time to prepare a full analysis an outline analysis will be better than nothing and may take less than an hour to prepare. For the preparation of an elemental cost analysis we shall require a priced bill of quantities, a drawing showing plans and elevations, and a list of elements. It would also be very helpful to have the sorted slips or dimensions although these are not essential.

Each item in the bill has to be allocated to one or more elements until every item has been dealt with and the elemental totals will equal the total of the tender. Once an office has adopted a certain form of analysis it will be possible to prepare the bill with subsequent analysis in mind; this will ease the task of the analyser considerably and will make it unnecessary to refer to slips or dimensions. It is not necessary to depart radically from the usual order of billing as long as the main elements can be kept separate within each trade or section of trade.

For instance the 'sawn softwood' section of woodwork could be billed under headings of roof timbers, upper floors, stud partitions; and the 'reinforced concrete' section of the concrete work could be separated into sub-structure, frame, upper floors, roof, and staircases. This should not involve lengthening the bill greatly as there will be very little duplication of items between different elements; the concrete work is the only section where this should occur to a significant extent.

A bill prepared in this manner will be useful not only for analysis but will be convenient for interim valuations and for the contractor's use in bonusing and site organisation generally. The so-called Northern system of taking-off, where the bill is written straight from the dimensions, lends itself to the preparation of a sub-divided bill of this sort.

Any further breakdown of the tendering bill into a completely elemental format is very unpopular with builders' estimators, since design cost planning elements are of little significance to them and a good deal of re-arrangement into trade order has to be carried out by them.

This is a convenient point to mention the ingenious bill which was pioneered by Hertfordshire County Council, in which each trade of each element was printed on a separate sheet and coded so that the bill could be shuffled into either elemental or trade order at will. The usual arrangement was for the bill to be sent out in trade form for tendering, and put into elemental form for subsequent use.

Another development is the Master bill for tendering, which can be used when a contract consists of several buildings each of which requires a separate bill which in turn is divided into trades and/or elements. In order to avoid inundating the contractors who are tendering with masses of paper the section bills are re-abstracted into a master bill which is used only for tendering. This master bill is in strict trade order and contains no sub-headings or sections apart from those required by the Standard Method of Measurement. The additional expense of preparing this bill may be worthwhile because busy contractors naturally prefer a two hundred page bill to a pile of bills containing up to a thousand pages, and keener tendering will probably result. By using this method it is possible to have the section bills divided into elements without bothering about trade divisions, which are only really required for tendering purposes.

Further refinements are possible when bills are prepared by computer. Providing that all the dimensions are given elemental code references the bill can be sent out to tender in traditional form and after tenders have been received the computer will be able to print an elemental version of the bill. In some systems the bill rates are also fed into the computer so that the elemental breakdown will be completely priced as well.

Before going any further there are three terms which must be defined in order that the processes of cost analysis and cost planning may be understood.

Elemental cost is the cost of the element expressed in terms of the superficial area of the building.

Elemental unit quantity (sometimes also called *quantity factor*) is the actual quantity of the element, expressed in square metres for such elements as floors, roof, walls, finishes or in terms of number of elements where this is not practicable.

Unit cost is the cost of the element expressed in terms of the element unit quantity, e.g. 1,000 m² of internal walling costing £1,200 gives a unit cost of £1.20 per m².

Elemental ratio is the proportion which the unit quantity of one element bears to that of another. A commonly used example of this is the ratio of external wall area to gross floor area.

When analysing a bill of quantities there are three different ways in which items may be allocated:

1 The description of the item may indicate the element to which it belongs, in which case there is no necessity to refer to slips or dimensions.

2 The item may be too trivial to spend time on, in which case it may be allocated as seems most obvious or as is most convenient.

3 It may be necessary to refer to the dimensions in order to allocate the item correctly. Since it would take far too long to refer to the taking-off for every item in the bill, the analyser must use a good deal of discretion about this. In cases of doubt he would be influenced very much by the cost importance of the item.

Element unit quantities

Unfortunately a simple analysis of cost into elements will not satisfy all our requirements. It will give us a money total for each element and we can divide each total by the floor area to get the elemental cost per square metre, but it does not give us the unit quantities or the unit costs. We need these if the analysis is to be of much use. Some commonly used element unit quantity factors are set out below as a guide; where 'none' is marked the elemental cost per square metre is the only basis of cost comparison.

Work below lowest floor finish. Area of lowest floor.

Frame. Area of floors relating to frame.

Upper floors. Area of upper floors.

Roof. Area on plan of roof measured to external edge of eaves, but excluding area of rooflights.

Rooflights. Area of structural opening measured parallel to roof surface.

Staircases. Number and total vertical rise of staircases, or area on plan.

External walls. Area of external walls excluding window and door openings. Basement walls given separately.

Windows. Area of clear opening in walls.

External doors. Area of clear opening in walls.

Internal walls and partitions. Area of internal walls excluding openings.

Internal doors. As for external doors.

Ironmongery. None.

Wall finishes. Area of finishes.

Floor finishes. Area of finishes.

Ceiling finishes. Area of finishes.

Decorations. None.

Fittings. Often none, but where appropriate the total length of benches or other details might be given.

Plumbing and hot water. Number and type of sanitary fittings, number of hot and cold draw-offs.

Heating services. Heat load in kW, cubic capacity of accommodation served.

Gas services. Number of outlets.

Electrical services. Number of points, total electrical load.

Special services. Such information as will indicate the extent of each service (e.g. for lifts the number, capacity and speed of each and number of stops should be given).

Drainage. None.

External works. None.

As well as giving the consolidated unit quantities and costs for each element it is common practice to sub-divide them into categories of different construction or finish and give quantities and cost for each. For example, for a building of 2,400 m^2 of total floor area:

ELEMENT: UPPER FLOORS

Total cost of element:	£85,525
Cost per m^2 of total floor area (elemental cost)	£35.63
Unit quantity and cost	
Element unit quantity	1,000 m^2
Unit cost per m^2	£85.00
Sub-division	
550 m^2 of 150 mm r.c. slab at £82.50	45,375
350 m^2 of 225 mm r.c. slab at £105	36,750
100 m^2 of 25 mm softwood boarding	
on 175 × 50 mm joists at £34	3,400
	£85,525

While these sub-divisions are very useful as explanations of how the total cost is affected by specification it must never be forgotten when using them that the greater the detail in which a priced bill of quantities is analysed the less reliable are the results. It might well be found that a bill for the same job priced by another builder would give completely different figures at this level of breakdown, though quite similar in total.

The overall unit costs however, will be found particularly useful when preparing early estimates of cost before specification details are available.

Obtaining unit costs from elemental analyses

In order to obtain unit costs as well as elemental costs the analysis has to be more elaborate than if elemental costs alone were being recorded, particularly if we are going to keep separate costs for the different forms of structure or finish within each element.

There are two problems to solve: the separation of the costs within the element and the recording of areas and other quantities. Such things as areas of walls, floors, and finishes can be obtained most easily from the bill, but sizes of window and door openings may be difficult to get from this source (particularly where there are fanlights, sidelights, or windows glazed directly into frames) and it may be necessary to refer to the dimensions or the drawings. There is also the difficulty of areas which occur more than once, for instance the areas of concrete in floors will be duplicated by the formwork areas and these must not be added in again, but areas of concrete floors and hollow pot floors will require to be added together – so either the analysis needs to be done by somebody technically competent or else very clear procedures have to be laid down. No attempt need be made to separate constructions which differ only in detail (such as 100 mm, 125 mm and 150 mm floor slabs), because the aim is to obtain overall unit rates for basic constructions, not 'Bill Rates' for individual items.

The analysis in its final form

This is one of the places where a standard form, preferably the SFCA form, should always be used.

It is not necessary to fill in too much specification detail if the analysis is for office use, as it should be possible to refer to the contract papers if anything more than a very broad outline is required.

There should be a reference number for the analysis so that a list of analyses can be kept, and there should also be a cost index value so that allowance can be made for changes in market prices when comparing with past or future jobs.

Note that elemental costs continue as a running total but unit costs cannot be carried forward, as a total of them would be meaningless.

The BCIS on-line system

As well as providing printed elemental cost analyses the Building Cost Information Service offers an on-line link whereby their data base of analyses can be examined from a distance, and any interesting examples down-loaded to the cost planner's own computer for use in the cost planning of new projects. This service also provides the facility to amend the BCIS analyses for the cost implications of a change in tender date or UK region of construction, so enabling comparison between jobs carried out at different times or in different places.

It is then possible to download a complete detailed analysis and amend this as required to form the cost plan for a new project, using the methods described in Chapters 16 and 17.

Chapter 13
Resource-based Cost Models

In Chapter 6 and Chapter 9 we looked briefly at the cost of actual building operations, and it might be worthwhile to re-read these chapters before starting on this chapter. The detailed cost models which we have examined so far have been largely based upon the measurement of finished work in place, and its valuation from bills of quantities or other market-price-orientated data. Quite apart from the general fallibility of bill rates, these models have embodied two major fallacies:

1 That the production cost of a building element is proportional to its finished quantity.
2 That the cost of a building element has an independent existence which can be considered separately from the rest of the building.

We have been aware of these drawbacks almost from the start and are able to come to terms with them through the exercise of professional skill. But it is worthwhile considering whether it would not be better to base our estimates and cost control procedures on production cost criteria, as is done in most (perhaps all) other industries.

One major reason why we do not normally adopt this approach has already been given, which is that under lump sum contracting arrangements it is the market price, rather than production costs, which concerns the client and his advisers. This reason, however, is not valid if we are considering a cost-reimbursement or management type of contract, or if we are looking at a situation where the cost planning is being carried out inside a design-and-build organisation.

There is, however, another and quite different ground on which resource-based cost planning can be criticised. Much of the resource cost of modern building is attributable to plant and organisation, and both the initial estimating and the subsequent refining of the estimate require the envisaging of technological solutions. Whether the building will be steel-framed, precast concrete frame and panel, or in-situ concrete, will demand totally different approaches with probably significant cost differences for any particular configuration and set of user requirements.

This last paragraph in fact sounds like an argument in favour of a resource basis rather than a criticism, but the big disadvantage is that it moves the design considerations of a building into the production field much too early in the process. An economical structural system may be postulated and the configuration of the building developed, for example, to suit the radius of action of the tower crane (or cranes) placed in the most efficient positions, or to suit the repetition of precast units. The architect's traditional approach, as we have already seen, is the opposite one – the form of the building is evolved primarily as a set of user-oriented spaces, and the best technological solution for that configuration is then investigated. Quite clearly, in both cases, some compromises may have to be reached, but the basic issue is whether user needs and environmental considerations come first and the construction follows, or whether production efficiency should be the consideration from which design develops.

There are too many existing buildings which remind us that construction optimisation lasts for months but the consequences last for years (although not always as many years as the designers intended). Unfortunately production managers are no less lazy than the rest of us and prefer to postulate easy solutions rather than applying themselves to devising an efficient way of building a user-oriented design. It ought to be possible to do this provided the details of construction have not been developed too far before the builder is called in.

However, whilst it is true that the quantity surveyor's traditional elemental cost planning approach enables the early design process to proceed with some semblance of cost control prior to construction decisions, such decisions do have to be faced sooner or later. Once this point is reached a resource-oriented approach has much to recommend it in cases other than the competitive price-in-advance situation, particularly where the builder who will be undertaking the work is a party to the cost planning exercise. If this is not the case, then the method is of dubious advantage; there are many different ways of organising a building site and each builder has his preferred methods and equipment, so that there is unlikely to be a 'best' solution that any builder would automatically accept. Much of the advantage of a resource-based estimate is that it can be used for production-cost control purposes, and this benefit will be lost if the chosen builder throws it out of the window.

A resource-based estimate will deal quite separately with the different cost components of labour, plant, materials, and sub-contractors, rather than amalgamating them into a series of 'all-in' rates as is the quantity surveyor's custom.

However, having said this, a very substantial part of the total cost will comprise sub-contractors' work, and the resource-oriented estimator will have to base much of his estimate on the specialists' all-in prices in exactly the same way as his product-based colleague. (The builder's estimator thus knows no more about the so-called 'real cost' of building than the independent professional cost planner.)

The resource-based estimator's immediate concern will be with that part of the work which is usually undertaken with the builder's own resources, normally the structure of the building including excavation.

But his real expertise (and the area where the product-oriented estimator is weak) is in determining and pricing the site organisation and the project duration, including the management of all the specialists' programmes and the integration of their work with the building structure and with each others' efforts. If he is working in a design-and-build, contracting, or construction management organisation he will almost certainly have the assistance of a planning engineer or project manager.

He may well be looking at alternative configurations for the proposed building, as well as alternative means of obtaining the same configuration.

As far as estimating the cost of the builder's own work is concerned the materials present few problems, since unlike labour and plant their cost is reasonably related to the quantities of finished work. Anybody preparing a resource-based estimate will, therefore, have to start off by measuring (or assuming) quantities of finished work in order to determine material requirements and also as a first step in looking at the scale of the project and the distribution and inter-relation of its parts. The person doing this will tend to work in bulk quantities (e.g. cubic metres of concrete) split into categories and locations which seem to him to be organisationally significant. Before he can proceed much further however, he will need to prepare an outline programme for the works.

Resource programming techniques

The estimator, in respect of each different estimate, will need to decide the principal operations to be undertaken and their methodology and duration. The operations cannot be considered in isolation since they will be inter-related by two factors:

1 The need to use labour and plant effectively, so that neither men nor machines stand idle for long periods between tasks, nor are required to be working in two different places at once, nor spend too much time moving from one part of the site to another.
2 The inescapable sequence of building, so that, for example, the walls and columns cannot be built until the foundations are completed, and the first floor cannot be placed until the ground floor walls and columns have been built.

There are various techniques in common use to assist him in this task, which tend to rely in the first instance upon graphic methods.

The traditional method is the 'bar chart'. On this chart a horizontal time scale is used, often divided into weeks, and the various operations comprising the project are listed vertically down the left-hand side. The timing and duration of each operation is then indicated by a horizontal bar spanning the relevant period of weeks and shown on the same line as the operation it refers to. An example is shown in Fig. 13.1 and it will be seen that the bars follow the classic pattern of moving diagonally from the top left-hand to the bottom right-hand of the chart.

The bar chart is simple and easy to follow. It gives quite a good picture of the way in which the various operations fit into the total contract period and is very popular on building sites for the purpose of monitoring progress and forward ordering. It is, however, a better communication tool than a planning tool; it does not help in determining the duration of operations (it cannot be drawn until this is decided) nor does it bring their interdependence, or otherwise, to the notice of the planner. It tends to show the results of planning which has been undertaken by other means.

At one time these other means were usually informal, but in present-day practice the network diagram is commonly used. 'Network diagram' is a generic term and also includes 'precedence diagrams' and 'critical path diagrams', which are special uses.

The network consists of a series of nodes joined together by lines or arrows, and normally moves from a start at the left-hand end to a finish on the right. In one system (Fig. 13.2), the lines or arrows represent operations (or activities) and the nodes represent the interdependencies which occur at the start and finish; a network prepared in this manner has the advantage that it may be drawn to scale with the lines or arrows of a length proportionate to their time requirements, and can develop into something that is almost as easy to read as a traditional bar chart. However, this option is open to the same objection – that it cannot be drawn until the durations are known, and that once drawn, it is very difficult to amend if durations change during the progress of the work. It is really a bit of a halfway house. The system generally is open to the objection that it can be difficult to follow, particularly as it is usually necessary to draw a number

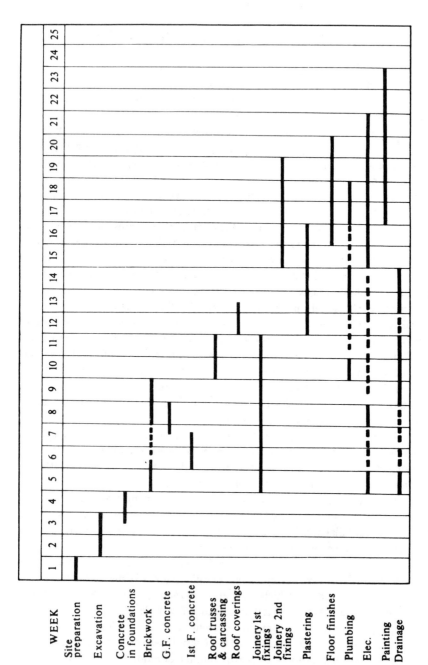

Fig. 13.1. A bar chart.

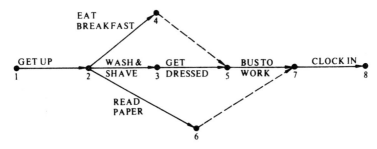

Fig. 13.2. Simple 'arrow = operation' network to illustrate going to work in the morning. Each node represents the beginning or end of an operation. Note the need for dummy (dotted) arrows. You cannot catch the bus to work until you have dressed and had breakfast. The paper need not be read until the bus ride.

of 'dummy' arrows in order to close the network, and there is always a tendency to see the length of the arrows as having some significance even if they are not drawn to scale.

For planning purposes, the alternative approach is to be preferred (Fig. 13.3) in which the nodes represent the activities and the lines or arrows joining them simply illustrate dependencies. The nodes are usually drawn in the form of sequentially numbered circles which are large enough to contain some numeric information; the length of the lines joining them is of no significance and is chosen arbitrarily to suit a clear layout of the diagram. The diagram can thus be drawn without any idea of the length of time which any operation will take and in this form reflects the earliest planning stage, in which the planner is identifying those operations which fundamentally depend upon each other and those which do not. Once this has been done it will never need to be redrawn (unless the project itself alters); any changes will be made only to the numeric information within the nodes.

A simple network in an unquantified form is simply a 'precedence diagram', showing which operations must precede which. No operation can take place until all operations arrowed into it have been completed. Remember, on a building project, it is

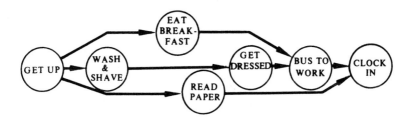

Fig. 13.3. The same 'project' on the 'node = operation' system. The arrows simply indicate precedences. Note that in neither case does the diagram mean that breakfast is eaten whilst washing and dressing, simply that it is not essential for you to be dressed before eating – you might have your breakfast first. Similarly, the paper need not be opened until you got on the bus, but you have the option of looking at it at any time after you rise.

fundamental construction dependence which counts and not convenience or the efficient use of resources, which are considered at a second stage.

Two important points are worth noting. Firstly, the network can be expanded to include not merely the construction but also the whole of the planning and design process and can thus become a tool in the hands of the client's professional representatives, particularly useful where overall time requirements are important or where design and construction are to run in parallel on a management, design-and-build or fast-track contract.

Secondly, the operations may be shown at a strategic level ('build walls') or at a much more detailed level ('build ground floor external walls', 'build in sills and lintels', and so on). For eventual control purposes a fully detailed network may be used, but this is expensive to prepare and would be inappropriate at planning stage, before the actual detailed design has been finalised. On the other hand, a simple 'planning' network cannot show the interdependence of meshing operations. For instance, if erection of precast beams can commence after only some of the columns have been completed this would require the work of each group (or even pair) of columns to be shown as separate operations. This problem occurs throughout the project, but at planning stage is only likely to be considered in those parts which the planner feels to be of crucial importance – probably major structural work rather than finishes.

Once the network has been drawn and the precedences and dependencies settled it is time to quantify the problem. The procedures are sufficiently standardised for there to be a number of computer packages available, and the work then entails filling in the estimated durations on a schedule, but if manual methods are being used a suitable method is to mark durations in the node circles which are divided into four quadrants for the purpose of quantification (see Fig. 13.4). In the top left-hand quadrant is written the reference number of the operation concerned (full description of the operation probably being written out in a numbered list) and in the top right-hand quadrant the estimated duration in days, weeks, or whatever unit is being used.

When all the individual durations have been inserted the total project time can be calculated. Beginning at the 'start' and working along the arrows the shortest possible elapsed time to the completion of each operation is inserted in the bottom left-hand quadrant. This is arrived at by adding the duration of the operation concerned to the highest elapsed time of any of the other operations which precede it and which are therefore arrowed into it. By the time 'end' is reached the total time for the project will have been calculated.

In respect of each operation, we will now know the earliest time at which it can be completed. Where the operation concerned is a crucial one this will also be the latest

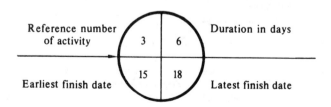

Fig. 13.4. Activity node (showing division into quadrants for reference).

time at which it can be completed if the project as a whole is to be finished on the calculated date. However, many less important operations may be able to be delayed without delaying the completion of the whole project.

In order to identify the crucial operations the bottom right-hand quadrant of each node may be used to show the latest possible date at which the operation can be completed without causing delay. This is calculated by working backwards from the finish and in the case of each operation calculating from the earliest start time of any of the operations which immediately follow it.

When this task is completed it will be found that for some operations the earliest and latest dates for finishing are the same. These are the critical operations, and the sequence of arrows which joins them forms the so-called 'critical path' through the project. In the case of a non-critical operation the difference between the earliest and latest finishing date represents the 'float', or the period for which that operation can be delayed without affecting the completion date for the project. All these figures can of course be set out in the form of a schedule, instead of on the diagram itself.

It should be noted that if part of the 'float' for a particular operation is actually utilised this may affect the float available for succeeding operations. If all the float is used up then the operation, and its successors, will have become critical and the critical path will have changed. It is, therefore, necessary for the project controller throughout the progress of the job to keep a close eye on operations which have a very small float, as well as those which are already on the critical path.

Resource smoothing

There may well be two or more operations in a network which do not depend upon each other in the construction sequence – for example, foundation excavation and drainage trenches. It will not matter which is done first, and they could in theory be done simultaneously (whereas the foundation excavations and the concreting of them could not). However, their relative timing may depend upon resource utilisation. To dig the foundation and drainage trenches simultaneously would involve bringing an unnecessary number of diggers to the site, whilst to leave several weeks between the two operations would either involve machines standing idle or else taking them away and bringing them back. If both operations were on critical or near-critical paths then the additional expenditure would have to be accepted (or the time for the job altered), but otherwise their floats would be adjusted to give a more economical end-on sequence. The adjustment of a programme in this way to make the best use of expensive plant, and to give gangs of men a consistent and balanced work pattern, is an important part of the exercise, and is often called 'resource smoothing'.

Use of techniques in relation to design cost planning

As already explained, the cost planner who is intending to use resource-based methods will need to start off with some idea of the configuration of the proposed building – its size, shape, height, and preferred technology. He may well suggest a suitable combination of these requirements, based on systems already rationalised in his organisation and which he knows can be built quickly and efficiently, and for some types of developments (e.g. industrial premises) this could be a valid approach. Alternatively, he may begin with a solution proposed by the architect, test it in terms of resource use, and then investigate alternatives.

At this stage he will be principally concerned with those things which he can identify as major time constraints and production cost constraints, by using the skills derived from his experience. He will be looking for a solution which gives effective use of labour and equipment by providing smooth and continuous work-flows, and which enables the most economical type of plant to be used. For example, if one single heavy component has to be lifted at the extreme reach of the tower crane this will dictate a much heavier and more expensive type of crane to buy and run, probably for the whole duration of the contract. The crane 'cycle' will also dictate the intervals at which components, or skips of wet concrete, can be lifted into position, which in turn may determine the size of gang and the method of working. Alternatively, an efficient method of working may require a different cranage configuration to that first thought of. In turn, formwork usage will have to be considered and the concreting programme planned to allow for optimum reuse, with sufficient time allowed for any necessary alterations. This is where the builder may look for savings, for example, by keeping column sections constant all the way up the building, as he may decide that the extra cost of the extra material required in the upper storeys will be more than saved on labour, plant, and formwork. Decisions of this kind can often only be made in the light of the actual construction programme which is envisaged, and whether site-mixed or ready-mixed concrete is to be used.

The Property Services Agency of the DOE in the UK produced an inter-active computer system (COCO) on an experimental basis, which is intended to assist a cost planner in making the right cost decisions on some of these matters, although this does not appear to have emerged from the experimental stage.

A further matter to be considered is the staffing of the site, and the number and type of supervisory and control staff.

The builder will also be very concerned, even at an early estimating stage, with the integration of the engineering services work with the construction programme. Some of this integration concerns the way in which the various contractors will have to work together, as the installation of pipes, conduits, and components may be incorporated into the design in a way which means that they will interfere with the builder's own work sequence. He will also be on the look-out for major items of equipment whose placing may cause difficulties. Equally he will want to know whether, and if so when, he can use the permanent engineering facilities for the purposes of his own work – lighting, toilets, and passenger and goods lifts in particular. The duration of the complete project and of major stages in it, will have a most substantial effect on plant, supervisory, and establishment costs, and this duration may in the end be determined as much by service sub-contract requirements as by his own work. In highly serviced buildings in fact they may be the prime determinant.

The criticism is sometimes made that in assessing and balancing all the above matters builders' estimators and planners tend to attach too much importance to optimising their 'own' part of the work, whereas it is often the work of the specialists and their co-ordination which in the end proves to have dictated the time for a project. This is perhaps where the independent professional cost planner can take a more balanced view.

Resource-based estimates are certainly no cheaper to prepare than a quantity surveyor's elemental estimate. Much the same sequence will therefore be employed:

1 Strategic estimates taking account of major factors only, often those which are of importance in a comparison between alternative schemes. A resource-use plan at this level may also be used in the preparation of a competitive tender based upon bills of quantities.

2 A detailed production plan and estimate, probably incorporating a full critical-path network. This would be intended for use as a production control document, and would be unlikely to be prepared until the design itself had been finalised and until the builder was expecting to be chosen to construct it. It would almost certainly involve consultation with major sub-contractors, and the use of a computer package.

Obtaining resource cost data for building work

Keeping an accurate record of costs split into:

1 Site labour
2 Materials and sub-contractors
3 Plant
4 Establishment charges

and allocating them to the right contract, is comparatively easy, but the lump sums obtained do not really contribute very much to the understanding of how costs are incurred. A detailed site costing system is much more difficult to arrange.

In theory it might be possible to cost each contract in terms of finished work, attaching costs to the measured items (or groups of items) in the bills of quantities. One disadvantage of trying to do this is the difficulty and expense of keeping records of time and material for a multiplicity of items and subsequently processing them, but more important is the fact that many costs are not directly related to specific quantities of work. We can identify four types of cost, as follows:

1 *Quantity related.* These are the costs which bear a straight relationship to quantity of finished work – many material costs fall into this category as do some components of labour cost. With this type of cost, twice the quantity of work costs twice as much.

2 *Occurrence related.* These costs are related to a particular event, or occurrence, such as the bringing of excavation plant to the site, or the moving of a plasterer's equipment from room to room. While the scale of the occurrence is obviously affected by the scale of the work which is anticipated, its costs will not vary in proportion to actual quantity of work executed.

3 *Time related.* Some costs (such as the hire of a major item of plant) are related to a length of time, and not to the amount of work done in that time.

4 *Value related.* Fire insurance, for instance, will be related to the value of the project; establishment charges are also often allocated on a basis of project value and can be included under this heading.

In addition to these four types of cost, it must be remembered that some of the work done will relate to temporary items (such as erection of site huts and conveniences) which do not form part of the permanent works. On many projects the shuttering of concrete work is a very large item in this category.

Identification of differing variable costs

It is all very well to make these rather academic distinctions between different types of cost, but in practice they may be difficult to separate. For instance, the plasterers are unlikely to keep separate the time involved in moving from room to room, and the charge for a major item of plant may well incorporate several cost types, e.g.:

Quantity related	Fuel, and part of hire charge.
Occurrence related	Bringing to site and removal, moving, assembling and dismantling.
Time related	Part of hire charge.

However, even an arbitrary division into these different types will be better than trying to allocate total cost pro rata to quantity of finished work, and would form a more suitable basis for using cost information for estimating or cost planning purposes. Recent revisions of the Standard Method of Measurement have taken a step towards identifying costs under these various categories.

Costing by 'operations'

In practice, costs are rarely kept to any lower level than an 'operation', which has been defined by the Building Research Station as 'a piece of work that can be completed by one man, or a gang of men, without interruption by others'. A typical operation might be the whole of the brickwork to one floor level, or the whole of the first fixings of joinery. It is comparatively simple to record costs on site for such overall parcels of work, and to allocate materials to them, and it is much more meaningful to attach the time-related and occurrence-related costs to total operations than to try to split them up among quantities of measured work to which they bear little relation. If the contractor's estimate can be similarly sub-divided he will be able to compare cost with estimate at each stage of the job.

Chapter 14
A Cost Planning Strategy

A complete system of cost planning must comprise both:
1 cost planning and control of the design process, and
2 cost control of the construction procurement stage.

It has been the custom in the past to consider these quite separately, since under traditional contracting arrangements they have been discrete consecutive operations with very little data transfer between the two. Under the various systems of fast-tracking which are now popular this is no longer the case but there are still advantages in looking at them separately in the first instance.

Cost planning and control of the design process comprises:
1 Establishment of the brief
2 Investigation of a satisfactory solution
3 Cost control of the development of the design

It is unfortunately true that the amount of time and effort spent by the cost planner on each of these three aspects is often in reverse proportion to their relative importance, being almost entirely concentrated on the third aspect with perhaps a little advice being given on the second. Sometimes the excessive effort required to keep the design development within cost limits actually stems from unwise decisions made at the earlier stages without the benefit of cost investigation. In fact, if the strategy is right the design development will usually present few problems.

The correct strategy can be summed-up in one word – 'zeroing-in'. Estimates given at an early stage carry a fair degree of risk, and as further information becomes available they are almost certain to become subject to amendment. What is important is that whenever this happens the relevant decisions can be taken (to amend the cost, to amend the scheme, or even to abandon the scheme altogether), without a lot of abortive expenditure having been incurred. As a corollary of this, the estimates themselves should not be prepared in any more detail than is relevant to the current stage of progress of the design.

Although the establishment of the brief and the investigation of a satisfactory solution are shown as two separate and consecutive functions, there should be a certain amount of iteration – design investigation may suggest modifications to the brief which in turn will need to be investigated. This is all to the good, and will probably result in improved performance. Even if it involves, as it will, a good deal of abortive work by the cost planner, his work at this stage is relatively cheap compared to the potential benefits. The cost control of design development, on the other hand, demands considerable resources, and should not be carried out until a satisfactory solution has been defined and agreed. Substantial iteration between the second and third stages of the process brings nothing but disadvantages.

In formulating the brief the cost planner should be co-operating, preferably on a team basis, not only with those professionals concerned with the actual building design but also with representatives of the client's organisation or his project managers as well as valuation surveyors, accountants, and possibly planners.

As the investigation of a satisfactory design proceeds further into the realms of building configuration the cost planner will become increasingly involved with the designer and his consultants, including perhaps a construction planner, to the gradual exclusion of the other parties. At a fairly early point in this process the cost models are likely to adopt an elemental format, but may subsequently (if appropriate) change to a format based on production operations when basic design decisions have been ratified.

As has been stated, the basic principle to be adopted is one of moving from the estimating of outline design proposals to the detailed costing of production drawings in a series of steps, at each of which previous assumptions can be checked in the light of further design development, and any necessary modifications made to the estimates or to the design before proceeding to the next stage.

As the brief is developed in more detail, and as the design itself develops, it may become apparent that the building which the client wants, with a proper balance between economy and quality, will cost either more or (in theory at any rate) less. It is possible to reduce or increase the quality of the specification to get back to the original figure, but the cost planner should not automatically assume that the client will want this to be done. He may be more concerned with getting the building he wants than with 5% on the cost; on the other hand he may not, but the choice is his.

The difference between this approach, and the situation (without cost planning) of a large and unexpected gap between estimate and tender is that the client makes his choice consciously with a knowledge of the amount of money which his decision is costing (or saving him). In this way the time and expense involved in abortive detailed design work can be avoided.

A programme for cost planning and control – traditional procurement methods

The sequence which follows is specifically oriented towards a project which is to be designed prior to contractor selection and which is to be let on the basis of main contractors tendering for the whole project on a bill of quantities. It requires little modification where the traditional contract arrangements do not apply and the contractor is appointed by negotiation at an early stage, provided that the above principles are clearly understood and adopted. The main difference is that the contractor will become a party to the cost planning arrangements much earlier, and the equivalent of the tender stage will therefore be less traumatic.

1 BRIEF STAGE – *Preliminary estimate based on floor area*

This is the minimum contribution that the cost planner should be making to the initial formulation of the brief. No drawings exist at this stage, and only the vaguest of information, so it is pointless to start constructing complex models.

2 SKETCH PLAN – *Outline elemental cost estimate(s)*

These are produced in order to evaluate the designer's first sketch designs, or even to help the designer select a configuration. Drawings of some kind are needed, so that wall areas etc. can be calculated, but they can be very simple. These are sometimes called 'Preliminary Cost Plans', but the title is misleading as they are most unlikely to be used for subsequent cost control purposes, which is what distinguishes a cost plan from an estimate.

$224 - 7$

3 APPROVED SKETCH DESIGN – *Elemental Cost Plan*

Eventually the client will approve one of the sketch designs, and this is the correct time to produce the formal cost plan. This will probably be based upon the most recent outline elemental cost estimate but developed in as much detail as a full elemental cost analysis, complete with outline specification (agreed with the designer) for each of the elements, and this will form the basis for the system of design cost control. It will make everyone's life much easier, and the system of cost control more efficient, if the design development then proceeds as closely as possible in accordance with the details of the cost plan.

This expensive document should not be produced prior to approval of sketch design, since it might have to be done again a second (or third) time if the design changes, and in any case the designer will find it difficult to make decisions on detailed specification matters while the design itself is still undecided.

Neither should it be left until the design is further developed – it may be easier to do, but it will become a record of what has been decided rather than a plan for controlling design – just a rather fancy cost estimate in fact.

4 PRODUCTION DRAWINGS – *Cost Checks*

As the designer produces the working drawings they will have to be checked against the cost plan to see whether the assumptions that were made still hold good. This is a time-consuming business, and the adjustments which will have to be made if major discrepancies emerge are even more so. The process will be minimised if the cost plan was prepared in considerable detail and if the designer has adhered to it closely.

The task of cost checking can also be minimised by not doing it thoroughly because of the vast amount of work and time involved in evaluating all the changes which have been made and making and evaluating the further changes which would be necessary to get the project back on target. If this point is ever reached then the money which the client has spent on cost control can be regarded as wasted, and if the débâcle is not of his making he will have every right to feel cheated by his professional advisers.

Sometimes, however, substantial changes are unavoidable during design development, and in such cases it may be better and cheaper to go back to one of the previous stages and start again rather than trying to adjust a detailed cost plan to fit a scheme which has changed into something quite different.

5 RECEIPT OF TENDERS – *Tender reconciliation*

When the tenders are received they should be reconciled with the cost plan. If all has gone well this will be a matter for the cost planner alone – simply putting the records straight. However, if some reductions have to be made because of cost over-run this process will assist the making of the most economical suggestions. If the foregoing procedures have been properly carried out these should only be of a minor nature.

6 CONSTRUCTION STAGE – *Post-contract ('real-time') cost control*

This is an essential part of the process. The cost planner must continue to monitor the cost of the project whilst construction proceeds, in the light of the provision which the contract makes for adjusting the contract sum. This stage is at least as important as those which have gone before.

A programme for cost planning and control – management-based procurement methods

The above six stages change very little, the main difference being that stages 4 – 6 no longer form discrete time stages but will proceed concurrently. The cost plan referred to in stage 3 should preferably have been prepared in an operationally-based format rather than elementally, as this will be required if it is going to play its part in the cost control process.

Stage 5 (Receipt of tenders) will become much more complicated, since the team will be seeking separate tenders from all the specialist 'works contractors', and each of these has to be dealt with separately.

It might seem a good idea to try and obtain the most important tenders at more or less the same time, since it would then be possible to consider their total effect. However, this is rarely advantageous in practice since the whole idea of fast-track systems is that design and construction proceed simultaneously, with the design of later stages of the work going on whilst earlier stages are being executed. Specialist work should not be tendered for until the design work on that particular specialism has been completed, otherwise unnecessary variations are likely (and this will be no help whatever to cost control).

So it is more usual to carry out the cost control of such projects with the budgets for the specialist works contractors being turned into firm prices one by one. This does have the advantage that it is possible to amend the budgets (and specifications) for later specialists if there are cost over-runs on the earlier ones, so that tenders can be sought on the revised basis. This gives better cost control than having to amend works contracts to obtain savings, as may happen if tenders have been called too early.

Resource-based estimates are certainly no cheaper to prepare than a quantity surveyor's elemental estimate. Much the same sequence will therefore be employed:

1 Strategic estimates taking account of major factors only, often those which are of importance in a comparison between alternative schemes. A resource-use plan at this level may also be used in the preparation of a competitive tender based upon bills of quantities.

2 A detailed production plan and estimate, probably incorporating a full critical-path network. This would be intended for use as a production control document, and would be unlikely to be prepared until the design itself had been finalised and until the builder was expecting to be chosen to construct it. It would almost certainly involve consultation with major sub-contractors, and the use of a computer package.

Chapter 15
Cost Planning the Brief

It might be thought that this title is misleading – since the cost plan is not going to be prepared at this stage a better title would surely be 'costing the brief'?

This might be so if that was all that the cost planner was going to do – passively estimate the cost of carrying out the client's ideas.

But this will not be good enough. We have seen that this is the stage where the big cost decisions are made, consciously or unconsciously, and design work should not start until the brief itself has been got right. There is little point in spending large sums of money meticulously and successfully cost planning and controlling a design which is not in fact a very good answer to the client's needs.

In the case of profit projects such as speculative offices and flats it is often possible to charge higher rents or get a higher price for leases if a better standard of amenity and finish is provided, so that the budget is flexible to a limited extent and the cost disadvantages of larger and more luxurious public spaces, more and faster lifts, better fittings, and so on will have to be evaluated in relation to income. The acceptable final cost may be quite different to the original rough estimate due to such decisions.

A client for a private residence may find that his ideas become more ambitious as the drawings progress, and again provided that he is kept informed of the consequences of his decisions he may prefer to have the house he wants rather than the cost he first thought of.

It is much better that these matters should be decided at brief stage rather than during design development or, even worse, during construction. Many authorities consider that the traditional practices of the British building industry make it too easy for the client to make changes during the construction stage and that this is partly, or even largely, to blame for the comparatively high cost of building in the UK compared to some other countries.

The cost planner therefore must be helping to develop the brief and examine its cost implications critically (unless he has been specifically told not to do so – but clients rarely do that these days!).

Nevertheless, the first task is indeed to cost the brief as presented.

Preliminary estimate based on floor area

This will be a very approximate estimate; it could hardly be otherwise in view of the lack of information at the cost planner's disposal. There are exceptions to this, for instance where the project is one of a series (such as a chain of service stations) and information has been accumulated on a number of almost identical buildings, or where the first estimate has to be calculated according to a government formula and the building will be tailored to fit the estimate.

In the preparation of the preliminary estimate based on floor area previous experience of the designer is most valuable, as there are some whose designs always come out above average costs just as there are those who can be relied upon to achieve an economical design. At an early stage the cost planner is quite likely to be told 'This will be an economy job' and only a knowledge of the people concerned will indicate how far this statement can be relied on in framing the estimate.

The cost planner should never allow the amount of the estimate to be influenced by opinions expressed by interested parties, or worse still by the figure that the client would like the job to cost, unless these also coincide with the cost planner's own judgement. When an estimate has worked out at £3,500,000 it is all to easy to remember that it was supposed to be a £2,250,000 job. The danger then is that the cost planner may assume that the estimate is too high and reduce it a little in order to make the difference less breath-taking. Remember that it is the person who has prepared and possibly signed the estimate who will be held responsible for it, not the designer or the client. This is as it should be; cost planners are paid for their skilled evaluation of real probabilities, not for telling people what they want to hear.

The initial estimate, like all others, must make quite clear what is and is not included. It is quite good practice to use a form for preparing the estimate, with a definite section to show inclusions and exclusions; a computer spreadsheet program should similarly present a menu of possible inclusions and exclusions to jog the cost planner's memory and to print them out when preparing the estimate for the client.

Such items as complete internal decoration, joinery fittings, venetian blinds and curtaining, furnishings, office partitions, light fittings, carpeting in offices and flats, site works, specialist plant and services, shopfitting, and security systems, should be specifically dealt with, not just forgotten about or dealt with by some vague overall clause. In fact it is even more important to define these things properly in a preliminary estimate than it is later on, when the detail of the estimate itself may be sufficient to answer many of the questions.

Make sure that professional fees and VAT (where appropriate) are clearly excluded or included; many clients think that the estimate shows their total commitment unless they are told otherwise. If fees are shown it is important to get them right. Until the mid-80s this was comparatively easy, as everybody charged according to the fee scales of their various professional bodies. Today, when fees are subject to negotiation or competition, this is more difficult but unless something has already been agreed to the contrary it will probably be best to assume fees according to scale at this preliminary stage. Cuts can always be made later, if necessary.

Professional fees should be estimated to project completion, not just design completion, and should include the various specialists as well as expenses and document reproduction. In these days of escalating paperwork this last item can be a considerable sum.

The estimate must also be specific about inflation. If it is based upon current building costs it should say so, and also make clear that it is subject to revision in the event of increases or decreases in costs. There have only been one or two periods when building costs have been fairly static in the last forty years; the longest such period has been in very recent times so that there is a tendency not to get too worried about this, particularly among students and trainees who have never experienced major escalation in building costs. But even if things seem to be set fair at the time when the estimate is being prepared it is still wise to mention that it is based on current costs and may need adjustment for inflation.

If on the other hand the estimate is based upon anticipated cost levels in the future the basis of calculation should be set out in the estimate so that revision can be made if the prediction proves to be false. The person who could accurately forecast economic conditions two or more years ahead would be in demand in other and more lucrative fields than construction cost planning!

A final problem relates to whether the tender or the final cost is being forecast. Clearly the latter is the preferable alternative, since it is this sum of money which the client is going to have to find in the end, and in any case there is little alternative in the case of projects which are to be built under some contractual arrangement that does not involve a tender as such.

However, particularly in the public sector, the custom is often to estimate the tender amount, and it is important in both public and private sectors to be specific about which alternative has been used. At times of high inflation there can be a considerable difference between the two figures, especially if the cost planner makes some allowance for the claims which often arise during the course of the work and which will be reflected in the final cost but not in the tender amount.

As the preliminary estimate cannot be related to the subsequent cost plan (except in total) it is not worth preparing it in any greater detail than is necessary for its primary purpose, that is to give an approximate estimate of cost. And, although cost planning is very much more than the preparation of an estimate, it gets off to a bad start if the first estimate is badly wrong. This stage is therefore extremely important.

Finding the right rate at which to price floor area is not easy, and the most reliable starting point is a similar job from office experience. Failing this, or as a check on it, a wide selection of costs from a published source such as the BCIS system should be used, resisting the temptation to choose only from the lower part of the cost range. The BCIS 'on-line' computer system is a particularly good source, not least because of the way in which it facilitates the adjustments which are about to be described.

To adapt the floor area rate from one job to another when shape is not known requires consideration of seven factors, as follows:

1 Market conditions
2 Size, number of storeys, etc.
3 Specification level
4 Inclusions and exclusions
5 Services
6 Site and foundation conditions
7 Other factors

A most important point is to check what is actually meant by floor area. This might be:

1 The total area at each floor level measured over the external walls including all staircases, lift wells, and similar voids, and including all internal and external walls.
2 The gross floor area measured according to SFCA/BCIS rules, which is as (1) but excluding the thickness of external walls.

3 Something rather less than (1) or (2), perhaps excluding voids, circulation spaces. or plant rooms.

4 The commercially lettable floor space (often still given in square feet to complicate the issue).

Obviously for the same building these would each yield quite different cost factors, and it is important to check that the same option is used for all the data-source examples and the new estimate, or that conversion factors are duly applied.

Further points for care include factors which may not be properly reflected in traditional floor-area formulae, and which may occur in either the project being estimated or in the cost data examples. The most troublesome of these is the atrium, which was not a common design feature when these floor area formulae were devised, and which if large cannot be treated as an ordinary void if comparison is being made with non-atrium buildings. A somewhat similar problem arises with buildings which have an unenclosed ground floor used as part of an open car parking area.

We can now proceed to work through an example in which the above factors are taken into account where applicable.

The project used as an example will be a new 2,400 m^2 six-storey office block in a provincial city, with shops on the ground floor and offices on the other five. The previous job on which the estimate will be based is a project of similar quality in outer London, but of four storeys only again with shops on the ground floor. The total floor area was 1,235 m^2 and the tender price 12 months previously was £680.00 per m^2.

1 Market conditions

This adjustment deals with the changes in building prices and in tendering conditions between the time when the previous job was priced and the anticipated date of tender, or tenders, for the new project.

One of the published cost indices previously described would probably be used, but where the estimates which are being compared are not too far apart in time (up to a year or so) it is quite possible to make the adjustment for market conditions by simply adding a percentage based on known labour and material fluctuations. It may indeed be necessary to do this for short-term estimates because of the time-lag in publishing official indices.

For this purpose the proportions of labour, material, and plant content given in Chapter 11 should be used, with a heavier bias towards materials if the contract is for an expensive building with good finishes and high-class materials. In making this calculation the figures should be taken as fractions of the gross contract amount without deducting overheads and profit since these will rise more or less in proportion to increases in prime cost.

Let us assume an increase of 10% on this contract, ascertained by comparison of indices; £680.00 per m^2 plus 10% = £748.00.

2 Size, number of storeys, etc.

The new provincial block will have a larger plan area than the London example; this will tend to reduce the proportion of external wall area to floor area (assuming a similar plan shape) and hence should reduce the cost per m² of floor area a little.

However, the new block will have six storeys instead of four, and this will have the effect of increasing the cost, as it is a general rule that higher building costs money. But the difference between four and six storeys is not critical, and the probability is that this and the previous factor will cancel each other out. The less that is known about the new building the less point in adjusting for hypothetical trifles. We are working with a broad brush at this stage, trying to get a ball-park figure as a starting point for the cost planning process.

However, a more important point to consider is that on the outer London block one-quarter of the floor area was set aside for shops whereas on the six-storey building the figure will be one-sixth. As shop areas are normally left in an unfinished state for the shopkeeper to fit out according to his needs and tastes this part of the building will be cheaper than the rest, and by reducing the proportion of the building allocated to shops we shall be increasing the overall price per m². Assuming that the saving in floor, ceiling, and wall finishes and services will be £72.00 per m² of floor area:

Outer London block

Shops	25%	309 m² at £748.00 minus	
		¾ of £72.00 = £694.00 =	£214,446
Remainder	75%	926 m² at £748.00 plus	
		¼ of £72.00 = £766.00 =	£709,316
			£923,762
Check – total 1,235 m² at £748.00		=	£923,780

Provincial block

Shops	16.67%	400 m² at £694.00 =	£277,600
Remainder	83.33%	2,000 m² at £766.00 =	£1,532,000
			£1,809,600
£1,809,600 divided by 2,400 m²		= £754.00 per m²	

Similar adjustments may require to be done wherever the buildings are divided into areas with different cost profiles and where the proportions are different on the two buildings. Car parking areas in office buildings frequently lead to adjustments of this sort.

3 Specification level

It is important to remember that the building cost has already been adjusted for inflation and therefore any specification adjustments must be done at prices ruling at the date of the new scheme, and not those ruling when the previous or example job was carried out.

Alternatively, of course, all specification adjustments could be made at the earlier pricing level and the inflation adjustment for the total scheme done at the end. But the two methods must not be mixed up.

It is assumed that the specifications of the two buildings will be of the same general standard, but the designer has decided that the cheap floor covering used in the offices in outer London was a mistake, and wants to allow an extra £10.00 per m² for this. The client also wants natural stone dressings in lieu of the cast stone used on the previous job.

The adjustment for the flooring is easy. Allow an extra £12.00 per m² over the area of the offices, say two-thirds of the total floor area. 1,600m² at £12.00 = £19,200 which is equal to £8.00 per m² of gross floor area when divided by 2,400.

Calculations like this can be done by simple proportion if preferred (1600/2400 of £12.00 is £8.00), but it is quite easy to make a mistake doing this, inverting the ratio for example, and the longer calculation is more easily checked.

The stone will be a little more difficult to allow for (in the absence of any drawings, remember). The area of external walling in a medium-rise office building of this sort should be about 10% more than the floor area (again, check this with several analyses if possible). Not much of the wall area will actually be stone – most of it will probably be windows. We will assume that 25% of the area is stone and that natural stone facing will cost £130.00 more per m² than cast stone.

So the extra allowance for natural stone is:

$$(110/100) \times 2,400 \times \tfrac{1}{4} \times £130.00 = £85,800$$

which is £35.75 per m² of gross floor area. The total cost after allowing for changes in specification is therefore £754.00 + £8.80 + £35.75 = £798.55

Note that a number of assumptions have been made about the London example which could have been avoided if a full SFCA/BCIS cost analysis had been available, even though it is not intended to produce this first estimate for the new building in elemental form.

4 Inclusions and Exclusions

The client has found that office tenants prefer to do their own partitioning and decoration, whereas on the outer London job these were provided by the developer.

This adjustment will most easily be carried out by extracting the relevant lump-sum figures from the bill of quantities for the London job.

	£
PC sum for office partitioning	26,250
Add for profit and attendance	1,313
Decoration in offices	4,300
	31,863

This figure must be adjusted for inflation, also for five-sixths of the building being offices in lieu of three-quarters.

£31,863 plus 10% inflation allowance (from above) = £35,049.30

Divide by floor area of offices on London building (926 m²) gives a cost of £37.85 per m² of office floor area.

The saving on the provincial block by omitting these items is therefore £37.85 × 2,000 m² = £75,780.00 in total.

£75,780.00 divided by 2,400 m² gross floor area = £31.57 per m².

By the (alternative) proportion method the calculation is:

£31,863 divided by 1,235 m² = £25.80

£25.80 plus 10% inflation allowance (from above) = £28.38

Omission of partitions on new building per m² of gross floor area is therefore £28.38 × $\tfrac{4}{3}$ × $\tfrac{5}{6}$ = £31.53 per m².

The total after allowing for exclusions is therefore £798.55 – £31.53 = £767.02

5 Services

Services form a major part of the cost of any modern building. On a project of any size there will probably be consulting engineers for the heating, air-conditioning, plumbing, and electrical work, and they will normally provide an estimate of the cost of the services for which they are going to be responsible. The building cost estimate would be adjusted as follows:

Provincial office block

	£
Estimated figures received from consultants:	
Space heating	165,000
Plumbing	52,000
Electrical installation	108,000
Lifts	118,000
	345,000
Builder's work, attendance, profit 7.5%	25,875
	370,875

£370,875 divided by 2,400 m^2 = £154.53 per m^2

Outer London block

	£
Comparable figures extracted from bills of quantities	186,000
Add 7% for inflation as main estimate	13,020
	199,020

£199,020 divided by 1,235 m^2 = £161.15 per m^2

Reduction in cost of services compared to London block

£6.62 m^2

The total after allowing for services is therefore
£767.02 minus £6.62 = £760.40 per m^2

In view of the early stage at which this estimate is being prepared the engineering consultants may not feel able to give an estimate for services, but it is important that they should do so. This will avoid the situation which might arise if the cost planner were to assess the cost independently, and the amount required by the engineering consultants when they prepared their detailed scheme proved to be very different. The resulting arguments would do little to improve either the harmony within the professional team or the client's confidence in the team as a whole.

Nevertheless it is quite possible that the consultants, having been asked to establish a preliminary cost target before they have fully established what is required, will play safe and allow themselves a sum of money sufficient to cover any eventuality. (It is, after all, in the nature of engineering design to allow for the likely 'worst case').

The situation where engineering services appear to be taking rather too large a slice of the whole cake is common enough to be met by most cost planners from time to time, and how it is dealt with depends very much on the cost planner's terms of appointment. If these include a measure of executive authority then the cost planner might well call a meeting to thrash matters out; otherwise it may be only a matter of pointing out the problem to the architect, project manager, or whoever is directing the project.

The cost planner should bear in mind however that prices per square metre for services are extremely difficult to estimate (there is still a lack of cost data in this important area), and should try to co-operate with the various consultants rather than raising difficulties. It may, for instance, be very reasonable for the engineer to insist on client requirements being defined with more precision before he can give even a rough estimate.

If there are no consultants appointed, or if at this stage they are not prepared to commit themselves, the cost planner would be better not making any adjustment to the estimate for the services, unless there are glaring differences between the outline requirements for the two buildings. It is, however, becoming increasingly common for professional cost planners to be closely involved in the cost planning of the engineering services which, after all, represent a considerable proportion of the budget for which they are responsible.

Special services, especially transportation services such as lifts, should always be adjusted, as their cost bears little relation to the area of the building and they are much more easily priced on a 'per unit' or 'per installation' basis. The adjustment should be done as a comparison of lump sums, as shown above.

6 Site and Foundation conditions

These are best adjusted on the basis of the so-called 'footprint area' or site area of the building. On most buildings this should be more or less the same as the roof area.

Both the sites under consideration are in congested urban areas, but the new provincial job will require piled foundations which were able to be avoided in London. Either the cost of piling per m² of footprint area can be obtained from a suitable analysis, or a certain number of piles can be allowed for, as follows:

Say 90 No. piles average 6 m long at £75 m = £40,500

£40,500 divided by 2,400 = £16.87 m²

Note that the complex of pile caps and ground beams in connection with the piles will cost little less than conventional foundations, so there is no saving to be set against the cost of the piles.

The total after allowing for site and foundation conditions is therefore £760.40 + £16.87 = £777.27

7 Other Factors

The official wage rates in the provincial city will be slightly lower than in London, but differences in economic climate between one part of the country and another may mean that what actually has to be paid in the two cases bears little relationship to official figures. It may be difficult to get workers and sub-contractors to come on the site at all, and necessary to bribe them to keep them there, or alternatively people may be glad of a job and highly co-operative. The same applies to delivery of materials.

This sort of situation often tends to be relatively short-term (because an overheated local construction sector just has to cool down somehow), and thus is not always easy

PRELIMINARY ESTIMATE OF COST

PROJECT: Office Block, Midtown
DATE OF ESTIMATE: 12.2.90   ASSUMED DATE OF TENDER: Sept 1990
GROSS FLOOR AREA: 2,400 m²

BASIS OF ESTIMATE:	£ per m²
Office Block, London (Sept. 1989)	
1235 m²	680.00

ADJUSTMENTS:

1. Market Conditions:
      RICS Cost Information Index. Add 10%

	£ per m²
	68.00
	748.00

2. Size, number of storeys, etc.:
      6 storeys in lieu of 4 - see notes

+	6.00	
	754.00	

3. Specification level:
      See notes. Wood block flooring/
      Portland Stone

+	44.55
	798.55

4. Inclusions and exclusions:
      Office partitions and decorations excluded

−	31.53
	767.02

5. Services:
      See consultants and notes

−	6.62
	760.40

6. Site and foundation conditions:
      Piling (90 No 6 m long)

+	16.87
	777.27

7. Other factors:
      Tendering (as notes) add 2%

+	15.54
	792.81

Professional fees	excluded
	£792.81

Total estimated cost 2,400 m² @ £792.81 = £1,902,744

INCLUDED (and as above)	EXCLUDED (and as above)
	Floor, ceiling, and wall finishes in shops.
	All furnishing, fittings, blinds.
	Electric light fittings (except in public spaces).
	VAT
	Professional fees.

Fig. 15.1. Preliminary estimate of cost.

to pick up from published building cost indices. Local enquiry may be worth while in a case such as the present one if the new project is in an unfamiliar region. After such an enquiry, a cost reduction of 5% has been assumed.

The client, however, is anxious to have the building erected as quickly as possible and the contract period will be very short. This will tend to increase the cost, both because the site labour force will need to be larger than is economical and because the number of local contractors and sub-contractors who can be relied upon to work to a tight timetable is small and competition will therefore be limited. The contract price may therefore have to include a great deal of overtime. This is expected to add about 7% to the cost, so after allowing for the reduction in wage rates a net addition of 2% will be made to the estimate.

The total after allowing for other factors will be £777.27 plus 2% = £792.81

Fig. 15.1 shows how the estimate would be set out in an appropriate format. It must be assumed that the question of professional fees is set out in a covering letter.

Cost reductions

It may be that the estimated cost is higher than the client can afford, or more than he wants to pay. It would be very dangerous to reduce the amount of the estimate for this reason, or on some vague grounds such as a general reduction in standard, although of course adjustments could be made for specific items such as (on the basis of the above example) cheaper lifts, or a return to the cheaper floor finish. But if the total of the estimate is much too high it is best to face facts, and either reduce the size of the building or get the brief substantially modified.

It is the easiest thing in the world to cut an estimate by 10% (or by any other percentage) under the influence of the client's pleadings and the designer's optimism, but it must be made clear that an overall 10% cut in costs means much more than a 10% cut in standards as much of the structural work, for example, cannot be reduced in cost. It must also be remembered, as stated earlier, that the estimate will carry the cost planner's signature, not those of the people who want to modify it.

It can be assumed therefore that if the cost planner amends the estimate from its original figure there must be some tangible reason or reasons. These reasons need to be spelt out if a modified figure is presented.

Finally mention must be made of the threat 'the job won't go ahead on this figure, do you want it to go ahead or not?' This may be the response of a profit-oriented client, or an architect who does not want to lose a commission. Alternatively, it may be the attitude of a public body which knows that a major project will never get the go-ahead if its true cost is known beforehand but will be very difficult to stop at a later stage when the enormity of its cost becomes apparent. Nobody ever seriously believed, for example, that the Sydney Opera House would only cost one or two million dollars.

We are now well into the realms of professional ethics, about which people must make up their own minds in the light of the case before them, but whatever the ethical position the responsibility of the cost planner for whatever figure goes forward must be remembered. The other parties will adopt a very low profile if things go wrong!

Data sources

The example of the two office blocks assumed that there was a comparable scheme already in the office. In the later stages of estimating, when a cost plan is prepared on an elemental basis, an exact similarity of use between data source and proposed project is not vitally important – costs per square metre of external walls, windows, doors, floors etc., need not vary much between a library or an office block. The costs per m² of floor area of the buildings themselves, however, will vary considerably because of the differing proportion of these items in them. If there are no truly comparable schemes in the office it may be possible to get information from a professional colleague, or as a last resort to use published information.

When working from one's own information it is best to use one building as an example because the variables in design and tendering are known, but it is often helpful to supplement this with published information for similar buildings, in which case as many examples as possible should be obtained to try and average out all the variable factors. Published information is always useful for checking and comparing an estimate prepared from one's own sources.

Mode of working

It may be objected that the above method of working is far too crude. For instance, instead of guessing a percentage for an accelerated programme it would surely be better to compare the consequences of a normal programme and an accelerated one by preparing network diagrams and costing out the two sets of resources. Well, of course it would, except that no design exists for the building at this stage, and although it might be possible to study the effects in detail for a known building of fairly similar characteristics the answer would not be sufficiently valid to justify the amount of work involved (and in any case there was a 'market forces' component to the figure based on restricted tendering which such an exercise would not disclose). And the workload would be quite heavy; even if computer simulations are used they require a fairly considerable data input.

To repeat what was said at the beginning. What is being looked for is simply a ball-park figure for a hypothetical development – ('how many noughts') – and there may often not be enough firm information even to carry out the very modest adjustments set out above. There will be plenty of opportunity to use more sophisticated techniques as soon as there are some tangible proposals to examine, and this process begins with the preparation of tentative sketch designs.

Chapter 16
Cost Planning at
Scheme Development Stage

Cost parameters

A parameter is 'a quantity which is constant in a particular case considered, but which varies in different cases.' (*Concise Oxford Dictionary*) It is now necessary to look at the factors which influence the quantity of components in terms of area, number, and size, and also their quantity in terms of cost. Unfortunately insufficient research has been undertaken to date to give clear indications of the degree to which changes in the parameters of the building (or by implication its model) will affect the cost of that building. There is however a very great depth of knowledge gained by practitioners which provide us with some general 'rules of thumb'. In some cases we can be quite specific about how cost varies. For example, if we change the shape of a single-storey building so that the area of external brick cavity wall is increased, then we can be sure that, all other things being equal, the wall cost will probably have increased in direct proportion to the increased area. Similarly, if the quality of facing bricks is increased and the shape of the building remains fixed, then the wall cost will have increased by the extra material cost of providing the better specification. Whilst we can probably rely on this type of simple wisdom for small brick buildings it may not be adequate for dealing with more complex multi-storey concrete clad structures.

If we change the shape or height of the building, it may not be just the extra quantity and quality that we have to pay for, but indirect costs such as different lifting equipment, improved fixings to deal with increased exposure, access, and manoeuvrability and dispersal of plant on site, and so on. Some useful studies have been undertaken in such areas, but every research project opens up a whole series of new research proposals. It may be many years, if ever, before we obtain a reliable body of scientific knowledge on which to base our judgments. A particular difficulty lies in producing rules of general application, rather than in relation to one constrained set of circumstances, and indeed there is no real agreement that such rules exist. Very often the answer seems to depend upon the methods that a given builder normally uses.

Using our existing knowledge we can however establish some starting principles which could be the foundation for any further cost research. We can view the parameters at two levels:

1 The form of the building itself (or as it is sometimes called its 'morphology'), where we can study the effect of shape and height on cost,
2 The major components of the building, and the factors which influence their size, quantity and cost.

Building shape

The building shape has its major impact on the areas and sizes of the vertical components such as walls, windows, partitions, etc., as well as the perimeter detailing such as ground beams, fascias, and the eaves of roofs. It would seem obvious that the building that has the smallest perimeter for a given amount of accommodation will be the cheapest as far as these items are concerned. However, the shape than has the smallest perimeter in relation to area is the circle and this does not very often produce the cheapest solution for the following reasons:

1 The building is difficult for the constructor to set out.
2 Curved surfaces, particularly those incorporating timber or metalwork (e.g. in joinery or formwork) are expensive to achieve.
3 Circular buildings seldom produce an efficient use of internal space, as inconvenient odd corners are generated between partitions and external walls. There is a tendency for them also to generate non-right-angled internal arrangements.
4 Standard joinery and fittings are based upon right angles and will not fit against curved surfaces, nor into acute-angled corners.
5 The circle does not normally allow efficient use of site space.

In these circumstances it would appear that the right-angled building which has the lowest perimeter will provide the best answer. This shape is of course the square. Now although a compact square form is generally recognised as the most economic solution because of its reduction in cost of external vertical elements (and the lowest area of external wall for heat loss calculations) there are some important qualifications to be made:

1 Where the square plan produces a very deep building requiring artificial ventilation, air-conditioning and lighting which would not otherwise have been needed, then the shape may become uneconomic, It is only the single-storey building with its opportunities for natural top-lighting and ventilation which may be regarded as a partial exception to this rule, although really satisfactory arrangements for top-lighting and venting are likely to be expensive in first cost and (particularly) maintenance.
2 Where there is a high density of rooms the use of the external wall as a boundary to the room helps to reduce the amount of internal partitioning required. It is therefore sometimes preferable to elongate the building so that rooms can be served from either side of a spinal corridor rather than have a deep building resulting in a complex network of corridors to serve all rooms plus the possibility of artificial ventilation to those that are internal. A real-life case concerns a building where the internal offices with no natural lighting were so claustrophobic that no-one would work in them and they were used as stores for rubbish; a small saving in cost per m² of floor area had produced a lot of floor area that was unusable, and a poor bargain.
3 A given amount of accommodation housed in a square, but multi-storey, block, may be much more expensive than the same accommodation housed in a less compact two-storey block, for reasons to be discussed later.
4 On a sloping site involving cut-and-fill it may be more sensible to provide a long

building running with the contours rather than a square building which would cut more extensively into the site.

These are just a few of the qualifications which need to be made when talking about the efficiency of shape. Like a number of rules-of-thumb, this one was developed for traditional buildings in the UK. Two exceptions, one relating to more modern buildings and the other to overseas experience, show how dangerous it is regard such rules as universal laws.

1 High-tech office buildings require large floor areas, air-conditioning, and artificial lighting as norms. None of the objections under (i) above therefore apply to such buildings.

2 In Britain external walls are expensive partly because they need to have a good thermal performance. In warm countries this is not the case, and the ability to obtain natural ventilation cheaply may lead the cost planner to try to *maximise* the perimeter of the building.

In any case, the shape of the building is often dictated by the site boundaries, topography, and orientation, and the degree of choice is therefore rather limited. If the national construction programme tends towards redevelopment rather than the exploitation of green field sites, as has been the case in the UK in the 1980s, then ideals of building shape become fairly meaningless.

There have however been a number of attempts to measure the cost efficiency of a building shape, and some simple examples are listed below:

1 Wall/floor ratio

This is perhaps the most familiar of all the efficiency ratios but it can only be used to compare buildings with a similar floor area and does not have an optimum reference point such as those below.

2 $\dfrac{P - Ps}{Ps} \times 100\%$

(J. Cooke)

Where P = perimeter of building

Ps = perimeter of square of the same area

This simple formula relates any shape to a square which would contain the same area thus providing a reference point for shape efficiency.

3 Plan compactness or POP ratio (Strathclyde University)

$\dfrac{2(\pi A)^{\frac{1}{2}}}{P} \times 100\%$

Where P = perimeter of building

A = area of building

In this case the point of reference is the circle (a square would have a POP ratio of 88.6% efficiency and yet it is probably the best cost solution in initial cost terms).

4 Mass compactness or VOLM ratio (Strathclyde University)

$2\pi \dfrac{[(3V/2\pi)^{\frac{1}{2}}]^2}{S} \times 100\%$

Where V = volume of hemisphere equal to volume of building

S = measured surface area of the building (ground area not included)

This formula chooses a hemisphere as the point of reference for considering the compactness of the building in three dimensions.

5 Length/breadth index (D. Banks)

$$\frac{p + \sqrt{(p^2 - 16a)}}{p - \sqrt{(p^2 - 16a)}}$$

Where p = perimeter of building

a = area of building

In this index any right-angled plan shape of building is reduced to a rectangle having the same area and perimeter as the building. Curved angles can be dealt with by a weighting system. The advantage here is that the rectangular shape allows a quick mental check for efficiency. As these formulae are only for guidance purposes this index is probably sufficient for early design advice.

6 Plan/Shape Index (D. Banks)

$$\frac{g + \sqrt{(g^2 - 16r)}}{g - \sqrt{(g^2 - 16r)}}$$

Where g = sum of perimeters of each floor divided by the number of floors

r = gross floor area divided by the number of floors

This is a development of the previous index to allow for multi-storey construction. In effect, the area and perimeters are averaged out to give a guide as to the overall plan shape efficiency.

While the above indices are useful they obviously have severe limitations as they consider only those elements that comprise the perimeter of the building, or in the case of VOLM the perimeter and roof. However, the repercussions of shape on many other major elements are considerable. For example, wide spans generated by a different plan shape may result in deeper beams, which in turn demand a greater storey height to give the same headroom, and thus will affect all the vertical elements. These implications need to be represented in any advanced model of building form, and an awareness of knock-on cost effects of this kind must be part of the cost planners knowledge.

Size of building is another important factor in cost efficiency. The larger the plan area for a given shape, the lower will be the wall/floor ratio. For example, a single-storey square building of 100 m² plan area with a height of 5 m would have a wall/floor ratio of 2, whereas a square building of the same height with a floor area of 1,000 m² would have a wall/floor ratio of 0.63. Although the previous points which have been made, concerning artificial lighting and ventilation in particular, may counteract this saving in envelope area, other savings that result from larger plan-area buildings make the grouping of functions into a single large building cost-attractive.

External walls and their finishes need to have a high performance, as they are exposed to the weather and tend to be expensive, so that the savings to be obtained from a reduction in external wall area are considerable.

Terraced housing is a good example of cost savings due to grouping; by taking three detached houses and making them into a terrace you save two brick-faced flank walls, two lengths of foundation for these walls, two roof verges, downpipes, and so on. You would also replace two other faced brick walls with cheaper party walls. Further savings could be made by having deep narrow-fronted terrace houses rather than wide-fronted ones.

Another advantage of large buildings containing grouped accommodation is the sharing of common facilities. The more accommodation units that can be grouped around the central service and circulation core of a building the less will be the cost of the service and access facility for each unit. This applies both to residential buildings and to offices and the like, though over-centralisation can involve long circulation routes and corridors, and problems with fire escape, and in the case of offices can make the sub-division of spaces for letting difficult.

However, given the choice most of us would prefer to live in a detached house rather than a terraced one, or to work in an office which did not involve a long walk to an over-used lift. Here we come face-to-face with a nasty fact: once any obvious extravagances in a scheme have been cut out further savings nearly always involve a reduction in the quality of the building environment. A conscious choice has to be made; sometimes the saving is clearly worth-while while in other cases the net benefit is much more questionable. The experience and knowledge of all members of the design team must be called upon here, but most importantly the client should be closely involved in the process of choice, and the resultant trade-offs.

Height

Here it is possible, and indeed desirable, to be dogmatic. Tall buildings are invariably more expensive to build than low-rise buildings offering the same accommodation, and the taller the building the greater the comparative cost. The only exception to this rule is that the addition of a further storey or storeys to a tall building in order to make the best use of lifts or other expensive services may slightly decrease the cost per storey, but this is an exception which does not invalidate the general rule.

What are the reasons for this?

1 The cost of the special arrangements to service the building, particularly the upper floors. Apart from the necessity of providing sufficient high-speed lifts it is necessary to pump water up and to break the fall of sewage and other rubbish coming down. Complete service floors often have to be provided at intervals of ten or fifteen floors to deal with these problems. Special ventilation and lighting arrangements are needed because of the impossibility of providing adequate light wells in a tall building, together with a high standard of fire-resistant construction and practicable escape arrangements.

2 The necessity for the lower part of the building to be able to carry the weight of the upper storeys, which obviously makes it more expensive than if it were carrying its own weight alone. Also the structure of the building and its cladding will have to be designed to resist a heavy wind loading, a factor which hardly affects a low building at all. Experience with many of the tall buildings of the 1960s and 1970s has shown how demanding is the required standard of windows, wall panels, etc., at high levels, and how expensive is the failure to meet these standards. One is talking about a very different price range indeed from similar components for low-rise construction.

3 The cost of working at a great height when erecting the building. Apart from the cost of hoisting all materials and operatives to the required level, and of the time spent by operatives going up to their work and down again at the beginning and end of each day and at break times, there are the extra payments for working at high altitudes and all the safety requirements which this entails, and again the bad climatic conditions for working at many times of the year.

4 The increasing area occupied by the service core and circulation. As the height of a building increases so it needs more lifts, larger ducts, wider staircases, etc., and these installations take up more and more of the lower floors so cutting down on the usable area. It is possible to imagine a building so tall that the whole of the ground floor is occupied by vertical services; adding extra floors to such a building would produce no increase in usable area at all!

Many of the above factors will also influence the running and maintenance costs – such items as window cleaning, repainting, and repairs to the face of the building will all be much more costly than similar work to a low-rise structure.

Therefore high building should never be considered favourably on cost grounds unless the saving in land costs due to the smaller site area in relation to accommodation will pay for the considerable extra building costs. Land values must be high for this to occur, since a tall building needs a lot of space around it and the reduction in land requirements is not proportional to the reduction in plan area. An increasing problem today is the near-impossibility of dealing with the car-parking needs of a tall building within its plan area.

Just as high building is not usually economic neither is single-storey building. The exceptions here are much more numerous and important, for instance:

1 Where large floor areas free from obstruction by walls or columns are required it is more economical to build horizontally rather than vertically, since to provide loadbearing floors over such areas would be far more difficult and expensive than a roof.

2 Similarly where very heavy floor loadings are required it is cheaper to build floors resting on the ground than high-performance suspended floors over other storeys.

3 Temporary or sub-standard buildings where the low-cost foundations and structure cannot be made capable of supporting a further storey.

However, both for retailing and for manufacture there are so many user advantages in single-storey accommodation that this type of building is becoming very common. One of the problems in some areas is in finding a reasonably level site for a large building of this kind.

The reason for the relative economy of two- or three-storey building is that one roof and one set of foundations will be serving two or three times the floor area, and the walls or frame will be capable of carrying the extra load with little or no alteration. In domestic construction it will be possible to use cheap timber-framed upper floors which will help the comparison still further.

Once the building exceeds this number of storeys various factors contribute to higher building costs. It is less often possible to dispense with a separate frame, the frame itself must be more substantial, and the necessity for lifts, fire-resisting construction, and other expensive measures makes it difficult to attain the low costs possible with two-storey construction.

Optimum envelope area

The 'envelope' of a building, that is the walls and roof which enclose it, forms the barrier between the inside and the outside environments. The greater the difference in these environments the more expensive this envelope will be, but it is always at least a significant factor in the cost of constructing and running the building. We have already seen that a square building is inherently economical in wall area, but the total envelope/floor area ratio will also depend upon the number of storeys that are chosen for the accommodation.

For example, imagine a large single-storey building. If we arranged the same accommodation on two floors we should reduce the roof area by more than the consequent increase in wall area, so that the total envelope area would be reduced. The same thing might happen if we arranged the same accommodation on three floors, but if we continue making the building higher and higher but retaining the same gross floor area the process eventually reverses and the increase in wall area becomes greater than the roof area saving. This reversal happens quite slowly, so that the envelope area changes very little over a range of several possible storey arrangements close to the optimum. It is obviously useful to know what this optimum is as a design guideline.

We can use a formula to calculate this optimum for a square building thus:

$$N \checkmark \quad N = \frac{x \checkmark f}{2s}$$

N = optimum number of storeys
x = roof unit cost divided by wall unit cost
f = total floor area (m^2)
s = storey height (m)

If the desired width in metres (w) is known the formula for a rectangular building is:

$$N^2 = \frac{xf}{2sw}$$

More complex formulae involving several elements have been developed to optimise the shape of the whole building, but although interesting as research tools these result in over-simplification which renders them of little practical help.

Elemental estimates at sketch plan stage

As soon as the first sketch drawings are available an outline elemental estimate should be prepared in order to see whether the architect's initial solution is possible within the cost limits which have been set. This will be particularly true if the shape or design is rather unusual, or looks inherently expensive.

This is sometimes called a 'preliminary cost plan', but this is a little misleading, since it is really only an approximate estimate, and the real cost plan will be prepared afresh. It is probably the elemental format which leads to this terminology.

There are some advantages in using a 'short list' of consolidated elements for preparing the preliminary estimate, as the only measurable items will be the main floor and wall areas and there is unlikely to be a specification as yet.

For the purpose of this example a short list is assumed, and four main areas need to be measured, as follows:

1 Ground floor area.
2 Total floor area.
3 Area of external walls, including windows and doors.
4 Area of internal walls, including windows and doors.

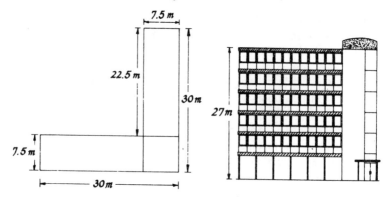

Fig. 16.1. Architect's first sketch design for provincial office block.

Let us look at an example, based on the architect's first sketch design for the six-storey office block as shown in Fig. 16.1:

Ground floor area 30 + 22.5 = 52.5 multiplied by 7.5 = 393.75 m².

				m²
Total floor area	6	× 393.75 m²	=	2,362.5
plus tank room	8 m ×	6 m	=	48.0
				2,410.5
External wall area	4/30 m ×	27 m	=	3,240
Tank room	28 m ×	2.5 m	=	70
				3,310
Internal wall area, say				1,200

Already we can see that the ratio of external wall area to total floor area looks rather high (33:24), and we may feel that this is going to prove an expensive building. However, the elemental estimate will confirm or reject this theory better than any amount of conjecture. The best rates for pricing would be those obtained from an analysis of the outer London block, since we based the first estimate on the same job. We must remember to add the 10% price increase and the 2.5% tendering differential so as to keep the two estimates on the same basis, unless circumstances have changed in the interim, in which case of course the total of the estimate will have to be revised. Again a standard form should be used.

The form shown in Fig. 16.2 is an example of the sort of thing required. Some of its features would vary according to the preferences of the individual office – for instance some people might prefer to show the preliminaries as a separate item, or to make a separate element of the frame. On the other hand it would be possible to combine the finishes with the relevant structural items. However, the selected arrangement should be standard within the office so that the recorded information can be kept in a suitable manner. The overall rates for elemental estimates can be obtained when doing the ordinary analysis and filed separately, and being large omnibus items there is perhaps less risk than usual in poaching prices from more than one job.

Let us return to the particular example. The establishment of rates for the various

ELEMENTAL ESTIMATE No. 1

PROJECT: Office Block, Midtown
DATE OF COST PLAN: 3.4.90
ASSUMED DATE OF TENDER: Sept 1990

		Total Cost £	Cost per m² of floor area (divide by 2411)
GRD.FLR. AREA (A) 394 m²			
TOTAL FLR. AREA (B) 2411 m²			
EXTERNAL WALLS (C) 3310 m²			
INTERNAL WALLS (D) 1200 m²			

			£	Total Cost	Cost per m²
Work below lowest floor finish	(A)	394 x 306.00		120,564	50.01
Upper floors inc. frame & stairs	(B-A)	2017 x 123.50		249,100	103.32
Roof inc. frame	(A)	394 x 164.50		64,813	26.88
External walls	(C)	3310 x 201.00		665,310	275.95
Internal walls	(D)	1200 x 50.00		60,000	24.89
Floor finish	(B)	2411 x 44.00		106,084	44.00
Ceiling finish	(B)	2411 x 27.50		66,303	27.50
Wall finish	(C+2/D)	5710 x 13.00		74,230	30.79
Decoration	(B)	2411 x 4.50		10,850	4.50
Fittings				2,750	1.14
Services				485,000	201.16
Drainage				9,000	3.73
Site works				27,500	11.41
Contingencies				30,000	12.44
				1,971,504	817.71
Professional fees				-	-
				1,971,504	817.71

(Inclusions and exclusions as approximate estimate unless otherwise stated)
VAT not included

Fig. 16.2. Preliminary cost plan.

elements will have been complicated in this instance by the different proportion of unfinished shops compared to the outer London job. The rate for floor finishes for example will have been obtained as follows:

		£
Rate per m² for floor finishes on outer London offices		27.78
Add 12½% increased costs and tendering		3.47
		£31.25

	£
1,235 m² at £17.10	= 38,594
Shops 25% = 309 m² at say £10.00	= 3,090
Balance 926 m²	= 35,504
Which is £38.34 per m²	

	£
New provincial block 2,410 m²	
16⅔ shops = 402 m² at £10.00	= 4,020
83⅓ remainder = 2,008 m² at £38.34	= 76,987
Total 2,410 m²	= 81,007

	£
which is £33.61 per m²	£33.61
Add say ¾ of £7.50 for more expensive floor finish in offices (as approximate estimate)	10.00
	£43.61
Say	£44.00

From an examination of the elemental estimate shown in Fig. 16.3 we can see that the scheme is likely to prove more expensive than the amount of the first approximate

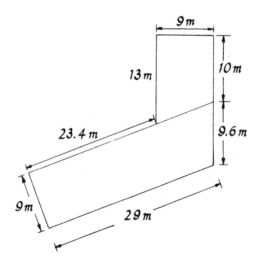

Fig. 16.3. Revised sketch plan for provincial office block.

estimate (£1,902,744). A little of this increase is no doubt due to the slightly increased size of the building (2,410 m² instead of 2,400 m²) but the cost per square metre of floor area is also increased from £792.81 to £817.71. While the increase is not very large (it could be got rid of by a small reduction in the level of specification) it does make us wonder whether this is the most economical solution to the design problem. At this point the figures in the right hand column (cost per square metre of floor area) come in very useful in enabling us to make comparisons with other contracts.

It certainly seems as though the external walls are costing a lot, which confirms our first thoughts on looking at the plan, but we now have figures to prove the matter. Although it would be practical to build this scheme within the budget it would be worth the architect's while to see whether an alternative shape would be possible.

The rather unsatisfactory appearance of the first elemental estimate should lead to the consideration of alternative schemes. It may not always be possible to improve wall/floor ratios very much, due to site restrictions or other unalterable factors. But let us suppose that in the case of the provincial office block the architect has been able to produce an acceptable alternative design (Fig. 16.3):

$$\text{Ground floor area } \frac{33 + 29}{2} + \frac{14 + 10}{2} = 43 \text{ multiplied by } 9 = 387 \text{ m}^2.$$

$$
\begin{array}{lr}
 & \text{m}^2 \\
\text{Total floor area } 6 \times 387 = & 2,322 \\
\text{plus tank room} & \underline{48} \\
 & \underline{2,370}
\end{array}
$$

$$
\begin{array}{lr}
 & \text{m}^2 \\
\text{External wall area } 23.4 + 13 + 9 + 19.6 + 29 + 9 = 103 \times 27 = & 2,781 \\
\text{Tank room as before} & \underline{70} \\
 & \underline{2,851}
\end{array}
$$

Internal wall area say 1,200 m² as before.

The approximate cost for this scheme is shown in elemental estimate no. 2 (Fig. 16.4).

This is a much more hopeful design with an estimated cost of £1,877,219, slightly below the approximate estimate. The allowance for site works has been increased, as additional treatment to the front areas would be required owing to the front elevation being set back on the splay.

It should be realised that outline elemental estimates will not be completely accurate, although the comparisons obtained from them will be valid enough. Each of these breakdowns would only take a few hours of a senior assistant's time, and each hour spent at this stage is worth days spent on cost checks of minor elements later on. This is where the process of cost planning really pays for itself.

This is the place for a word of warning. Because we are dealing in such large figures at elemental estimate stage it is vital that all arithmetic, and if possible all measuring, should be checked. A silly mistake in calculating the principal areas would have dreadful consequences later on, and just because they seem (and are) simple the elemental estimates must not be taken lightly. The same remarks apply to the preliminary estimate, which we dealt with in the previous section.

It is also important to remember that there are many other considerations in

ELEMENTAL ESTIMATE No. 2

PROJECT: Office Block, Midtown
DATE OF COST PLAN: 10.4.90
ASSUMED DATE OF TENDER: Sept 1990

		Total Cost £	Cost per m² of floor area (divide by 2370)
GRD.FLR. AREA (A)	387 m²		
TOTAL FLR. AREA (B)	2370 m²		
EXTERNAL WALLS (C)	2851 m²		
INTERNAL WALLS (D)	1200 m²		
Work below lowest floor finish (A)	387 x 306.00	118,422	49.97
Upper floors inc. frame & stairs (B-A)	1983 x 123.50	244,901	103.33
Roof inc. frame (A)	387 x 164.50	63,662	26.86
External walls (C)	2851 x 201.00	573,051	241.79
Internal walls (D)	1200 x 50.00	60,000	25.32
Floor finish (B)	2370 x 44.00	104,280	44.00
Ceiling finish (B)	2370 x 27.50	65,175	27.50
Wall finish (C+2/D)	5251 x 13.00	68,263	28.80
Decoration (B)	2370 x 4.50	10,665	4.50
Fittings		2,750	1.16
Services		482,050	203.40
Drainage		9,000	3.80
Site works		45,000	18.99
Contingencies		30,000	12.66
		1,877,219	792.07
Professional fees		-	-
		1,877,219	792.07

(Inclusions and exclusions as approximate estimate unless otherwise stated)
VAT not included

Fig. 16.4. Revised preliminary cost plan.

designing a building besides cost, and that the cost planner must not be tempted to try and usurp the architect's function when the shape and type of building is under consideration. The cost planner must be prepared to give the architect all possible assistance at this time, but should avoid making detailed suggestions about matters of planning and architecture which are entirely the architect's responsibility.

Chapter 17
Cost Planning at
Production Design Stage

Elemental cost studies

Cost research or cost studies are usually called for in the earliest design stages of the project, when the architect is considering design alternatives. It is at this stage that cost models such as those described earlier will be most effective. However, at the present time these studies would normally be done by means of approximate quantities, about which sufficient has already been said.

Cost studies if done properly are expensive in professional time, and therefore cannot usually be undertaken in respect of the majority of items in the project. They should therefore be reserved for comparisons of important components, or for projects which are themselves unusual and for which ordinary elemental cost data are of little use. Large-scale cost studies are inappropriate where the component is not a significant part of the cost structure, or where the difference in cost between the alternatives is obviously going to have little effect on overall costs.

Where there is little cost difference between one solution and another the decision should be taken on grounds other than cost and not on the possibility of a marginal saving; if the contract is to be awarded in the traditional manner with competitive tendering on bills of quantities the cost planner cannot forecast which of two alternatives with a marginal cost difference would actually be priced the cheaper by the (unknown) successful contractor. In connection with this the cost planner must remember the natural tendency of the contractor to play safe when pricing work involving unknown or experimental materials and techniques.

Another point to bear in mind in connection with cost studies is that all constructional systems have certain optimum conditions (of loading, span, etc.) in which they are at their most economical. There is usually a reasonable spread of conditions on either side of the optimum in which the system remains reasonably economic, but once the conditions get outside this range costs will start to rise steeply. Even though the particular system may still be perfectly feasible in these unsuitable conditions it is likely that there will be a cheaper and better solution.

It is also worth noting that a high standard of fire resistance, thermal insulation or sound insulation normally costs money, and that it is usually wasteful to employ methods of construction or materials which offer these advantages if they are not required.

We will now look at some of the principal elements in the building, and the particular aspects of design costs which affect each.

Foundations

The type of foundations required will normally be influenced by the type of structure which they are to carry; obviously a structure which imposes heavy point loads cannot be carried on conventional strip foundations. Piling is almost always a very expensive solution and is normally employed only where conventional foundations are impossible because of the depth at which a bearing formation exists, combined sometimes with the waterlogged nature of the ground. Different systems of piling have particular advantages and disadvantages but the specialist piling firms are usually more than ready to offer their services in connection with early cost investigations. As previously noted the cost of the piling itself will usually be a complete extra, because the complex of pile caps and ground beams associated with this form of construction is often as expensive as conventional foundations.

Frame

Cost studies may be necessary to determine the type of frame to be used or indeed whether there should be a frame at all. External walls of habitable buildings need certain qualities of weather resistance and heat insulation, and such a wall capable of bearing quite substantial loads costs no more than a non-loadbearing wall with similar weathering and insulation performance. In conditions where normal external walling would be capable of carrying the weight of floors and roof the cost of a frame is a complete extra, and it follows that for small buildings of not more than three storeys a frame is likely to be an expensive solution, even though it may be desirable for other reasons. A frame for small buildings only becomes economically justified where it is necessary to have a very high proportion of the external walls glazed, or in farm sheds and the like where the walls do not need good insulation or aesthetic qualities and cheap non-structural materials such as asbestos-cement sheeting can be used.

A frame, if required, is likely to be of steel or reinforced concrete, with light alloys and timber as recent and rather specialised competitors. As with other material comparisons, if either steel or reinforced concrete held all the cost advantages the other would long since have disappeared. For multi-storey construction there is a general tendency for reinforced concrete to be cheaper in normal conditions of loading, although the difference may often be small. The structural steelwork industry is continually improving its techniques of design and construction, particularly when the cost competition from reinforced concrete becomes severe. Although steel is inherently cheaper than reinforced concrete the saving can sometimes be wiped out by the measures necessary to give the structure an adequate fire-resistance. This is almost certain to be the case if in-situ concrete cladding is required to beams and columns, so that the type and standard of fire resistance is therefore an important factor in any such cost comparisons.

Steel therefore gives its best comparative cost performance as a frame to single-storey buildings or in roof framings where fire resistance is not normally required by the building regulations and where the steel may be used without any protection other than paint. For such applications reinforced concrete is not usually competitive, except in some proprietary systems for constructing factories or sheds out of standard precast units.

It is worth noting that some contractors (particularly small or medium sized firms) prefer working with a steel frame to erecting a reinforced concrete structure of their own, and this preference would be reflected both in tendering enthusiasm and in completion time for the project.

In designing cost-effective steel frames attempting to save weight of steel can be counter-productive if the result is increased complexity – by using compound welded members, for instance, in lieu of simple sections. A steel fabrication firm's advice should be sought as to the point where savings are swallowed up by fabrication costs.

Most constructional steel firms and speciality reinforced concrete contractors are very helpful in giving cost information for a new project. If a consulting structural engineer is appointed for the project the cost planner should work with him in providing cost information; in many cases the consultant would provide most of the cost information himself but as alternative frame designs may affect other elements as well he may not be able to ascertain the full cost implications without assistance.

As well as a choice of materials, frame design will involve a choice of frame shapes and spacings. The most uneconomical solution will occur where heavily loaded beams have to span long distances especially if they are also restricted in depth. The spacing of columns and beams will have an important effect on cost, particularly if the columns are expressed on the elevations and covered with expensive cladding. If the frames are spaced too closely the savings in sections will not pay for the additional frames, since although frames spaced at 2.5 m centres will each be carrying two-thirds of the weight of the frames spaced at 3.75 m centres they will cost much more than two-thirds of the latter. This is because the choice of section for a beam is affected by its own span as well as by the load it carries, and the span of the beams will not have been changed (see Fig. 17.1).

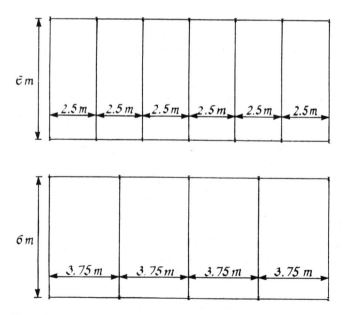

Fig. 17.1. Alternative grid layouts.

Since the most economical spacing will depend upon the span and upon the loading it is necessary to consider each case individually. For normal floor loadings however it is unlikely that spacing as close as 2.5 m will give the most economical solution, while spacings much in excess of 5 m will begin to produce additional costs on the floors and roof owing to their increased span, even though the frame design itself may still be economical.

Staircases

Since the structure of a concrete staircase represents less than a third of the total cost (the remainder being caused by finishes and by balustrades) and since the whole elemental cost is only a minor one it is doubtful whether the staircases of a building are the most fruitful field for cost studies. However, the true total cost of a staircase would include the surrounding walls of the staircase together with foundations, windows, wall finishes, doors, roof, etc., and would be quite considerable. The most rewarding approach is therefore likely to be the reduction in number of staircases rather than in the details of their design.

Upper floors

In contrast to staircases upper floors are an important element and require comparatively little work in the evaluation of alternative designs, which may therefore be well worth doing. As with some other elements, it is difficult to arrive at true cost comparisons without considering the 'frame' element as well.

Roofs

The architect may require cost studies of alternative roofs to which most of the preceding remarks apply. In addition the comparative costs of flat roofs and pitched roofs may be involved; these will depend more upon the level of the comparative specifications than upon the basic difference in roof type. However, for medium to large spans a satisfactory pitched roof is likely to be rather cheaper than a flat roof of comparable quality, partly because of the simplicity of spanning large areas with roof trusses rather than deep beams.

Pitched roofs also tend to be more durable than flat roofs which are sensitive to poor workmanship and poor design detailing, unless very expensively built. Bad maintenance experiences with cheap flat roofs has led to the avoidance of these wherever possible.

Rooflights

Rooflighting is not normally the most economical method of lighting rooms where windows are a possible alternative, and individual domelights or small lantern lights are probably the least satisfactory from a cost point of view. However, there are many reasons why the architect may choose this means of lighting and the element is not important enough for this to affect the total building cost very considerably.

On some current buildings steeply pitched roofs effectively form the walls of upper storeys, and these require the same level of fenestration as a wall. This is one of the areas where the theory of elemental cost planning breaks down, and it will almost certainly be necessary to consider the walls and the roof together for cost planning purposes.

External walls

This may be the most important structural element, particularly in a multi-storey building, and being one in which a tremendous range of constructional methods and finishes may be used is a suitable element for cost studies. In the type of building where the external walls comprise a series of repeating panels it will be sufficient to study a single panel in detail. It will often be difficult to consider this element in isolation from windows, internal finishes, and frame. Note that ordinary brick or concrete block walls finished with facing bricks externally are likely to be far cheaper than any other construction of comparable performance and durability; so that a considerable area of plain walling on a building may enable smaller areas of luxury construction or finish to be used while still keeping the overall elemental cost quite reasonable.

Internal walls and partitions

The comparative costs of traditional partitioning methods are well enough known for cost studies to be unnecessary, but if it is proposed to use a special type of partition on a large scale then a cost investigation would be worthwhile. Comparative costs of small areas of special partition or glazed screens on the other hand would not have much effect on the total building cost and would not usually be worth investigating.

If *The Architects' Journal* or similar list of elements is being used it is important to remember that the cost of finishings to traditional partitions are included in other elements, but the complete cost of self-finished proprietary partitioning is included in the 'partitions' element, so that in this instance elemental costs will not be strictly comparable.

Windows

These are often a major element of modern buildings and have a very large cost range. Cost differences will normally be due to performance requirements rather than the actual materials used in manufacture; standard metal or timber windows for housing schemes are competitive in cost as are high class metal or hardwood windows for prestige buildings, but the latter category may be four or five times dearer than the former. Apart from the actual material and section of framing the cost of windows is substantially affected by performance, particularly as regards double-glazing and weather-stripping, but there are also factors affecting individual windows or groups of windows rather than the fenestration of the building as a whole:

1 *Size of window*. Small windows tend to be high in cost per square metre, both because of the greater intricacy of the window itself and the cost of forming the opening in the wall; these two costs vary according to perimeter rather than area.

2 *Size of panes*. Very small panes increase the cost per square metre, so also do

panes which are so large that it is necessary to glaze them with stout float glass instead of sheet glass. It is unlikely that any saving on window frame or glazing bars would counterbalance the cost of plate glass unless an extremely expensive type of window was being used.

3 *Opening lights.* This is probably one of the most important factors in window cost, particularly where a high standard of weather resistance is required. The heavy comparative cost of opening portions of windows (as against areas of fixed glazing) makes it very difficult to compare window costs on an overall square metre basis unless the pattern of opening and fixed areas is very similar. Unfortunately it is not even possible to compare the costs of opening portions on a square metre basis as so much of the cost (hinges or pivots, fasteners, framing to angles) varies according to the number of sashes rather than their size.

Because of the above factors it is often necessary for any cost studies of windows to require individual consideration of all the window types in the building, rather than being confined to a typical window and the results being applied to the remainder on a square metre basis.

Doors

Except on buildings which are divided into a large number of very small rooms, such as hostels or flats, the doors are not usually a significant element. Because of this, because they are one of the components most subject to heavy mechanical wear, and because they are also one of the most noticeable features of the building it is probably a mistake to be excessively concerned with cost when designing this element.

Floor, wall, and ceiling finishes and decorations

Cost studies of these elements are likely to be comparatively simple, and where large areas are involved are certainly worth doing.

Joinery fittings

Of all minor building components single joinery fittings involve the most work in arriving at a cost, and so cost studies of such fittings should not usually be attempted. Where a fitting has been designed for repetition (such as a bedroom fitting for a large hostel, or a typical bench to be used as a basis for furnishing a set of laboratories) the position is of course different.

Generally

We have seen many ways in which cost studies can contribute to the design of a building. But it is important to realise that it would not be economically possible to employ more than a few of them on any particular building, as a complete cost study of every element, major and minor, would not produce savings commensurate with the amount of time expected. The only occasion where such a course might be practicable is where a large number of similar buildings are to be erected, as for instance a standard house design for a large housing authority.

The cost plan

We now come to consider the cost plan itself. This may take many forms, but all the different systems have much common ground. The plan will almost certainly be prepared by elements and the cost will be expressed per square metre of floor area even if it is calculated in a different manner, in order to be able to make comparisons between other schemes of different size on a common basis.

If there is a cost target it will by now have been set up with some degree of finality and the architect will want to go ahead with the working drawings. Since cost planning involves cost control the architect must design in accordance with the cost plan, which must be available before the working drawings have proceeded very far. What do we need in order to prepare such a plan?

1 A drawing and a standard of specification to which the cost plan can be related.

2 A cost analysis of a comparable project.

Opinions differ on how detailed the specification information should be when the cost plan is prepared and this is a suitable place to discuss the issues involved.

By training, and often through bitter experience, the cost planner likes to deal in solid facts. His natural inclination therefore is to try and get as much detailed information as possible from the architect before preparing the cost plan, and if this is not completely forthcoming to fill in the blanks himself. When the cost plan is complete he can supply the architect with a detailed list showing the type of construction and finish on which the cost is based, and the architect knows that although he may alter anything in the specification at will, any such deviation may have an effect on the planned cost. The cost planner may also feel that this specification gives him some protection in the possible event of an architect failing to design within the agreed limits, since these limits are written into the cost plan and are not just assumed.

However, this procedure is open to the objection that the cost planner appears to be taking the whole of the responsibility for design out of the architect's hands and into his own. The alternative way of dealing with this problem is to leave the responsibility for design entirely where it belongs, with the architect. When the cost plan is prepared the architect's ideas on specification are incorporated into it, but if these are not forthcoming the cost planner does not provide them for himself. Instead he devises a target cost for the element based on standards of cost performance achieved elsewhere and it is left to the architect to design as he wishes within that cost. If the element, when designed, costs more than the sum of money allocated to it then it must either be redesigned or the elemental cost increased at the expense of other elements.

That the philosophy behind this approach is sound few will deny, but the first alternative is the one which was always preferred by most cost planners, and today is almost universally adopted. Apart from the factor already mentioned (of protection against uncooperative architect) it is said that most architects prefer the discipline of a specification to the imposition of a cost figure for the element which is unrelated to specification and may necessitate several attempts at design before it is achieved. The 'detailed specification' method also has the advantage that if the architect works to the agreed specification then subsequent cost checks will be unnecessary.

In order to prepare an elemental cost plan the analysis of a comparable project will be required. It is not necessary that the buildings should be for a similar use as long as there is some reasonable compatibility as buildings. For instance a police-station and a health clinic are both public buildings with fairly similar storey heights and divided into both small and large rooms; the elemental analysis of a police-station could therefore be sued to prepared the cost plan for a health clinic if nothing better was available whereas between say a church and a block of multi-storey flats there is no resemblance whatever.

The cost plan must be prepared on a standard form, for reasons that have already been emphasised in connection with analyses and preliminary estimates. The form should show the chosen list of elements, contingency sum, preliminaries and insurances (if separate), and fees. It is common practice to allow a 'design margin' of between 1 and 5% as a design contingency in addition to the contract contingency, and this margin may come in useful during the cost check when it can be absorbed into any element that needs it. The cost plan should also show the cost index value which is being used.

Although specialist computer software is available for cost planning work most cost planners prefer to use an ordinary spreadsheet package adapted to the practices of their own particular office. Whichever alternative is used the software will guide the user through the processes of preparing the cost plan, making sure that all necessary adjustments are made and that nothing is forgotten.

Prior to the large-scale adoption of micro-computers for the purpose much the same control was obtained by the use of standard manual forms and procedures.

The first page of a typical manual form is shown in Fig. 17.2.

Method of relating elemental costs in proposed project to analysed example

There are two ways of doing this, either by using unit costs of the elements or by proportion. In spite of what may be thought to the contrary these are really two ways of doing much the same thing; in both methods the unit quantity of the element (e.g. area of external walls) is first measured from the sketch drawings of the proposed project.

Using the first, or 'unit cost' method this quantity is priced out at the unit rate per square metre for the same element obtained from the analysis, where this is applicable, but if specifications are different then approximate quantities are more likely to be used. This gives the total cost of the element, and the total is then divided by he floor area of the proposed project to give the elemental price per square metre.

When the second, or 'proportion' method is used the ratio of unit area to floor area in the proposed project is compared to the corresponding ratio in the analysed example, and the elemental cost per square metre is adjusted in proportion to the difference in the two ratios. With this method one is not concerned with the actual unit cost (either in total or per square metre) at all, although the total cost for the element will be required for comparison purposes at cost check stage.

In order to illustrate the systems in use, two examples are worked out on page 200 using both methods. It will be seen that either method will allow differences in quality as well as quantity to be adjusted.

Methods of relating proposed project to analysed example: 1

Analysed example
    Total floor area 5,000 m²
    External wall area 4,000 m²
    Wall:Floor ratio 0.80
    Cost of external wall per m² of floor area (elemental cost) = £75.00
    Cost of external wall per m² of wall (unit cost) = £93.75

Proposed project
    Total floor area 4,800 m²
    External wall area 3,600 m²

Elemental cost of external walls on proposed project

UNIT COST METHOD

3,600 m² of wall at £93.75 = £337,500

£337,500 ÷ 4,800 = £70,312

Total cost 4,800 × £70.312 = £337,500

PROPORTION METHOD

Wall: floor ratio $= \dfrac{3,600}{4,800} = 0.75$

$£75.00 \times \dfrac{0.75}{0.80} = £70,312$

Method of relating proposed project to analysed example: 2

Analysed example
    Total floor area 5,000 m²
    Internal doors 35 no.
    Costs of doors per m² of floor area (elemental cost) = £2.10
    Cost of doors each (unit cost) £300

Proposed project
    Total floor area 4,800 m²
    Internal doors 50 no.
    Cost of doors each £330 (better quality)

Elemental cost of internal doors on proposed project

UNIT COST METHOD
50 doors at £330 = £16,500

£16,500 ÷ 4,800 = £3.44

Total cost 4,800 m² × £3.44 = £16,500

PROPORTION METHOD

          Quantity   Quality
       adjustment adjustment

$£2.00 \times \overbrace{\dfrac{5,000}{35} \times \dfrac{50}{4,800}} \times \overbrace{\dfrac{33}{30}} = £3.44$

What are the advantages and disadvantages of the two methods? The proportion method is the logical choice for use where the elemental cost per square metre is regarded as an index of performance rather than the result of a specification, and it tended to be preferred by central government departments in the early days of cost planning.

But the unit cost method is almost exclusively used today. It is more in line with the traditional quantity surveying approach and most cost planners would feel less likely to make mistakes when using it, particularly if the workings become complicated. The unit cost method is more easily related to approximate quantities either at cost plan or cost check stage, and we must remember that at one or other of these stages fairly detailed quantities have got to be introduced.

If the analysed example will not yield all the necessary information for preparing the cost plan it must be obtained elsewhere:

1   By taking a price from another analysis. This is potentially dangerous; we have seen how a priced bill of quantities will probably contain compensating errors and that these will be incorporated in the analysis. However, if we take items from more than one analysis we may get cumulative instead of compensating errors; it would be possible to take a number of elements from different analyses all of which are priced too low and this would make the total of the cost plan quite unrealistic. This cannot happen where a single analysis is used throughout.

   If prices are obtained from other analyses it is wise to collect as many examples of the same item as possible so that any errors are likely to cancel out in the average so obtained.

2   By the use of approximate quantities. These are likely to be very approximate indeed (remember that we have no working drawings) and herein lies the danger; one is much more likely to leave things out than to put extraneous items in. Therefore when using approximate quantities the surveyor must be very careful to include all the items which are included in the description of the element (e.g. skirtings and their screeds with 'floor finishes', or rainwater pipes with 'roofs'), and also to include the percentage for preliminaries, insurances or contingencies if these are included in the elemental rates. Needless to say, the quantities must be priced at the same level of market rates as the cost plan itself, although the rates will be higher than ordinary bill of quantities prices because of the need to include the cost of secondary work which has not been measured separately.

The approximate quantities themselves will be similar to those normally employed by surveyors for estimating purposes except that the elements must be kept separate; so that the plastering on the underside of a concrete roof slab, for instance, could not be included in the overall rate for the roof but would be worked out separately as part of 'ceiling finishes'.

Examples of use of information from analysed building

This information is given below.

Analysis

    Total floor area 2,000 m²
    Element: internal partitions
    Elemental cost: £27.00 m²

1,400 m² of 75 mm breeze partition at £13.00	£18,200
900 m² of 225 mm and 300 mm brick walls at £33.00	29,700
120 m² of glazed partitions at £50.80	6,096
2,420	53,996

      Average unit cost = £53,996 ÷ 2,420 = £22.32 m²

Proposed project

    Total floor area 3,100 m,²
    Total area of partitions (measured roughly from drawings) 3,500 m²

In some offices it is the custom to keep a complete schedule of prices for use when pricing approximate quantities; this certainly gives a consistent level of pricing but it is doubtful whether any but the largest organisations would find the preparation and maintenance of such a list to be an economic proposition.

If we do not know the type of partition required, or if the proportions of various types are likely to be similar to those of the analysed example we can use the average unit rate, or work by the proportion method:

    £3,500 m² at £22.32 = 78,120 total cost: 3,100 = £25.20 m² elemental cost

But if even at this stage we can see that the proportions of the various sorts of partition are quite different to the analysed example we must use the individual unit cost as shown below.

20% 75 mm breeze partition = 700 m² at £13.00	£ 9,100
10% 225 mm brick wall = 350 m² at £33.00	11,500
60% glazed partition (cheaper type)	
= 2,100 m² at £45.00	94,500
10% timber stud partition	
(no rate for this: price built up by	
approximate quantities) = 350 m² at £23.00	8,050
	123,200

£123,200 ÷ 3,100 m² = £39.74 elemental cost

When severely altering the proportion of items like this it is necessary to watch out for any freak rates. For instance, it is possible that the rates for glazed partitions in the analysed example may have been much too low, and being a very small part of the whole job this will not have had much effect on the total price. On the proposed project however the proportion of glazed partition has been increased from 10% to 60% and such an error would be much more serious. Again this emphasises the point which has been made before, that the smaller items in the analysis may well have the least reliable prices.

Elemental costs

By whatever means the total cost of the element has been obtained it should always be expressed per square metre of floor area, for two reasons.

Firstly, for comparison of reasonableness with other buildings. If the partitions (for instance) are costing far more per square metre of floor area than on other buildings of the same type it is worthwhile investigating the reasons. These may be perfectly sound; the cost may include built-in cupboards, there may be a greater degree of insulation required between rooms, but it is also possible that there is insufficient planning of the accommodation and this would also be reflected in other elements.

Secondly, the elemental price per square metre enables the importance of any extravagance or economy to be judged in a way that is not possible when considering unit prices only. Let us consider the unit costs for three different specifications of internal doors:

1 Hardboard faced skeleton framed flush doors hung on steel butts to softwood frames, no architraves. *Unit cost £45.00 per m².*
2 Plywood faced semi-solid core flush doors hung on steel butts to softwood frames with softwood architraves. *Unit cost £82.50 per m².*
3 Mahogany veneered plywood faced solid core flush doors hung on solid drawn brass butts to mahogany frames with mahogany architraves. *Unit cost £175.00 per m².*

From a comparison of the unit costs it would seem that specification (1) is prohibitively expensive for normal work compared with the more usual specification (2). Yet to adopt the better specification in a typical school of 3,000 m² with 70 internal doors would only increase the elemental cost per square metre by about £2.16, and to use the completely inadequate specification (1) would only save £0.88. Considering that the total cost per square metre of the school would be around £600 these increases and decreases are quite insignificant. On the other hand a marginal difference in some other elements, such as the particular choice of hardwood for wood block flooring, which may only alter the unit price by 10% will have as great an effect upon the cost of the school per square metre as a 100% difference in door prices.

It will soon become obvious to the cost planner which items need the most time spent on them at cost plan stage – in other words, which elements have the greatest cost significance. It would be worthwhile to use approximate quantities for a major element such as external walling where a difference of one or two pounds in the unit cost per square metre would have an important effect on total cost, and to check the cost plan very closely when drawings subsequently become available; such elements as internal doors or ironmongery on the other hand can be dealt with much more rapidly.

Presentation of the cost plan

The cost plan should be presented as neatly as possible; it will almost certainly be going to the client and should reflect credit on the firm which has prepared it. To what extent

COST PLAN

PROJECT: Office Block, Midtown

DATE OF COST PLAN: 12th March 1990

ASSUMED DATE OF TENDER: July 1990

TOTAL FLOOR AREA: 2390 m²

Note: This Cost Plan is based upon the attached outline specification, and both documents should be read together.

	Unit quantity	Unit cost £	£	Total cost £	Elemental cost per m² £
1. WORK BELOW LOWEST FLOOR FINISH. Grd. Flr. Area 390 m²	390 m²	306.00		119,340	49.93
2. STRUCTURAL FRAME 2390 m²	-	-		125,600	52.55
3. UPPER FLOORS 225 mm Hollow pot Floor 150 mm in-situ r.c.	386 m² 1585 m²	60.00 41.00	23,160 64,985 88,145	88,145	36.88
4. STAIRCASES R.C. staircases One 25 m rise One secondary staircase 21.5 m rise	25 m/ rise 21.5 m/ rise	1225.00 900.00	30,625 19,350 49,975	49,975	20.91
continued				£383,060	160.27

Fig. 17.2. Front sheet of typical cost plan.

OUTLINE SPECIFICATION

in connection with Cost Plan

for

Office Block, Midtown

WORK BELOW LOWEST FLOOR FINISH	12th March 1990
Foundations to walls and/or columns	In-situ concrete piles average 6 m long (approx. 90 No.) reinforced concrete pile caps and ground beams.
Basements, walkway ducts, etc.	Semi basement 1.5 m approximately 8 m x 4 m for boilers. Also lift pits approx. 0.75 m deep in waterproofed concrete.
Rising walls	Clay common bricks in cement mortar built off ground beams (approx. 0.75 m high generally to d.p.c.). Reinforced concrete walls (0.3 m thick) around stair well and lift pits.
Ground floor slab	Generally 150 mm slab reinforced with fabric on building paper and hardcore. Note thickness of hardcore to make up levels (0.6 m average).
D.p.c's. and membranes	Lead-cored bituminous d.p.c. in walls. 3 coat Synthaprufe or similar on g.f. slab. 2 coat asphalt tanking to semi basement.

STRUCTURAL FRAME	
Generally	Reinforced concrete beams (at 3.5 m c/c) and columns similar to Outer London House.

Fig. 17.3. Specification notes to accompany cost plan.

expense should be incurred in order to give the plan an impressive appearance is a matter for individual preference, but if the document has been prepared by computer a high-quality printer must be used for the purpose.

Another matter for personal decision is the extent to which the detailed build-up of the plan should be made available to the architect. Whether this should be done in all cases will depend upon the degree of mutual confidence which exists and also upon the available facilities for copying a large quantity of rough working.

With the cost plan should go the specification upon which it is based; this will necessarily be more in an outline form than the type of specification which will eventually be embodied in the bill of quantities. It may either be included as part of the cost plan itself or prepared as a separate document; a typical example is shown in Fig. 17.3. If a separate document is used it is important that there should be a space for each specific item so that it can be quickly seen whether everything has been dealt with or not. The specification should preferably be filled in by the cost planner in consultation with the architect rather than being sent to the architect for him to fill in himself.

Cost checks on working drawings

After the cost plan has been prepared and agreed it will be the cost planner's task to see that the work shown on the working drawings and measured in the bill of quantities corresponds with the plan, as otherwise the tender figure is likely to be quite different from the amount of the estimate. Any necessary adjustments must be made before the bill is prepared, not afterwards.

For this purpose all drawings should be cost checked as they come into the office; this must be a matter of routine and a rubber stamp should be made so that the drawing can be marked when the check has been done.

Unfortunately this stage in the quantity surveyor's work (the receipt of working drawings and subsequent revisions and details) is usually a hectic one and there is not time to spare for things like cost checks. Again it is unfortunate that these checks can best be done by a senior person, who is usually very busy in other directions.

Where the cost plan has been based upon approximate quantities tied to a fairly detailed specification, and where the design has proceeded in accordance with it, the work involved in cost checking will be minimal. If, however, the design has been radically changed for some reason it may be better to start the process again and prepare a fresh cost plan at as early stage as possible rather than attempting to check the detailed drawings against a cost plan prepared for a somewhat different building. Cost checking procedures are quite the most time-consuming and error-prone part of the whole process and the more they can be legitimately minimised the better. Should it prove to be necessary to undertake a thorough cost check of the whole design (perhaps because the cost plan did not give the architect sufficient guidance), then the drawings will have to be prepared by elements if the feedback from the cost checks is to be of any real use.

The extent to which cost checking should be carried out will depend upon:

1 The amount of extra time which the cost planner can be allowed. This will vary from job to job, but it must be realised that if lack of time prevents essential cost checks from being carried out then the whole attempt at cost planning will largely be a waste of time.

2 The amount of apparent alteration from the cost plan. If the cost plan was based upon a very similar project for the same architect and client and the details of specification and design were therefore known fairly well, the cost check may entail little more than a quick glance at the drawings to see that nothing is substantially different. Even where the circumstances are not quite as ideal as this it may be possible for the cost planner to satisfy himself by inspection that the drawings show what the cost plan envisaged. The drawings should be stamped and initialled, even so.

3 The amount of detail in the cost plan. If it was possible to take out fairly full approximate quantities the cost check may simply consist of comparing these quantities with the working drawings.

4 The degree of confidence which exists between cost planner and architect. If the architect is known to be cost conscious, capable of and enthusiastic about cost design, the cost checks will be much less important than where this confidence is lacking.

5 The familiarity of the project. Most cost planners would feel fairly confident of cost planning a school, whereas a planetarium would be a very different proposition. In the latter case the cost plan would probably incorporate rather a lot of assumptions and it would be necessary to check these in detail.

6 The importance of the element. It would be a very self-confident cost planner who would not bother to check the external walling, or the roof of a single-storey building, even under the most ideal conditions. On the other hand if time is pressing some of the smaller elements can often be ignored.

Carrying out the cost check

Again this is a suitable occasion for using standard forms. Where the architect is designing by elements the procedure will be considerably simplified, but even if he is not doing so the check must still be done by elements.

The design of the cost check form may be left to the tastes of the firm or other organisation. It should show at least the following information:

1 Number of check. The checks on a particular project should be numbered consecutively, starting at 1.

2 Total estimated cost of project after completion of previous check. (This amount will be got from the previous cost check form, or if this is the first check it will be the unaltered total of the cost plan.)

3 Reference number of drawing or other information being checked.

4 Element being checked.

5 Amount allowed in cost plan in respect of element (or as amended by previous checks).

6 Estimated cost of element as calculated from the drawing, or other information, being checked.

7 Difference (plus or minus) between (5) and (6).

8 Estimated cost of project after (7) has been added to or deducted from (2).

A typical example of a cost check form is shown in Fig. 17.4.

As each cost check is carried out a copy of the cost check form should be sent to the

```
                    CONTRACT  Office Block, Midtown

                    COST CHECK NO.  3

                    ELEMENT  Structural Frame


                                         Total Cost Carried Forward
Date 6.7.90                              from Cost Check No. 2 £1,873,038
Gross Floor Area 2390 m²
```

					per m²
Total Elemental Cost from Cost Plan			£	125,580	52.54
Elemental Cost Check					
(See attached dimensions) 380 m³ Reinf. conc. in beams & cols	£90	£34,200			
2950 m² Formwork to beams and cols	£16	£47,200			
31 tons reinfct (as Mr Smith of Consulting Engineers 28.6.90)	£1200	£37,200			
Sundries		£10,000			
Total revised elemental cost	£	128,600		128,600	53.80
Amount of ~~saving~~/addition				£3,020	1.26

```
            Revised total cost carried forward
            to next Cost Check           £1,076,058
```

Fig. 17.4. Example of cost check form.

architect so that he has an up-to-date running total of the project. If any check shows a substantial increase or decrease it would be as well to discuss the matter with the architect rather than merely sending the results of the check to him. Similarly he should be warned if the general standard of detailing appears to be more lavish than was allowed for.

If a drawing shows changes in more than one element a separate form should be used for each. The rough workings in connection with each check can be done on dimension paper and stapled to the office copy of the form.

A list should be kept of all drawings or other information received with columns for marking:

1 That the cost check has been carried out.
2 The reference numbers of the cost check form or forms.
3 The elements involved. The purpose of this is to enable the checker, on receiving a drawing showing (say) part of the roof to look back through the list to see whether any adjustments have already been made to this element.

If it appears that a detailed cost check of any particular drawing or information is not necessary the list can be marked accordingly. But all drawings and information (such as sub-contractors' quotations or replies to queries) must come to the cost checker in the first instance.

In connection with cost checking it is as well to remember the total cost of the project and also the inherent margin of error in cost planning; a cost planner whose estimates consistently get within plus or minus 5% of the accepted tender is doing well. In these circumstances it is obviously not worthwhile to spend much time cost checking isolated details of a large project, as the possible differences in cost are so small compared to the probable overall error.

Prices for cost checking may be obtained from the same sources as have been used earlier; although as there will be a tendency for the items to be more detailed than at cost plan or estimate stage the cost planner is likely to be using either built-up rates, price book rates, or actual rates from the bill of quantities from which the example analysis was prepared. Again it must be remembered that the builder's permission must have bene obtained for this latter course.

It will be very tempting to lift prices from bills of quantities for other projects at this stage, but the errors that can arise from this practice have previously been pointed out.

If prices are built-up or obtained from price books the level of market prices on which the estimate is based must be remembered. It will also be necessary to include a realistic percentage for preliminaries and insurances; this will not necessarily be the same percentage as the builder used in the analysed example.

Cost reconciliation

When tenders have been received the tender which is most likely to be accepted should be reconciled with the cost plan. If, as should occur, there is little difference between the totals this is still worth doing; there are quite likely to be large compensating discrepancies in the various elements and these may provide information for future use.

The reconciliation is in fact a comparison of the final cost plan (as amended by cost checks) with a cost analysis of the tender. This also ensures the cost analysis getting done at the earliest possible moment instead of being left until 'someone has time to do it'.

There may be all sorts of explanations for a considerable difference between the cost plan estimate and the actual cost of a particular element; perhaps the cost planner made a mistake, quite possibly the builder's estimator made a mistake or did a deliberate price adjustment. But if there is a definite one-way divergence between cost plan and tender right through all the elements it is likely that either the cost planner misjudged market levels or that the tenders themselves are abnormally high or low. If the cost planner feels that the latter is the case then he could report accordingly.

Note that the contractor's permission is never required to analyse his tender for cost reconciliation purposes, it is only if the analysis is to be published, or if the individual prices are to be used for cost checking other projects, that the question arises.

Post-contract cost control

With the successful reconciliation of cost plan and tender the design cost planning of the project is completed. It will remain to control the cost during the progress of the works as described in the next chapter. If the tender is within the client's budget as a result of cost planning then variations ought not to be extensive, partly because the cost requirements will have been satisfied by the job as tendered for, and partly because the architectural pre-planning which should accompany cost planning will have produced workable solutions to the design and construction problems. Thus the extra time spent in cost planning may enable the total time spent on financial control of the job to be actually reduced.

Office organisation

With regard to the amount of work involved in cost planning, any sensible person would want to keep this to a minimum. The cost planner will learn from experience which items are of cost significance, where short cuts can be taken, where cost checks can be avoided. Experience will also show that the use of standard forms and procedures speeds things up as long as the cost planning staff are familiar with them. For this reason, and also because of the acceleration in working brought about by familiarity with cost and prices it may be found worthwhile in a large office to concentrate cost planning in a separate departments rather than leaving it to be done by the takers-off on each project.

In coming to a decision on this point a powerful argument in favour of cost planning by the takers-off concerned must be considered; that it will give them familiarity with the project and thus make the actual preparation of the bill much easier. Whether this proposition is as valid as it sounds is open to question; certainly a number of authorities who originally used this method subsequently changed over to separate departments. There is scope for experiment here, but if the cost planner and the taker-off are two different people there must obviously be a close and friendly liaison between them.

A critical assessment of elemental cost planning procedures

It has been suggested that elemental cost planning techniques will be justified if they consistently succeed in forecasting cost within a margin of 5% up or down, and indeed recent research suggests that even this target may be over-optimistic, given the variability which exists in tender levels. It might well be asked whether all this work is worthwhile if at the end of it all the tender is liable to differ from the estimate by such an amount. Traditional and much easier methods of estimating have often come far closer than this.

The answer is threefold; firstly that most budgets are flexible by at least this percentage or that if they are not it is simple enough to make the requisite modest alterations in the scheme. Secondly it must be emphasised that traditional single-rate methods cannot be relied upon to achieve anything like this standard of accuracy, and that for every 'spot-on' square metre estimate there is another instance of such an estimate being up to 50% out. It must be made quite clear that where an estimate, however prepared, forecasts the cost to within 1% or 2% under traditional competitive tendering conditions, then luck as well as skill has played its part in the result. Thirdly, elemental cost planning achieves a balance and economy in the building which cannot be attained by any other method using traditional tendering procedures.

The author remembers an instance concerning a technical school which was estimated on a square metre basis and where the difference between estimate and tender was much less than 5%. Satisfactory though this was nobody was able to answer a very simple question from the architect during the design stage as to whether that estimate would cover a certain type of wall cladding. There was nothing in the estimate with which it could be compared.

In view of the alleged unreliability of estimates prepared by single-rate square metre superficial area methods it is as well to explain why these methods can safely be used for the preliminary estimate before the cost plan is prepared. It is because the cost plan will show up any error in the estimate before working drawings or bills of quantities have been prepared and the necessary adjustments can be made before any of the detail work is started. This is a very different situation to discovering the error at tender stage when time and money have been spent designing an impossible scheme in detail.

While elemental cost planning has made an outstanding contribution to the study, forecasting, and control of building costs, it nevertheless suffers from a number of basic limitations which have prevented it from developing much beyond the stage which it reached within the first few years. One might wonder whether it is doomed to be like the lead-acid electric battery which, although very useful, will never be capable of development into the kind of power-storage that was originally hoped for.

These limitations are:

1 Nobody has yet produced a set of elements each of which performs a single function but which can be easily cost-related, nor are they ever likely to. This means, amongst other things, that it is not really possible to compare the cost performance of the same element on two different buildings, nor, within one element, to compare two different technical solutions concerning one building. In order to attempt this task (which the cost planner is often asked to do) it is necessary to consider all sorts of extraneous matters, some of which are difficult to quantify.

It might be thought that it would be possible to overcome this limitation by the use of complex computer programs which would enable the interaction of the elements to be worked out exhaustively. This might well be possible, but it would be very expensive and at the present time is not worth doing because of the next limitation.

2 As we have already seen, but as is worth repeating here, the 'costs' which are being manipulated may bear no relation to fact; they are not really 'data' in any scientific sense. It is a waste of time and money to use sophisticated methods to manipulate inaccurate data, as is recognised in the computer world where the

maxim 'Garbage in, Garbage out' is well known. In fact, such processing is worse than useless, because the fact that sophisticated analysis methods have been employed leads people to attach some importance to the results. They would immediately recognise the original data as unreliable if it were presented to them in unprocessed form.

It must be remembered that there is no pressure on the contractor to insert rates in his bill of quantities which represent carefully estimated production costs for each of the items, but there are many reasons (as set out previously) why he does not. The rates in the bill of quantities are not 'retail' prices for which the contractor is offering to execute individual pieces of work in isolation, but are merely a notional breakdown of his total offer. However, even if it were possible to compel the contractor to show true cost estimates, and to enforce standard practice in defining and allocating overhead costs, there would still be the difficulty of setting production cost against square metres of design elements, as so many of these costs (e.g. the tower crane) are not directly related to element unit quantity.

3 Even if the two previous limitations could be overcome the whole system suffers because under normal conditions there is no contractual commitment to the cost plan. The quantity surveyor has no responsibility for the tender amounts and the contractor has no interest in the cost plan.

The exercise this has a great deal in common with weather forecasting. The cost planner is able to draw on a considerable body of past experience and can consider present trends, but he is in no position to guarantee the result. He will often be right, but if any unusual circumstances manifest themselves he can still be badly wrong.

The comparatively crude methods which have been set out in this chapter are usually adequate for forecasting within the wide limits of market pricing; any attempt to get much closer in these conditions is unlikely to be worthwhile. Therefore if the cost planner cannot offer any kind of guarantee to his client the latter is unlikely to agree to expenditure on expensive and sophisticated methods of 'control' which are not in fact controlling anything but are just hopeful forecasts. If a higher standard of cost control is required than is provided by the reasonable development of these traditional methods then it will be necessary to sacrifice some measure of market freedom to obtain it, for example by involving the contractor at an early stage. There is just no way round this.

Finally:

1 Elemental cost planning should ensure that the tender amount is close to the first estimate, or alternatively that any likely difference between the two is anticipated and is acceptable.

2 Elemental cost planning should ensure that the money available for the project is allocated consciously and economically to the various components and finishes.

3 Elemental cost planning does not mean minimum standards and a 'cheap job'; it aims to achieve good value at the desired level of expenditure.

4 Elemental cost planning always involves the measurement and pricing of approximate quantities at some stage of the cost plan or cost check.

Use of resource-based techniques in relation to design cost planning

See pages 160–61.

Use of the builder's cost information for design planning

The builder would appear to be in a much better position than the consulting quantity surveyor/cost planner as far as real cost information is concerned, yet he would be ill-advised to use the actual costs from a particular project for cost planning another quite different one. There are two reasons for this:

1 Many of the factors which affect site costs on a particular project have nothing to do with the design of the building and will not repeat from job to job. These include the weather, industrial and personal relations, the skill with which the work was organised, accidents, late delivery of materials, defective work, failure by sub-contractors.

2 Although the 'operation' is an excellent concept for collecting costs it is very difficult to re-use the information arising from it because each operation is unique; the only way in which first fixing of joinery on one project can be compared with another quite different one is by looking at the quantities of the different types of work involved, and the whole essence of operational costing is that the costs are not broken down in this way. If they were, we would be back to costing work items from the bill of quantities, and, as explained, this is impracticable. In some instances, the operation may embrace more than one cost planning element.

However, the contractor's operationally based estimates have neither of these disadvantages when considered as a data base for cost planning. Firstly, they should be based upon an assessment of 'average' costs, and as such will have a closer relationship to future tender prices than ascertained costs from a particular project. Secondly, they will break down each operation into measurable characteristics for pricing purposes (the estimated costs of which may be based either upon experience or work study methods) and these can be used for the purposes of design cost analysis.

In so far as the builder has access to this kind of estimate, and the consulting quantity surveyor has not, he may be said to have an advantage in cost knowledge. Against this the average quantity surveyor is much more experienced than the average builder in translating project cost data into terms which are relevant at early design stage, and this expertise is essential to cost planning and control – the quantity surveyor's need is for a bill of quantities which reflects more closely the way in which operational estimates are built up. For instance, if the 'Preliminaries' section of the bill of quantities were always priced in a consistently itemised way (like the other sections of the bill) the consulting quantity surveyor could analyse and manipulate this important part of the cost in detail instead of as a 'percentage of the measured work'. It must be remembered that the proportion of project cost represented by the execution of the measured builder's work is steadily declining, and the proportion represented by specialist work, fixed charges, and management costs is increasing.

Chapter 18
Real-time Cost Control

Cost planning and control as described so far is concerned with planning and monitoring costs during the investigation, planning and design stages, and it finishes at the point where tenders are received, or a contract entered into. During this stage nothing is irrevocable – drawings can be redone, the scheme can be reduced in size or its configuration altered, the whole thing can be started again from scratch or postponed or even totally abandoned at what will be (from the point of view if the total project) a negligible cost for abortive professional work. Even if the project has got as far as the submission of tenders by contractors, the scheme can still be radically changed or abandoned without any commitment to the tenderers or any need to recompense them for their trouble.

Whilst the accurate forecasting of a tender amount and the signing of a contract for that sum (or an amended one where the forecast hasn't quite worked out) is of considerable importance, the matter does not end there. The final agreed cost, after both the job and the various financial negotiations arising from it have been completed, is rarely the same as the original contract amount, and it is this final figure which the client actually has to pay and which will be the building cost for which he has to raise and service the finance. In many cases the 'contract sum' may be merely an intermediate stage in arriving at this final cost, and a system of cost planning and control which stops at this halfway point will only be of limited use (and possibly not worth the money which it costs).

It is important, therefore, that the cost control process continues to the point of completion and hand-over of the project, and this continuation is often termed 'post-contract' cost control for obvious reasons. It differs from design cost planning in many ways, and these will depend upon the exact contractual arrangements, but two differences will be inescapable, firstly that the interests of a contractor or contractors now form an important addition to the considerations involved, and secondly that major expenditure is currently being incurred and future options proportionately reduced. It becomes increasingly difficult to make major alterations of any kind at the stroke of a pen, once drawings and specifications are being translated into an expensive organisation of resources and finished work, and even delay caused by hesitation or reconsideration may involve the client in heavy costs. It is to emphasise this similarity with real-time computer control systems, and because in some circumstances there may not even be a contract, that we have chosen to use the term 'real-time' rather than 'post-contract' to describe this type of control.

The basis of real-time cost control is reporting at regular (often monthly) intervals or on a special occasion when a major decision has to be made. This cost control report will set out the client's likely final cost commitment in some detail and also the cost consequences of any remaining major options. In some instances this report may be made to the architect or other professional project controller but it is preferable that it should be made direct to the client. This service should be envisaged when the cost planner's fee is being negotiated, and written into the agreement. Its purpose is to enable the client to budget for his likely expenditure, to enable the cost effect of any major changes to be seen in the context of the project as a whole, and to enable avoiding action to be taken if the total cost appears to be escalating unduly.

One of the problems attached to this service is the extent to which sums of money should be kept in hidden reserve by the underestimation of savings and the overestimation, or at least the pessimistic evaluation, of additional costs. There is a natural tendency towards this type of caution, which in its most extreme forms may be taken to the extent of inflating the projections sufficiently to cover without disclosure any mistakes which may be made by the architect or quantity surveyor, and which will still leave the client with a pleasant surprise when the final account is inevitably settled at below the forecast amount. Such conduct would be as unprofessional and unhelpful as inflating the quantities in the bill of quantities for similar reasons.

Nevertheless, caution dictates that some allowance be made for things going slightly wrong, particularly where final costs will be subject to negotiation or where the full consequences of decisions or events cannot be exactly foreseen. The reasonable judgement of these allowances, avoiding both excessive pessimism or optimism, is one of the major factors in making this a truly professional task. In particular, the cost planner who is successful in this role will develop a feeling for the average project, where not everything is going to go wrong, and for the occasional project where it would be wise to assume that it might.

In carrying out real-time cost control it is essential for the cost planner always to be fully informed as to what is happening and what is intended, and ensuring this may well be one of the more difficult parts of the operation. Ideally he should be part of the decision-making process, in which case the difficulty will not arise, but this is not usually the situation. However, merely sitting in his office waiting for information to come to him is not going to be good enough. Most contracts contain no requirement for communication to and from the builder to flow through the cost planner's office. He is likely to be informed eventually of everything necessary for him to play his part in the final settlement of accounts, but this information may be far too late to be of any use for real-time cost control purposes. It must be remembered that cost control requires not a record of costs incurred to date but of likely eventual cost commitments arising from current proposals or from acts of commission or neglect. The cost planner must, therefore, instigate his own 'current awareness' procedures, not only seeing that he gets immediate copies of all official orders, drawings, and letters, but also attending all site meetings and generally looking around the site to see what is going on and what verbal instructions might have been given. It has often been alleged that on many projects the 'official' documentation is merely trying to catch up with the real but informal communications system. The preparation of interim valuations for payment purposes provides an excellent opportunity to monitor what is actually happening and should always be used for this purpose. The maintenance and updating of real-time cost control records is an ideal application for micro-computer spreadsheet systems such as 'Supercalc'.

Although, as shown above, many of the factors in real-time cost control are common to any type of contractual situation, it will be most convenient to look at the detail in the context of each of the different types.

Real-time cost control of lump sum contracts based on bills of quantities

In the real-time cost control system for this type of project the starting-point will be the contract sum, and the provisions which the contract makes for amending it.

The provisions will usually cover:

1 The adjustment of 'provisional' sums of money or quantities of work, which are embodied in the contract, in the light of the actual cost or quantity of work carried out. This type of provision is often made for work such as foundation excavation which cannot be foreseen exactly.

2 Payment for variations and for additions to, and omissions from, the work.

3 The adjustment of amounts included in the contract for work to be carried out by nominated sub-contractors and others, in the light of actual cost.

4 Correction of errors in bills of quantities and other contract documents.

5 Adjustment for the effects of inflation.

6 Adjustment for the effects of new legislation.

7 Compensation to the contractor for the cost of delays in the work caused by circumstances for which the client is responsible (or, in more generous contracts, by circumstances for which the contractor is not responsible). These circumstances may include delays caused by the ordering of variations or by delay in the provision of drawings, etc., or where the more generous provisions apply, delay caused by bad weather or industrial action.

In addition to the likely revised contract amount the cost projections should include updated additional amounts for professional fees, furnishings (unless it has been agreed to exclude these), and also any other items which the client wishes to consider as part of the total cost.

An item which can give rise to controversy is the adjustment of the provisional contingency sum. This is an amount included in the contract sum by the architect to cover unforeseen expenditure and the controversy centres around whether it is at his disposal to cover design development (or more bluntly, mistakes) or whether it is at the disposal of the client to spend on extra items. Some public authorities hold the latter view so strongly as to demand the omission of the contingency sum as the first variation on the contract, and to treat it as a saving. Except where the client holds such extreme views it is probably best to show the contingency sum separately in the cost control report, deducting any incidental extras of the appropriate kind and carrying the balance forward to cover further similar extras which can occur at any stage of the project. In any case a provision of some kind must be made for contingencies during the remainder of the project.

A general difficulty which faces the cost planner is that although the contractor is required by most contracts to give written notice to the architect of circumstances having arisen which will give rise to an extra cost, he is usually only required to give this notice within a 'reasonable time' of the occurrence, which in contractual terms has to be interpreted fairly generously, and which may well be too late for cost control purposes. The remarks already made about current awareness therefore apply strongly. A further difficulty under the British JCT Standard Form of Contract (although not always overseas) is that there is no requirement for the financial effect to be stated or agreed at the same time, and in fact this is positively discouraged both by the contract clauses and by traditional practice. The cost planner's estimates, therefore, have to bear in mind that his figures may be subject to negotiation, and if the builder also suggests a figure at this stage (which he is not bound to do) the cost planner will have to guess the point between them at which a deal is likely to be made. For this reason the cost planner's real-time cost projections on this type of contract should never be shown to the contractor, since the cost planner may have prudently assumed a higher figure for extras than he is prepared at that stage to concede.

Although the contract usually contains a similar provision for notification of reduced or omitted work by the contractor there is obviously less incentive for this to be done, and the duty is easily 'forgotten'. The cost planner again is in a position to see that omissions are duly notified and adjusted.

The adjustment in respect of an amount included for a nominated sub-contractor's work may require to be done on several different occasions. The first adjustment will be done when a sub-contractor's tender is accepted for the work; this tender is likely to contain the same type of provision for further adjustment as the main contract and similar subsequent changes in cost are likely to occur which will have to be reflected in the cost-control report.

The adjustment of the cost-control report for inflation is a difficult one, particularly at an early stage of the project, since it involves not only a forecast of rises in building costs over the contract period but also an estimate of the likely progress of the work in cost terms so that the forecast increases can be applied to an appropriate proportion of the cost. There are three different figures to be calculated for inflation, and this will apply whether the contract adjustment is on the basis of the difference between ascertained actual costs of resources and costs at the time of tender, or whether it is done on an index-linked basis.

1 Increased cost incurred to date.
2 Estimated additional cost for remainder of contract of increases announced to date.
3 Allowances for effect of future increases.

The figure which the client will require for budgeting purposes is the total of the three, although the total should be split into the three categories to emphasise the progressively more conjectural nature of the estimates. (Although it is theoretically possible for there to be a decrease in cost through the operation of this clause of a contract there has been no known instance of a reduction in the contract sum from this cause in the UK during the fifty years prior to 1989).

A most important point regarding cost control of lump-sum contracts is the perennial conflict between tight budgetary control and getting the best deal from market forces. The best budgetary control is achieved if a firm contractual commitment can be obtained at the earliest possible moment – in the case of extra work, for instance, before the order is given to go ahead. Even the British JCT Standard Form of Contract makes some provision for this in its use of the words 'unless otherwise agreed' in its set of procedures for valuing variations. But, as previously mentioned, the contractor is under no obligation to agree a firm figure beforehand, with its attendant risk of underestima-

```
                          OFFICE BLOCK, MIDTOWN

Cost Report No. 2: 10.6.91
                                                                    £
                                                   Contract Sum   1,834,369
Net Omissions                                      £
Partial substitution of 'Conglint'
panels for Portland Stone (V.O.3)                27,050
* Saving in foundations and piling                7,000
                                                 34,050                34,050
                                                                   1,868,419
Net Additions                                      £
Decoration and partitions in
offices (V.O.1)                                  34,000
Carpeting in offices in lieu of
wood block flooring (V.O.2)                       7,100
                                                 41,100                41,100
                                                                   1,909,519

Adjustment of P.C. Sums              Omit         Add
                                      £            £
Heating installation
(Midland Heating Co.)                            7,500
Electrical Installation
(Sparks & Co.)                                     975
*Curtain walling
(The Curtain Walling Co.)            2,065
                                     2,065        8,475
                                                  2,065
                                                  6,410
Profit and attendance                              500
Net Addition                                      6,910                6,910
                                                                   1,916,429
Adjustments for Fluctuations
(including sub-contractors)                        £
*  (i)    Increase on work to date               2,000
*  (ii)   Allowance for effect of current
          increases on remainder of work        60,000
*  (iii)  Allowance for possible future
          increases, say,                       50,000
                                               112,000               112,000

                         Total estimated cost                    £2,028,429

Amount included for contingencies                  £
Contingency sum in contract                     30,000
*Sundry minor variations (net extra)             1,000
                        Balance remaining      £29,000

Note * indicates changed or additional item since previous cost report
VAT & Professional fees not included in above figures.
```

Fig. 18.1. Real time cost report, June.

tion, and may prefer to leave negotiation until he has all his final costs available. It is up to the cost planner to advise his client on the advantages and disadvantages of agreeing extras at an early stage in the light of the client's particular budgetary problems and the apparent willingness or otherwise of the contractor to do a reasonable deal on this basis.

As an example of a system of cost reporting, two consecutive monthly cost reports on the early stages of a job are shown in Figs 18.1 and 18.2. There are many possible ways of setting out such reports, and the reader may well be able to devise what he

OFFICE BLOCK, MIDTOWN

Cost Report No. 3: 12.7.91

			£
		Contract Sum	1,834,369

Net Omissions

	£		
Partial substitution of 'Conglint' panels for Portland Stone (V.O.3)	27,050		
Saving in foundations and piling *(revised figure)	6,652		
	33,702		33,702
			1,868,071

Net Additions

	£		
Decoration and partitions in offices (V.O.1)	34,000		
Carpeting in offices in lieu of wood block flooring (V.O.2)	7,100		
* Extra work to Entrance Vestibule and Staircase (V.O.6)	6,000		
	47,100		47,100
			1,915,171

Adjustment of P.C. Sums	Omit £	Add £	
Heating Installation (Midland Heating Co.)		7,500	
* Additional Thermostats (V.O.7)		500	
Electrical Installation (Sparks & Co.)		975	
Curtain walling (The Curtain Walling Co.)	2,065		
* Lifts (Ascenseurs Ltd)	1,995		
	4,060	8,975	
		4,060	
		4,915	
Profit and attendance		450	
Net Addition		5,365	5,365
			1,920,536

Adjustments for Fluctuations
(including sub-contractors)

	£	
* (i) Increase on work to date	6,500	
* (ii) Allowance for effect of current increases on remainder of work	65,000	
* (iii) Allowance for possible future increases, say,	40,000	
	111,500	111,500

		Total estimated cost	£2,032,036

Amount included for contingencies

	£	£
Contingency sum in contract		30,000
* Sundry minor variations (net extra)	2,200	
* Likely claim for delay in connection with Ground Floor walling layout (Site meeting 5.7.90)	12,000	
	14,200	14,200
	Balance remaining	£15,800

Note * indicates changed or additional item since previous cost report.
VAT & Professional fees not included in above figures.

Fig. 18.2. Real time cost report, July.

considers to be a better format. What is most important is that any changes since the previous report should be clearly highlighted, either by the use of asterisks and notes as in the example or else by simply giving the previous figures as totals. Using this system the figures in the second example for 'Net additions' could simply have been shown thus:

Net additions	£
As before	22,500
Extra work to entrance vestibule and staircase (V.O. 6)	3,000
	£25,500

Points where other people's tastes might differ include the rounding-off of estimated figures (a figure of £18,577 implies a standard of accuracy which £18,600 does not), and the item of 'sundry minor variations' which could have been shown in detail. It is the authors' view that this document is prepared to assist budgetary control, not to start a premature witch-hunt against the architect. Consideration of such matters can usually be better left until the final account, when minor problems can be seen in the context of the finished job.

It will be noted that the report takes account of the variations ordered, and quotations accepted for work which is not yet done, or even started. The possible claim for delay has been taken out of the contingency fund, which will not be able to stand many such demands upon it.

Finally, it must again be emphasised that under normal contractual arrangements the contractor has no duty to participate in any cost control (or cash flow control) scheme of the client, and there are sound commercial reasons why he may not wish to do so. Any such requirement, to be binding, must be written into the form of contract with legal advice and cannot simply be stated in the preliminaries of the bill of quantitities.

The question of cash flow control has been raised in the last paragraph, this will be dealt with shortly.

Real-time cost control of negotiated contracts

A negotiated contract provides excellent opportunities for cost control, indeed this is one of is principal advantages for which, as usual, some measure of market-force benefit may have to be sacrified.

From the moment he is drawn into the scheme the contractor should be committed by the architect and cost planner to participate in the cost control process both during the design stage and subsequently.

This requirement should be made clear during the early stages of negotiation and duly written into the contractual arrangements, together with a provision for prior negotiation and agreement of extra costs.

Real time cost control of cost reimbursement contracts

The situation here is quite different. There is no contract sum as a starting-off point and no possibility of contractual commitment to estimates of original or extra costs, so that real-time cost control by the cost planner becomes of fundamental importance. On the other hand there is no price mechanism to cloud the issue, and thus no valid commercial reason why the contractor should not be willing to make all his own estimates and costings fully available to the client's representatives. It is preferable that cost control of the project should be carried out by the cost planner co-operatively with the contractor since two heads will certainly be better than one and full advantage can be taken of the absence of commercial secrecy which is one of the main benefits of this type of contract.

At any point in the project the cost planner's cost control report is likely to be based upon:

1 Cost planner's original estimate of cost, which should have incorporated an allowance for inflation.
2 Cost planner's estimate of cost of variations, ditto.
3 Adjustment of estimated cost for actual cost of work completed or expenditure committed.
4 Any adjustments to the estimated cost of current or future work in the light of this experience.

Towards the end of the project he may switch to a simpler system involving the ascertained cost of work completed and the estimated cost of remaining work required.

The cost planner's original estimate is likely to be based upon approximate quantities, as are his estimates of variations. In both cases, and especially the latter, he may instead adopt a 'resource-based' approach in co-operation with the contractor. Even where approximate quantities are used, however, all estimates should be discussed with the contractor before being given to the client. The purpose of this cooperation is twofold; to ensure that the cost planner has not made any false assumptions and also to enable a positive contribution to be made where the cost planner finds that the contractor is proposing to use methods or equipment which he feels are unnecessarily expensive or inefficient. It must be remembered that the contractor's usual incentive to cost-cutting will be absent, it is not his own money which he is spending. In such an event there may have to be a meeting between the architect, cost planner, and contractor to decide what is to be done; this could in some instances involve minor redesign to enable something to be built more easily. This aspect of cost control is an important part of the cost planner's contribution.

A most difficult part of the whole task will be the adjustment of estimated costs for actual ascertained costs. Under some forms used for cost-reimbursement contracts the contractor's responsibility for costs is limited to providing a periodic statement of total labour and plant costs, and invoices received, to date. Such a document is totally useless for cost control purposes both in format and in timing since invoices are often not available until weeks or even months after the cost commitment has been incurred, and the cost planner must, therefore, ensure that the contractual arrangements provide for:

1 Facilities for the client's representatives to participate in the contractor's estimating and work planning procedures and in the appointment and control of sub-contractors.
2 The use of the contractor's own detailed costing system for the benefit of the client's representatives, or alternatively the installation of a suitable system where the contractor's system is unsuitable or non existent.
3 Early disclosure of all papers concerned with ordering, costing, delivery, or pricing.

It must be made clear prior to the contractor's appointment that these requirements exist, and the must be incorporated in the contract. The arrangements must not rely entirely upon goodwill, since at the least they are inconvenient to the contractor and are certain to involve him in some expense. This will need to be reflected in his fee.

The costing system will have to separate the costs in respect of specific parts of the work which correspond to identifiable portions of the cost planner's estimate, otherwise any form of reconciliation between costs and estimate will be impossible until the whole job is complete. This means in turn that the cost planner's estimate will need to be structured suitably, an elemental basis being quite a good one. However, work executed by different trades at different times should not be telescoped into a single priced item (for exampe roof carcassing and tiling). A further point to watch is delay in reporting; cost of plant and materials may take some time to work through the contractors' office system and a method has to be devised for overcoming this delay. For example, materials can be priced off delivery dockets which should be instantly available. Reconciliation of delivered materials with measurements of major items of work should also be carried out to ensure that there are no unexplained differences up or down due to theft, excessive wastage, or clerical error.

Reconciliation of allowances for inflation poses some further problems, since these will often not be separately identified in the ascertained costs of resources. In most cases some form of approximation will be adequate in showing how much of the cost of some particular operation should be set against the allowance for inflation since expensive and complicated book-keeping arrangements which do not affect the total to be paid are unlikely to yield any commensurate benefit to the client. Allowances for future inflation may need to be revised from time to time, as was the case with lump sum contract cost control.

It is necessary to watch closely the arrangements for sub-contractors. It may often be possible to appoint these on a price rather than a cost basis, as the work may be more clearly defined by the time they are involved. (For example, the structural items to which finishes are to be applied may actually have been built). Competitive tendering might well be possible; again if the cost planner does not look after this then nobody else will. It is much easier just to ring up a friendly sub-contractor and ask him to come round and start.

Finally, great care is necessary when work on a price basis and work on a cost-reimbursement basis are both comprised in the same contract. This is only likely to be satisfactory where one or more of the following circumstances obtains:

1 The cost-reimbursement work is only a negligible proportion of the whole (e.g. incidental dayworks on a lump sum contract).
2 The two types of payment apply to two separate organisations (e.g. the main contractor and a sub-contractor).
3 The cost reimbursement work is carried out by a separate gang of people using special materials.
4 The two types of work are carried out at different periods or at different sites.

Unless one of these conditions is fulfilled there will be a tendency for labour and/or materials used in the price-based work to be charged to the cost-reimbursement work, thus being paid for twice, and this is almost impossible to check without continuous monitoring. An example of the type of project where this could happen would be a small extension to an existing building where the extension was to be paid for as a lump sum but the alterations in the existing building were done on a cost-reimbursement basis. It would be very difficult to ensure that men's time was charged to the correct part of the work if it was all proceding simultaneously, and even more difficult to check afterwards that this had been done. Such hybrid arrangements should therefore be avoided.

It will have been seen that the real-time cost control of cost-reimbursement contracts involves the skilful combination of builders' costing systems and client budgeting and cost strategies, and is likely to be expensive. There is no halfway house, however; cost control of this type of project can only be done properly or not at all. The client must make up his mind about this at the beginning. The internal cost control of a package deal contract would have many features in common with the above procedures.

Real-time cost control of management contracts

In many ways the procedures would resemble those on a cost-reimbursement contract, but probably at a more strategic level since the work will be undertaken using a number of different contracts, any of which in turn will require its own cost control procedures of the appropriate type already discussed. Nevertheless, the general pattern set out in items (1) to (4) on page 239 will also apply to the project as a whole.

Cost control on a resource basis

As with real time cost control on any other basis this fundamentally depends upon swift and reliable reporting of what is actually happening, and the rapid reconciliation of this with the control document.

The cost controller will be concerned with two different aspects:

1 The overall time scale for the project. The preparation of the network will have involved a large number of assumptions about the time to be taken by the various activities, and revisions must be made in the light of actual progress. In this context it is only the activities on the critical or near-critical paths which have to be considered, but any substantial change (for better or worse) may cause a change in the critical path itself and in the duration of the works. Decisions will then have to be made as to whether a different duration is acceptable, or whether the time for remaining activities must be adjusted somehow. In both cases cost will be substantially affected.

2 Cost of individual activities. A costing system has three objectives:
   (a) To measure actual expenditure against estimated cost.
   (b) To indicate whether performance needs to be improved.
   (c) To provide feedback which will improve future estimating performance.

Costs are recorded under 'cost centres'. These can be anything from the whole job to a tiny part of it, but to meet the above objectives costs need to be recorded under headings which can be identified with sections of the estimate, and the estimate in turn needs to be prepared in a suitable format to achieve this. There is a need to establish a level of breakdown which is neither so fine as to make the recording and allocation of labour and material costs unduly detailed, nor so crude that the 'cost centres' are too broad to mean anything. Individual bill of quantities items are normally at too fine a level, a whole trade section would be too crude. A single activity, or a group of activities, from the network is usually a suitable basis.

Costing of site work has two main difficulties which distinguish it from normal production costing as practised in factories and as described in most text-books on costing, and they spring from the one-off nature of most of the work (both in terms of the finished product and the conditions under which it is produced).

The first is that the usual system of 'standard costs' does not really apply. A standard cost is an ascertained cost of carrying out an operation which can then be built into an estimating system, and which should normally be achieved in future – any substantial deviation requiring to be investigated. Machining operations in a wood-working shop can be costed on this basis – once a standard sequence of events has been carried out on a machine a number of times (inserting the work, operating on it, withdrawing it) there is no reason why the same operation should not always cost exactly the same in the future. Site operations are not only less standardised than this, but the position of the work, weather conditions, etc., varies continuously. It is very difficult to get a sufficient run of identical work to establish standard costs and still have enough similar work ahead to make it worthwhile to apply them. Certainly it is not usually possible to apply such data from job to job, as a joinery works or engineering works would do.

The second problem concerns the difficulty of actually recording and processing costs of labour and materials. When men are dispersed about the site it is almost impossible to record accurately what they are doing except in fairly general terms, unless a quite unacceptably expensive staff is engaged for the purpose. Many sophisticated costing systems have failed through the time sheets being filled in on a Friday afternoon in the foreman's office from memory.

Materials to some extent are an equal problem, as nothing like a factory control system is operated on most sites when materials are drawn from stock. The invoice is often the main weapon in materials cost control, so it is convenient if the cost centres are sufficiently large and identifiable for particular purchases to be identified with them. However, some contractors' systems do not bother overmuch about materials, since as we have already seen, these are much less liable than labour and plant costs to deviate from estimate, given a reasonable level of management. This is another advantage, from the contractor's point of view, of not using 'all-in' rates.

This may sound very defeatist – difficulties exist to be overcome, not to be excused. They can indeed be overcome, but except in some limited circumstances (such as repetitive house-building) nobody has found that the cost and trouble of doing this has produced economically worthwhile results, because of the difficulty of deriving 'standard costs' from the data for reuse. All one is likely to get is an expensive post-mortem report. A further point to remember is that some of the people on a site, particularly labourers, are doing work which has to be done but which may not have been identified as an operation in the network – sweeping up, carrying odd materials, attending on the Clerk of Works.

It is, therefore, preferable to use fairly broad cost-centres 'fixing formwork to second-storey', and devote one's attention to getting the feed-back rapidly and with a fair degree of reliability. It should then be possible to measure actual expenditure against estimated cost with reasonable accuracy, and also to a somewhat lesser extent fulfil the other two objectives of a costing system.

In the reconciliation of costs and estimate, the difficulties concerning inflation, and alterations to the scope of the work, which were mentioned in the previous chapter will again be encountered, although the former problem can be minimised by working in man-hours and machine-hours rather than money, for comparison purposes. A contractor's costing system will in any case need to highlight the cost of alterations, because very often they will form an extra to be charged to the client. It will also be necessary to ensure that the cost performance of sub-contractors is kept up-to-date in order to arrive at the projected total for the job, complicated by the fact that these may often be on a price, rather than a cost-reimbursement, basis.

Control of cash flow

So far in this chapter we have ignored the phasing of expenditure, except insofar as the effect of inflation on costs is concerned. However, in practice the client is usually very concerned about the sums of money he will be called upon to pay at any given time, and may require a cash flow estimate from the cost planner before signing the contract, rather of the type shown on page 62 of Chapter 5 in respect of the block of flats. He will need this in order to make adequate arrangements to raise the money, whether he is operating in the profit sector or in the social sector. Quite clearly he is unlikely to be able to place the whole contract amount in his bank trading account at the beginning of the project, ready to be drawn upon as required over the contract period, and even if he were theoretically in a position to do this he would still have more profitable uses for the money in the interim, which would probably mean that it would not be instantly available.

The cash flow estimate will require to be revised by the cost planner from time to time, preferably at the same intervals as the real-time cost control reports, and these revisions will take into account not only any changes in total cost but also the extent to which some aspects of the programme are running faster or slower than anticipated, or indeed whether the whole programme has slipped.

In many cases, a periodic report of this kind may be all that is required, giving the client an updated warning of cash requirements. (As well as performing this function it may also be very useful in monitoring programme slippage, because the level of cash flow required to complete on time may be seen to be quite impossible in the light of experience to date.)

However, in some instances a client will need not merely a report of what is likely to happen but some positive form of control. In the case of a cost-reimbursement contract it should be possible to order an increased or decreased tempo of work, and few problems arise. In management projects, also, the placing of contracts can be deferred or work accelerated, but serious difficulties occur in the case of lump sum bill of quantities contracts on the standard form.

Once a contract has been signed using the British JCT Standard Form of Contract the builder is not answerable for the phasing of expenditure. He does not have to provide an estimate of client's cash flow, and even if he agrees to do so as a favour he cannot be held to perform in accordance with it. His only duty is to proceed 'regularly and diligently' and to complete by the appointed date or a later date which may be agreed upon because of delays; it is in the contractor's interest to obtain high level of payment in the early stages of the job in order to finance the later stages and he may inflate the prices for earlier parts of the work, and arrange early deliveries of materials, to achieve this.

Therefore if a positive form of cash flow control is required, a suitable provision will have to be made in the actual contract, and not merely in the bill of quantities or in correspondence. The contract might state a month-by-month cash flow programme with a provision that the contractor would not be reimbursed ahead of this programme if the work was carried out faster than scheduled, and that if expenditure fell behind the programme there could either be a provision for damages to be ascertained or, more constructively, for the difference between the programmed amount and the actual expenditure to be paid into a trust fund. It might be thought that there would be no damage to the client's interests if expenditure fell behind the programme (provided the job was finally completed on time), but in the public sector and in some large private corporations construction is often funded on an annual basis and amounts unspent at the end of the financial year may be permanently lost to the authority or department concerned.

In the profit sector, however, quite apart from his general wish to reduce the cost of financing his cash flow, the client may find it difficult to raise the total cost of construction until nearer the time when the building will actually be producing an income. He might then have arranged for the contractor to bear part of the building cost during construction or even to take a share in the risk of the development. In such cases it would be essential for the arrangements for funding during the progress of the works to be formalised in the contract.

Cash flow control of major development schemes

Major development schemes may extend over many years and may include a large number of building contracts entered into at different times.

Cash flow control is necessary in the running of public development schemes to ensure that money is available to meet outgoings. If a project, or a development programme, is funded on an accrual basis (that is, unspent funds which have been allocated to it can be carried forward from year to year), and if it is not possible to lend out money at a profit (perhaps because the authority has not got the necessary powers), then nothing more than a 'passive' control of cash flow will be required. The authority will simply need to order its financial affairs to suit what is happening.

However, most public funding is on a yearly basis, in which unspent funds in one sector are not carried forward for the benefit of the project or authority for which they were allocated, but are used to meet overspending in other sectors or applied to a reduction in government expenditure. In addition, most major building projects involve an unavoidable commitment to expenditure for some years ahead, although the authority's budget for these years may be unconfirmed when the commitment is entered into. Trouble may occur if major construction projects have fallen behind schedule, moving much of the expenditure which should have taken place in the current year into future years, and leaving a cash balance which will be lost to the authority with no certainty of picking it up again. It is, therefore, necessary to manipulate cash flow in the light of changes so that the year's allocation of funds is completely but productively spent and so that a realistic revision can be made of cash requirements for ensuing years.

Similarly, if funds are reduced in any particular year, perhaps because of economic problems, the on-going programme may have to be amended. This could be called 'active' or 'positive' control of cash flow.

In any authority's development budget the expenditure will fall into three categories:

1 Old projects.
2 Current projects.
3 New projects.

With each category the discretionary factor in expenditure is increased.

1 *Old projects.* These are projects which have been completed but for which full payment has not yet been made (possibly depending upon the outcome of litigation, arbitration, or negotiation). Once payments are due they will have to be met.

2 *Current projects.* It may be possible for an authority to order a speed-up or slow-down of current projects in order to change its cash commitment during the year, even at the expense of an eventual increase in total cost. What more usually happens however is that circumstances beyond the authority's control cause such changes, and the cash flow budgeting has to be revised to suit. Almost invariably, expenditure tends to fall behind estimate because of delays in progress, although this may sometimes be counterbalanced to some extent by extras and by inflation. It might be thought that if an authority had a large number of projects on hand (such as a programme of 10 schools) the differences would tend to average out, but this does not happen in practice because the causes of delay are often national or regional in character – weather, industrial disputes, labour shortages, etc.

3 *New projects.* It is these which offer the greatest scope for control. If the start of a two year project worth £10,000,000 is brought forward or put back by one month the effect will be an increase or decrease of £400,000 in the current year's cash flow, and correspondingly greater for longer periods than one month. There are two major difficulties however – the starting of a major project may have a comparatively small effect in the year in which the go-ahead is given (but it will then join the ranks of the current projects and will be an inescapable and major commitment for the following two years or more), and expenditure on projects tends to follow the well-known 'S' curve where the rate of spending is at its greatest in the middle period of the project with a comparatively slow build-up in the early stages. (See Fig. 5.1 on page 47.) In many ways the best way to control the cash flow situation is by having a number of smaller short term projects ready to roll, and these can be largely completed during the financial year in question even if the order is not given until the year is quite well advanced.

A method of disbursing funds on a project which is falling behind programme is to pay for the work, or for materials, ahead of the legal liability to do so. There are obvious dangers in doing this and a public auditor would be unlikely to accept a straightforward attempt to do it. It might, however, be possible to pay money into a joint trust fund.

Up to now we have been considering cash flow as though we are simply concerned with the year as a whole, but this is rarely the case. Funds are not normally paid across to an authority in a single sum at the beginning of the year, nor are they paid as each cash commitment arises, but are usualy paid in quarterly instalments or something of the kind. Quite clearly, therefore, the authority must avoid too much cash expenditure in the early part of the year since funds will not be available, whilst if it is empowered to lend surplus short-term money profitably it may be able to arrange matters so that it can do this. Even where there is a smooth monthly outgoing throughout the year, the money for the third month of any quarter will be available for the first two months.

Although this section has been written in the context of public or social development much of it is equally applicable to profit schemes, except that the problem of 'spending all the money by the end of the year' is not so likely to occur and the pressure is likely to be in the opposite direction; the question of interest on money borrowed becomes of paramount importance so that any pushing back of payment dates will be welcomed, so long as final completion is not held up.

It is not unknown for this to happen in the public area; an overseas regional government, which was anxious to minimise the effects of a slump in the building industry, but had spent all its funds for the year, let contracts on the basis that no payment would be made until the following financial year, so that for the first six months or more the contractors were totally financing the programme. However, as would be expected, the tenders reflected this additional burden.

Chapter 19
Cost Planning of Refurbishment and Repair Work

Twenty or thirty years ago it was customary to think of building projects in terms of 'green field' projects, that is to say projects erected on an open site (preferably one which had never been built on previously), and the classical cost planning techniques which grew up in this era tended to be oriented towards such projects as the norm.

This was rather a convenient assumption, as it avoided a lot of complications, but it is doubtful whether it was ever really true. However, today it is clearly untrue, and a very substantial part of the current UK construction programme comprises refurbishment, renewal work, and major repair work.

As the last two categories nearly always includes some element of improvement, and as refurbishment will usually involve renewal and repair, they can all be considered together for the purposes of cost planning. And some of what is written will apply equally to quite a lot of the 'new' work which is being undertaken and which involves building on very restricted urban sites between existing buildings with difficult access and close proximity to the public.

The uncertainties of refurbishment work mean that it will be almost impossible, and certainly inadvisable, to undertake the project on the basis of lump sum competitive tenders for the whole of the works, and other more collaborative methods of procurement will have to be used – either cost-plus or some form of management contracting.

Although it is obviously possible to set out any estimates in an elemental format the normal process of elemental cost planning is not really applicable, since it will not be practicable to make the element-by-element cost comparisons with other projects which lie at the heart of this technique. The costs will depend on the state of the individual building in relation to what is proposed to be done with it, and comparisons with other projects on an element basis would be meaningless.

So the cost planning and control of refurbishment work needs to start from first principles.

The most important initial step is to recognise the conflict of objectives which is inherent in such work. There is of course a conflict in new work between cost, time, quality, and size, but the client's interest is simply to optimise the scheme in these terms, and cost planning allows him to do this. However, in refurbishment work these basic objectives are usually supplemented, or even outweighed, by major secondary objectives.

Occasionally a single objective, such as speed of completion, may be strongly dominant and this is little different to new work. But more often two or more conflicting objectives are perceived as dominant – for instance, continuing occupation required by a 'user' client may hinder the achievement of speedy completion which is the priority of the 'owner' client.

Even the emergence of a single dominant secondary objective, such as safety of the structure or of the public, may have consequences which cut across normal procedures, particularly those for efficient construction methods or financial control.

The likely extent of conflict can only be identified after a thorough assessment of risks and objectives, and the trade-offs between them, and frequent review of the relative benefits compared with the likely cost, time, and quality implications will probably be required.

Uncertainty is another characteristic of refurbishment work. Generally there is a high level of uncertainty not only in client objectives but in available physical data. Such uncertainties may extend into the construction period with a high likelihood that unforeseen events will occur.

Problems due to uncertainty may be mitigated by allowing a longer lead-in time and by assembling the work into discrete packages, but above all by bringing the contractor (or construction manager) into the team at an early stage. Even so there will remain a need for client, designer, and constructor to respond quickly to the discovery of previously unknown features of, or defects in, the existing building.

Safety is perhaps the most dominant feature of refurbishment work, and the problems in this area are intensified by the general uncertainty already referred to. Safety affects operatives, the users of the building, and the general public, and may relate either to the safety of the structure or the safety of the operations. Clearly where safety and cost conflict then safety must be paramount, so that it is vital that safety issues be identified at an early stage if estimates are to be relied upon.

However, safety is not discretionary so that there are no real choices to be made as regards the level of safety, only the means of ensuring it. But another major issue involves fundamental choice, and that is the extent to which (if at all) the building should remain in use during the works.

Here we can be dogmatic. Just as real estate agents are supposed to believe that the three most important things about a property are position, position, and position, so the three most important decisions in connection with refurbishment are undoubtedly occupation, occupation, and occupation!

The effect on costs, programme, and safety if the building remains in use are so enormous that a client who wishes to do this should be encouraged to think again and again to see if some other alternative can be found, and it is incumbent on the cost planner to keep pointing this out.

A problem in this regard is that seen in advance the difficulties of moving out seem much greater than the difficulties of staying put – it surely only means having to put up with a little dirt, noise, and inconvenience for a year or so, whereas finding alternative premises, moving in and out of them, getting customers and staff to come to terms with the new location, etc, appears to be far more of a hassle.

This is an illusion. Staying put involves much more than slight inconvenience.

British Rail recently found it worth while to close Crewe station completely for several weeks, and divert all trains to other routes, in order to rearrange the tracks and signalling – the job could have been done piecemeal without such drastic action but the time, cost, and safety aspects would all have been badly affected, and there would have been just as much inconvenience only spread over a longer time scale. And exactly the same situation applies to buildings.

If relocation is considered it may be worthwhile to make the timing of the whole project dependent upon the availability of suitable accommodation, rather than deciding on the programme first and then trying to see if anything is available at that time.

If relocation proves to be impossible or unacceptable then a very realistic allowance must be made for the additional cost and time involved. However, this will not represent the total penalty imposed by the decision – the wear and tear on staff, and annoyance to customers or residents, caused by dust and dirt, noise of jack-hammers, temporary access arrangements, etc.,will not appear on the job balance sheet but will be show itself in staffing difficulties and reduced patronage. Even if most of the work is carried out at weekends and at night the dust problem is difficult to overcome, and finding stock, papers, furniture, etc. covered in dust every morning is one of the greatest irritants to occupants and tends to lead to complaints about everything else.

This does underline the importance of having a very senior manager employed full-time to liaise between the building team and the occupants, with ultimate authority to override either party. His cost must be allowed for, as well as the cost consequences of some of his decisions.

It is almost impossible, within the bounds of reason, to over-estimate the total costs of refurbishing a building which is in occupation!

Even without occupation the costs are difficult to estimate, because of the factors to do with uncertainty which have already been mentioned. But before any figures are given the position of the building with regard to 'listed status' must be investigated, since the planning authorities may be able to impose extremely expensive requirements on the scheme.

The cost of obtaining, and working with, obsolete or obsolescent materials if this is either required or dictated is likely to be considerable.

A further point to notice is that the cost of temporary works – scaffolding, shoring, etc. is likely to play a much greater part than in the case of new works. Where a major gutting of a building's interior is taking place within retained external walls such works are likely to be very sophisticated and expensive. This is a type of work where the average cost planner's knowledge of both technology and cost is sometimes weaker than it is with new permanent works, and it is essential that a structural engineer be involved in the estimates at an early stage if any major structural work is contemplated.

The advantages of working with firms who have a good track-record of refurbishment work cannot be over-estimated, and this applies equally to design and management consultants as it does to construction staff and works contractors. Again however such firms are unlikely to be at the cheapest end of the range.

This really leads to the conclusion that if there is a choice between new-build and refurbish for a firm's premises the problems must be faced squarely at the start – tight, over-optimistic estimating has absolutely no place here. Nobody likes cost overruns at any time, but if they mean that with hindsight the wrong decision has been taken by the client (with professional advice) then the consequences are likely to be more than usually serious.

The cost control process should follow the usual real-time methods set out in Chapter 18. A particular point to watch however concerns the letting of work packages to trade or specialist contractors. It is always considered advantageous to firm these up fairly early in the project, so that actual quotations can go into the job cost in place of the cost planner's estimates. However, it is often better in refurbishment work to wait until the efforts of other trades have reached a point where the specialist's work package can be properly defined and a firm price obtained (perhaps even in competition). In the case of a large project the inconvenience of letting, say, the plastering work in several separate packages could well be justified by the ability to get firm prices for each stage of the work.

Finally, it hardly needs to be said that being aware, and keeping the client aware, of the current cost situation is even more important than usual on this type of project, where in spite of everybody's efforts the cost commitment is always liable to escalate at fairly short notice.

Chapter 20
Further Cost Modelling Techniques

Introduction

In Chapter 9 the concept of a building cost model was introduced and we looked at the way in which simple traditional cost models had developed, which allowed the quantity surveyor to improve his forecasting ability. At the same time their deficiencies were noted and we put forward a list of criteria for judging cost models. In this chapter it is intended to take this idea a step further and look at the possibilities of adopting other models, now being extensively used in other disciplines in the search for better cost information. Some of the models involve the use of statistical terminology, and these terms are defined in Appendix C for those readers who may not be familiar with them.

When we refer to better cost information what do we really mean? There are probably five major ways in which it is possible to advance. We can:

1. Provide cost information *quicker*.
2. Provide *more* information so that a more informed decision can be made.
3. Provide *more reliable* cost information which will introduce more assurance into the decision-making process.
4. Provide information at *an earlier stage in the design process*.
5. Provide information in a *more understandable form*.

By harnessing the power of modelling techniques it is hoped that each of these objectives can be achieved, or at least a step will have been taken to improve the chance of achieving them. However, the techniques on their own are unlikely to provide a very great advancement. The time involved in manual calculation and organisation inevitably means that only a very few alternative possibilities for a design can be explored. If a giant leap in cost knowledge is to occur, then it would appear that non-manual methods are essential. Of the methods available to us today the computer appears to hold the major potential for advancement. For lower cost than that of employing a man for one month in an office, it is possible to purchase a machine that will perform vast quantities of calculation a great number of times, and thus explore situations and possibilities which would be quite inaccessible if we were restricted to using manual methods. Future development is likely to make the above cost comparison even more favourable.

However, these machines have their own language and organisation and it is essential for any proposed use of the machine that its structure, constraints, and communication are well understood. These factors will affect its efficient use and, therefore, its profitability.

It is envisaged in this chapter that some form of computer will be used with the models to speed the calculations. The traditional models evolved in the way that they did because it was expected that manual labour would be employed to do the calculations. This has resulted in an over-simplification of the models, at all levels, for the sake of expediency. A new view of what is needed in terms of cost information, without reference to the constraints of manual computation, is required if the full potential of the new technology is to be tapped for the benefit of the design team. However, almost by definition, all models are simplifications of the thing they seek to represent and are consequently imperfect.

It is not possible to explore every type of model in a book of this nature. The approach will be to look at the classification, the principles behind the construction of the more important types, and the potential use, of models. Chapter 21 then looks at the way models can be used in optimisation.

When should cost models be used?

The simple answer to this question is that models should be used at all stages of the design and construction process. However, the particular interest of this book is in the cost planning of buildings and, therefore, the early forecasting and subsequent control of building cost. In Chapter 8 it was noted that probably 80% of the cost of a building had been committed before even the sketch design had been produced. This suggests that the major scope for modelling is to improve the flow of cost information at the stage prior to production of the scheme design. It is interesting to note that John Southwell, when a RICS Research Fellow at Bath University, suggested that an extra stage be introduced into the RIBA Plan of work (see Chapter 8, page 89). Between 'Inception' and 'Feasibility' stages, he saw the need for what he called a 'Viability' stage. This was defined as 'that period when a cost forecast may be required, but where no sketches of the building layout are available'. Feasibility is then defined as 'that stage when some form of elementary sketch layout is available, and the design team are therefore evaluating a definite proposal'. This is a useful concept as it allows a separate category for those techniques which attempt to give information based on abstract hypotheses (from which a cost strategy for design may be developed), and those techniques which take the result of a strategy in the form of a tentative proposal, and carry out an evaluation of that specific suggested design. However it is important to realise that most cost models are used by the cost consultant as a mere reference point or skeleton to which he adds his knowledge and experience. It is important therefore that the consultant can communicate and integrate with the model in a meaningful way.

When a cost model is being developed it is nearly always constructed to be as universal as possible, so that it can be applied to a large number of cost situations and building forms. The basis of the model is therefore nearly always an abstract hypothesis which is made real or firm by the application of quantities and cost to its parameters. Even at the viability stage some assumptions will need to be made about quantitative and qualitative aspects of the building.

Parameter can be defined as 'a quantity which is constant in a particular case considered, but which varies in different cases'. Each hypothesis, therefore, may involve the application of different quantities to a set of parameters which will themselves probably remain the same. It can be argued that the basic principles of any building cost model should be found in the simple models constructed before a design has been produced, and that any further development is merely a refinement of those principles into a more detailed model. To simplify this introduction, the models will be kept as straightforward as possible without detailed refinements, and will be orientated towards giving advice at the very early stages of the design process.

Classification of models

There are a large number of ways of classifying models. For example, classification may take place according to the function they perform, e.g. evaluating, descriptive, analytical, optimising, etc. On the other hand it may be according to the form of construction, e.g. iconic (physical representation of the item under consideration), analogue (where one set of properties is chosen to represent the properties of another set, for example, electricity to represent heat flow), or symbolic (where the components of what is represented and their interrelationships are given by symbols). The last is perhaps the more useful structure, but the definition lacks clarity as all models are to some extent symbols of the actual commodity under consideration.

In some ways it matters very little to which system or set the model belongs so long as the overall purpose and construction are well understood. Perhaps the most important knowledge concerning any model is an understanding of its limitations within the context of its use. Every model user should have a very clear idea of what the model is measuring, the data upon which it is based, and the factors that need to be applied to the results before a decision can be made. It is dangerous to ignore the simplifications which are inherent in the construction of cost models in particular.

A useful system for representing the role of models in the decision-making process is given below.

The process is as follows:

1 Raw data is collected on cost and its relationship to design variables.
2 The data is analysed and synthesized according to the requirements of the cost models that are to be used. The data is also arranged for quick access, updating, and use.
3 Analytical and predictive models are then used to explore cost situations and changes in the design variables.
4 Optimising models are used to select the best alternatives from the results of the previous models, based on criteria determined by the design team. It is extremely difficult to optimise in building design for reasons already given in Chapter 6. Even the optimisation of future and initial costs based on static assumptions of use is difficult because of the problems of cost prediction. To take into account aesthetics, comfort, efficiency, flexibility, etc., as well would be impossible (at least in any meaningful way) at the present time.

For this reason the models included are largely of the analytical type, which allow comparison between alternatives from a cost point of view. It is not proposed to deal with models that attempt to forecast the effect of, say, government and international decisions on the construction market, and, therefore, upon its price levels, as these are not yet sufficiently developed.

Three examples of primarily analytical models have been chosen which, it is felt, are likely to be typical of the major application of models in the immediate future. These are as follows:

1 *Causal or empirical models.* These are symbolic models which, in our case, are based on relationships between the design variables and cost, and which have been derived from observation and experiment. In previous chapters we have already seen that the quantity surveyor has been using a symbolic language to represent a building for a number of years, even from the very beginning of his professional existence, namely dimensions and descriptions. In the process of time this language has been refined and standardised and has allowed subsets of the language to be developed such as cost per square metre, elemental unit rates, etc. It is a small step therefore to translate this language into a general algebraic form relating to the morphology of the building and its components, to which cost factors can then be applied. By changing the value of the algebraic terms according to the requirements of a particular building, the change in cost related to the design variables such as shape and height can be discovered.

2 *Regression models.* In the empirical model it is usually assumed that there is a fixed relationship between the design variable and cost, and that the cost factor applied to the algebraic formula is constant. However, in the social sciences, and particularly where human performance is involved, exact relationships are not generally observed. Instead statistical relationships prevail, and only average relationships may be observed among the variables. Trends may be detected through the data that has been analysed and it may therefore be possible to determine a trend algebraically, and allow an estimate of cost to be made, knowing one or more of the design variables. A regression model is a method of determining the relationship between variables by the 'method of least squares' which seeks to minimise the sum of the squares of the difference between the observed values and the predicted values. For example by knowing the shape, height, or other variable of a particular building type we may, through this technique, be able to forecast the cost of the building more reliably than by using one of the traditional single-price methods.

3 *Monte Carlo simulation.* The problems that exist in forecasting the future or even investigating the enormous number of combinations of shapes, specifications, and use of buildings make it a task that cannot be completely solved by a simple mathematical expression. Some other technique is almost certainly required that will allow us to sample a large number of possible solutions to discover the best solution in that sample, and the range of costs that could be expected given the known probability of an event or cost occurring. The Monte Carlo technique is simply a way of sampling from a distribution, or distributions, in a random manner to provide a range of solutions which may be directed to finding the best solution (or simply to establish the range of possible solutions available). What is attempted by the method is to simulate a series of future events (production on site, maintenance characteristics of components, operating costs of buildings) or the design of a number of buildings. In doing so it may use causal or regression models to provide the forecast of cost for the sample it has chosen.

Causal or empirical models

In terms of building cost causal methods are in some ways just a refinement of traditional estimating techniques. If instead of measuring up a building and applying a unit rate to the derived quantities of each component we express the building in algebraic terms, then we can give values to those algebraic terms, apply cost coefficients and arrive at a cost estimate.

Let us take a very simple rectangular building as one example. We shall only look at the major structural elements and to retain simplicity we shall ignore the thickness of walls and slabs. (In a working model these would need to be accounted for if they are cost significant). Figure 20.1 shows the physical form of the model, but of course any realistic value can be given to each of the buildings dimensions. Let us approach the model in elemental form and deal with each element in turn.

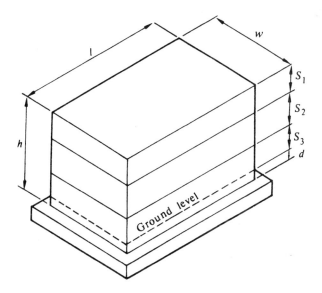

Fig. 20.1. Physical form of causal model example.

1 *Sub-structure*

To express this element in algebraic format we need to consider the cost significant components which need to be represented, together with their most suitable methods of measurement. Let us assume that from our observation and experience these are discovered to be:

  (a) *The ground floor slab.* The cost of this component can be considered to be a function of the ground floor plan area. The algebraic equation is therefore:

$$\text{Area of slab} = l \times w$$

To this equation must be applied a cost factor (CF_1) related to the cost per square metre of ground floor slab including additional items such as reduced level excavation and hardcore, which relate to this particular horizontal measurement. Already it can be seen that approximations and assumptions are creeping in to prevent the model becoming too complex.

$$\text{Cost of slab} = l \times w \times CF_1.$$

(b) *The perimeter strip foundation.* The cost of this component is a function of its total length. The cost factor (CF_2) for this item would include for excavation, concrete, brickwork, etc.

Length of strip foundations $= 2 \times (l + w)$
Cost of strip foundations $= (2 \times (l + w)) \times CF_2$

(c) *The bases for the frame.* If an average bay length (B_l) is assumed together with the average bay width (B_w) then a formula can be given as follows:

$$\text{Number of bases } (k) = \left(\frac{l}{B_L} + 1 \right) \times \left(\frac{w}{B_w} + 1 \right)$$

The value of the expressions (l/B_t) and (w/B_w) would probably be taken up to the next whole number to achieve an integer number of bays. This could easily be done by the machine if the model is computerised. The cost factor (CF_3) to be applied would include for soil conditions, loading, etc., that would affect the size of each base and would allow for excavation, concrete, reinforcement, etc., that would relate to this size.

$$\text{Cost of bases} = \left(\frac{l}{B_L} + 1 \right) \times \left(\frac{w}{B_w} + 1 \right) \times CF_3.$$

The total cost of foundations would therefore be the sum of these three expressions.

This type of build-up is summarised below for the remainder of the major components; these are taken in an unfamiliar order because some elements are dependent on the results from the calculation of another, e.g. area of solid external wall is dependent on area of windows. In a real computerised system a more familiar order could be adopted, either within the programme itself or by simply printing the results in a different order from that used for calculation.

2 *Window*

Height (h) $= (S_1 + S_2 + S_3) + d.$
Window area (WA) $= 2 \times (l + w) \times h - d) \times (g/100).$
Window cost $= WA \times CW_1.$
Where g $=$ percentage of external wall to be glazed.
   CW_1 $=$ cost per unit area of windows.
   WA $=$ window area.
   S_1 –n $=$ storey height

3 *External wall*

External wall cost $= ((2 \times (l + w) \times h) - WA) \times CE_1.$
Where CE_1 $=$ cost per unit area of external wall.

4 Roof

Roof cost $= (l \times w \times CR_1) + (2 \times (l + w) \times CR_2)$.
Where CR_1 = cost per unit area of roof.
$\quad CR_2$ = cost per unit length of perimeter detail.

5 Upper floor

Upper floor cost $= (l \times w) \times (N - 1) \times CU_1$.
Where N = the number of storeys.
$\quad CU_1$ = cost per unit area of upper floor.

6 Frame

(i) Column cost $= h \times k \times CS_1$.
$\quad$ Where k = number of column bases (see sub-structure).
$\quad CS_1$ = cost per unit length of column.
(ii) Beam cost =

$$N \times \left[l \times \left(\frac{w}{B_w} + 1 \right) \right] + \left(w \times \left(\frac{l}{B_L} + 1 \right) \right) \times CS_2.$$

Where (w/B_w) and (l/B_t) are rounded up to the next whole number.
CS_2 = cost per unit length of beam.

7 Staircases

Total rise in metres $= h - (S_1 + d)$.
No. of staircases $= \dfrac{l \times w}{f}$.
Where f is the maximum floor area to be served by each staircase and the result of $(l \times w)/(f)$ is rounded up to the next whole number.
Staircase cost $= [h - (S_1 + d)] \times \dfrac{l \times w}{f} \times CV_1$.
Where CV_1 = cost per unit rise of staircase.
(*Note:* It has been assumed that there is no staircase from the top floor to the roof.)

The above elements have been given to illustrate the build-up of an empirical model in algebraic terms. The rest of the building would continue in a similar way. Services would present a major problem but heating cost could be made a function of external envelope, 'U' value and building volume; lighting could be a function of lighted area; and plumbing and sanitary services could be a function of the number of sanitary appliances required. Preliminaries (i.e. supervision, plant, etc.) would be more difficult to forecast except as a percentage addition. However, with some research it may be possible to identify how these costs vary with changes in building shape, specification, and area so that a simple model could be included.

Obviously a great deal of simplification has taken place and some very broad assumptions made. No account has been taken of the inter-relationships between elements, and specification has been ignored except as a function of the cost factor applied to each formula. In practice, models of this type require fine tuning to ensure that they are sufficiently reliable for the point in the design process at which they will be applied. The cost information would probably come from the experience of the quantity surveyor, from an office data bank or from a manufacturer. The principles of the model would no doubt evolve over time as the user became more experienced in its operation and realised its shortcomings. It is also unlikely that a universal model for all building types could be achieved, as the quantity of various elements (e.g. stairs) would vary with the use and regulations pertaining to the particular type of building under consideration.

The great advantage of this particular model is the flexibility available in its use. By changing the length and width in our example it is possible to vary the area of accommodation of the building. In addition, different glazing ratios and different bay sizes can be assumed. If the cost is held in the store of the computer used for modelling, the process of exploration can be automatic with the machine's program altering the design variables to suit the possibilities that exist. If instead of using length and width we convert this to a universal plan shape index (see Chapter 16) then we can explore irregularly-shaped buildings.

The problems associated with the model are largely those facing the traditional estimating methods, and in particular, elemental estimates using elemental unit rates. These problems have been documented elsewhere in this book but it is worth reminding ourselves that one of the major problems of elemental estimating is choosing which cost factor is to be applied to the elemental unit quantity. The problem still exists in this model. However, provided that the right cost relationship exists between the components, that the cost factors do not themselves change with quantity, and that the costs are fairly reasonable, then the model will allow realistic comparisons to be made. If these comparisons are made within the design constraints then it should be possible to identify (providing all other facts are equal) which of the alternatives tested by the model represents the least cost. In this manner a point of reference is established for comparison with any other selection, and a step has been taken towards providing better value.

In addition to the cost factor problem there is the difficulty of representing a building in algebraic format. In our example, large simplifications were made to keep the formula simple. However, in reality the size, shape, and assembly of the building is far more complex. For example, if the bay size of the frame was changed it might result in a deeper beam which in turn might result in a greater storey height, a larger volume to heat, and so on. The repercussions would be much larger than just increasing the depth of the beam, and this would be very difficult to show algebraically. On the other hand the increased span might mean that it is more economical to choose another type of slab or frame altogether! It is possible to build into the computer program a check to see whether an alternative is a better proposition. However, the range of possibilities is so large that the simple model would soon have difficulty in coping, and a far more complex arrangement would be required.

When the model becomes very complex then updating, particularly of cost data, becomes a giant task. There is a strong case for keeping the model as simple as possible and allowing a high degree of human interaction with the computer. If the user acts as

The Comparative Cost of Components and their Fabrication for a Given Number of Floors

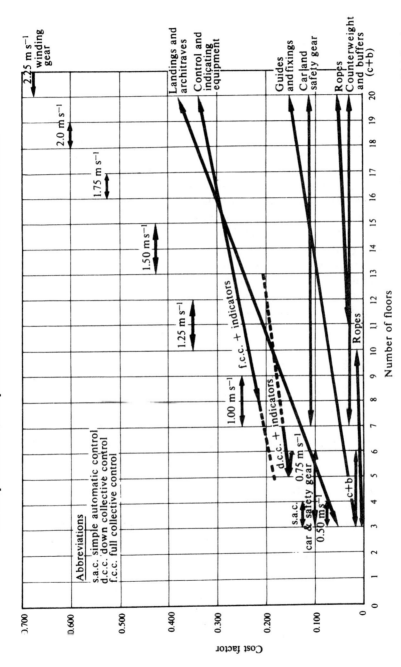

Fig. 20.2. Cost model for lift installation. (Note: this graph was originally prepared in 1974 when the actual cost could be determined by multiplying the cost factor by £10,000.)

the data bank and manipulates the choice of alternative specifications, and in some cases the size of components, then the experience of the quantity surveyor can be harnessed to the machine, and the judgement that makes a good estimate is not lost or handed over to a mechanical process.

Not all causal models are of the same structure. Take for an example the lift model shown in graphical form in Fig. 20.2. In this case several tenders for lifts were analysed to discover the important cost items and their relationship to one another in cost terms. The significant cost variables for the type of lift considered were taken to be:

1 Rope cost.
2 Counterweight and buffers.
3 Car cost.
4 Guides and fixings.
5 Landings and architraves.
6 Control system.

Costs were allocated under each of these headings by Mr R. Wiles, a student at Bristol Polytechnic in 1974, who followed the pattern of Messrs Knight and Duck in 1962. It is used here because it illustrates this type of model very well despite the fact that lift technology has advanced considerably since.

The tendency was for each component of the model to be linear in respect to the major variable of number of storeys served by the lift. (The linear trends could be produced by using regression analysis as described later). Although the tendency was linear this did not mean that the trends were continuous, and if you look at the winding gear in particular you see 'jumps' or 'thresholds' where the cost rises disproportionately to the number of storeys due to the increased speed required for efficient operation. In the same way the linear relationships of the control system costs are only applicable over a certain range, and moving from one control system to another produces a dramatic rise or fall in cost. To operate the original model in order to discover the cost of a lift it was simply necessary to add up the component costs above the number of storeys shown on the x-axis. The car size was assumed to be for eight people but could be increased in size at a special 'extra over' cost per person.

However, for this edition of the book we have translated these costs into comparative cost factors to show only the relationship. In a working model the actual costs would be used as in the original.

It is a simple matter to computerise the working of a model such as this. Each component can be expressed algebraically by the formula for a straight line $y = a + bx$ (a = intercept, b = the slope of the line, x = the number of storeys, and y = the cost of the component). If the program logic is used to select when a particular type of component is suitable then it is a simple matter for the computer to total the cost of all components it has selected and give the overall lift cost. If the prevailing National Association of Lift Manufacturer's index is then applied to the total, a reasonable current figure for the lift cost should be obtained.

However, there is a problem in using historic cost relationships to generate the equations, because the trends may change over time and the application of a cost index to the result may not reflect the current situation. A model of this type would need to be revised, by repeating the analysis procedure, about every three years, and this revision may well be a time-consuming procedure. In fact, if the model is not used very often the cost of writing and frequently updating the programs could exceed the cost of manual working.

In these circumstances it is a better proposition to express the relationships in terms of labour, plant, and material costs required for manufacture and fixing, in other words 'resources'. Any difference in the inflation rates will then be accounted for by the new resource costs that would be fed directly into a revised computer program. However, it is not always easy to find the resource costs, especially for engineering services, and in the case of the lift, historic cost data had to be used.

Regression models

Equations are used in mathematics to express the relationships among variables. Very often there are 'exact' relationships from which 'exact' values can be obtained when the equation has been solved. A familiar example is the formula for the perimeter of a circle which is normally given as twice the radius multiplied by pi. The results from this formula are considered to be accurate for any change in the value of the radius. In the symbolic model previously described, the derived formulae representing the building were assumed to be exact for the purposes of obtaining a result. However, it was recognised that broad assumptions had been made in the construction of the equations and that the unit costs to be applied were themselves subject to considerable variability.

This inherent variability is due in some measure to the fact that human performance has to be estimated to arrive at the unit cost, and this is not only very difficult to forecast but is unlikely to be consistent. In the social sciences exact relationships are not generally observed among the variables, but rather statistical relationships prevails. Average relationships may be observed, but these average relationships do not provide a basis for perfect predictions. For example the labour 'constants' used in contractors' estimating are guides obtained from experience and do not represent the actual time for completing a particular job. The actual time taken to complete a specific operation may be the result of a large number of factors including the incentive scheme offered, the location of the task, the degree of supervision and even the problems in the operative's home! Where relationships exist between a known performance or cost, and the major variables that explain why that performance or cost has occurred, then it may be possible to explain those relationships in the form of a mathematical expression. This expression can then be used to forecast future performance or cost, knowing the values assigned to the explanatory variables. A large number of examples of relationships can be quoted which apply to building cost such as the relationship of the cost of preliminaries to time, the relationship of the cost of heating installation to envelope area and U value, and the relationship of the cost of windows to area and shape. It may not always be possible to produce a reliable mathematical expression, or indeed to identify all the significant cost variables, but an improvement in estimating technique may be available nevertheless.

Regression analysis is a method by which estimates are made of the value of a variable (e.g. cost) from a knowledge of the values of one or more other variables (e.g. height and shape of building), and the errors involved in this estimating process measured. In its simplest form it describes the relationship between two variables by computing a straight line through the data obtained. This is known as 'two-variable linear regression'. The factor whose value we wish to estimate (e.g. cost) is referred to

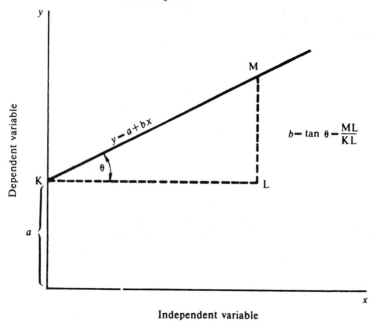

$$b = \tan \theta = \frac{ML}{KL}$$

Independent variable

Fig. 20.3. Formula for a straight line.

as the 'dependent variable' and denoted by y. The factor from which these estimates are made is called the 'independent variable' and is denoted by x. (Cause and effect is not necessarily implied by the terms 'dependent' and 'independent'.)

In two-variable linear regression the expression for the straight line is of the form

$$y = a + bx$$

where a is the intercept of the line with the y-axis and b is the slope of the line (Fig. 20.3).

To illustrate the use of the technique a hypothetical example has been chosen. It is

Table 11. Hypothetical example of *Building cost related to height*

Building	Height in metres	Cost/m² (gross floor area)
A	4	500
B	20	544
C	17	580
D	10	570
E	7	504
F	23	560
G	24	600
H	6	524
I	10	530
J	15	542

assumed that there is a relationship between building height and building cost per square metre of floor area. The first step is to establish whether or not a linear relationship exists and this is usually done in simple cases such as this by plotting the data on a graph to allow a visual examination. For the sake of simplicity, data from the superstructure of ten buildings of similar function and construction has been obtained and this is summarised in Table 11. (N.B. In practice this would be an inadequate sample and at least 30 data points should be found for a reliable result.)

This data is now plotted on a graph to produce what is known as a scatter diagram (Fig. 20.4). Standard practice is to plot the dependent variable along the y-axis and the independent variable along the x-axis. It can be seen that there is a trend for costs to increase with height and that this trend appears to be linear rather than curved or heavily dispersed.

A line is now required that will pass through the data and provide us with the most reliable estimate of cost related to building height. This line of best fit is established by a technique called the 'least squares' method. The method imposes the requirement that the sum of the squares of the deviations of the observed values of the dependent variable

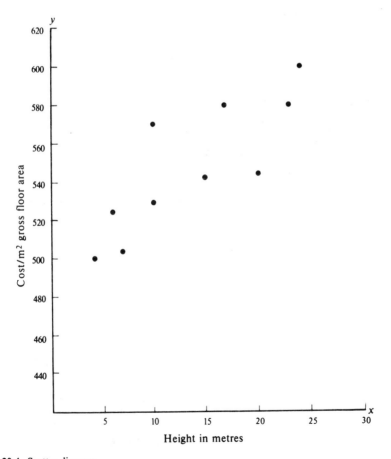

Fig. 20.4. Scatter diagram.

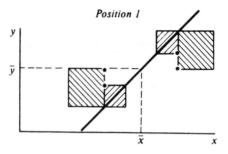

Position 1

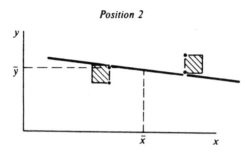

Position 2

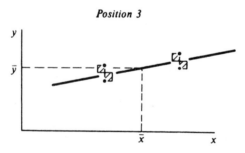

Position 3

Fig. 20.5. The principles of least squares. (Note: the smaller squared deviations from the regression line in position 3 makes this the line of best fit.)

from the corresponding values on the regression line must be a minimum. Figure 20.5 shows the principles of the technique in diagrammatic form using just four data points. It can be seen that a number of lines can be drawn through the data and also through the average value of x and y (a requirement of the technique) but only position 3 provides the least squared deviation from the line. Calculus is used to apply the condition that the sum of the squared deviations from a straight line must be a minimum and this results in two equations, one each for a and b in the straight line formula.

$$b = \frac{\Sigma xy - n\bar{x}\bar{y}}{\Sigma x^2 - n\bar{x}^2}$$

$$a = \bar{y} - b\bar{x}$$

where x and y are the arithmetic means of the x and y variables and n is the number in the sample

Using our previous data in Table 11 the equations can be solved as shown in Table 12.

Table 12.

Building	x	y	xy	x^2	y^2
A	4	500	2,000	16	250,000
B	20	544	10,880	400	295,936
C	17	580	9,860	289	336,400
D	10	570	5,700	100	324,900
E	7	504	3,528	49	254,016
F	23	560	12,880	529	313,600
G	24	600	14,400	576	360,000
H	6	524	3,144	36	274,576
I	10	530	5,300	100	280,900
J	15	542	8,130	225	293,764
	136	5,454	75,822	2,320	2,984,092

$\bar{x} = 13.6$ $\bar{y} = 545.4$

Therefore:

$$b = \frac{75,822 - (10 \times 13.6 \times 545.4)}{2,320 - (10 \times 13.6 \times 13.6)}$$

$$= 3.5$$

$$a = 545.4 - (3.5 \times 13.6)$$

$$= 497.8.$$

Placing these values in the straight line formula ($y = a + bx$) gives us $y = 497.8 + 3.5x$ and, this can now be plotted on the scatter diagram (Fig. 20.6).

A test can now be employed to discover the measure of the degree of association between the two variables, i.e. how well they are correlated. The measure usually adopted for this purpose is the 'sample coefficient of determination' (r^2) which is given by the following formula:

$$r^2 = 1 - \frac{\Sigma(y_E - y_A)^2}{\Sigma(y_E - \bar{y}_E)^2}$$

where y_A is the actual value of the sample data and y_E is the estimated value using the regression line.

The sample coefficient of determination may be interpreted as the proportion of variation in the dependent variable y that has been accounted for or explained by the variation in x. It is therefore a very useful assessment of how closely the regression equation fits the data. In our case this will mean how well it predicts the value of the cost

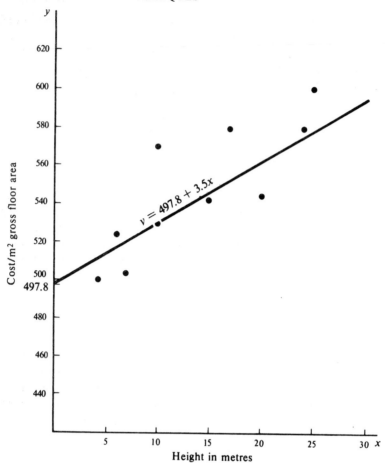

Fig. 20.6. Scatter diagram (showing best line of fit).

per square metre of the building, given its height.

Calculation of this expression can be quite tedious without a standard computer package and, therefore, short cut melthods have been devised for manual or self-programming use. One such formula using the values of a and b already calculated is:

$$r^2 = \frac{a\Sigma y + b\Sigma xy - n\bar{y}^2}{\Sigma y^2 - n\bar{y}^2}$$

(*Source*: Hamburg, M., *Basic Statistics* (Harcourt, Brace and Jovanovich Inc.))

By using this formula very little additional information needs to be found which has not already been included in the table to find a and b. Therefore, using the previous data in the formula we get:

$$r^2 = \frac{(497.8 \times 5{,}454) + (3.5 \times 75{,}822) - (10 \times (545.4)^2)}{2{,}984{,}092 - (10 \times (545.4)^2)}$$

$$= 0.61.$$

In other words approximately 60% of the total variation in the dependent variable is explained by the relationships between y and x expressed in the regression line.

The coefficient of determination estimates the proportion of the variation which the dependent variable shares with the independent variable(s). Another measure that is often used is the 'coefficient of correlation', which is the positive square root of the coefficient of determination. This measures the probability that there is a genuine relationship between the variables and that it has not arisen by chance. The closer the value of r is to plus one or minus one, then the narrower the range of the predicted value of the dependent variable.

The relative importance of the independent variable(s) can thus be determined, and a decision made as to which should be included.

Whether a value of r or r^2 is considered high depends on the field of study. In estimating r^2, values of over 0.9 should be expected which should preferably be higher still if a multiplier is to be applied to the resultant estimate. If the values are much lower than this, then either additional variables require to be introduced (as in the case of the crude cost/height relationship we have just looked at), or else the relationship is not a linear one.

The three major problems with the technique from the point of view of cost modelling are as follows:

1. Where historic data is being analysed the resultant coefficients are 'locked into time'. By this we mean that no matter how comprehensive the data, and how good a correlation is achieved, the line of best fit will relate to that sample of data at that point in time. The weightings given by the coefficient may well change as new production methods are developed, new materials are marketed, or cost relationships change. Unfortunately, the technique will not accommodate these changes unless a re-analysis is undertaken incorporating these new factors. It therefore has similar basic problems to the storey-enclosure method in that the weightings given may be satisfactory at the time of publication but quickly become outdated. One method of overcoming this problem is to undertake the analysis on resource quantities (e.g. the regression equation would predict the amount of labour required to build a high building) rather than costs. The current cost of that resource can then be applied to the estimated quantity and a total cost for it established at current levels. Although this overcomes the problem of a change in the relationship between the unit costs of resources it does not provide an answer to changes in, say, productivity. However, construction processes tend to change more slowly than resource costs and therefore error is not likely to creep in so quickly.

2. Unless there is a regular building programme it may mean that at certain times, for example, when the construction industry is in depression, very little data will be available with which to undertake the analysis. The method may not therefore be available to the cost adviser at times when the effect of his cost advice is probably most beneficial due to the economic constraints. A further difficulty is that lack of input data at times of recession may distort the index itself, unduly weighting it towards boom conditions.

3. It is possible to place a line mathematically through any set of data even though no correlation exists. Care must be taken therefore, to ensure that a linear

relationship does exist, and other statistical tests such as correlation should be used to confirm the strength of that relationship. Checks to make sure that there are no cost thresholds are also important. For example, in practice the cost per square metre will probably fall for the first four to five storeys and then rise again thereafter. Two linear regression models will therefore be necessary to accommodate these two slopes. If this point is missed the fitted line may show a continuous rise right from the very first storey. This will tend to throw all the estimates out to some extent and in particular those for the first four storeys. A careful look at the scatter diagram and a graphical analysis of residuals (see later) will assist in identifying any problem areas.

So far we have been dealing with two-variable linear regression, but in practice it is usual to find a large number of factors affecting the cost of a building or other major item. In the case of an office block for example, they may include the height, shape, density of partitioning, standard of finish, level of services, and so on. In these circumstances it is possible to use an extension of the technique called multiple linear regression. Here the same principles apply but unless there are only three variables (where a two-dimensional plane can be placed through the data points) it is not possible to draw the resultant regression plot on a scatter diagram. It is usual for a computer package to be used to establish the coefficients (b) to be used in the regression formula. This formula now becomes:

$$y = a + b_1x_1 + b_2x_2 + b_3x_3 \ldots + b_nx_n$$

where x is the value of each respective independent variable.

The number of 'b' coefficients included will depend on the number of variables chosen. Coefficients of multiple correlation can be determined which are similar in their interpretation to the simple expression used before. The technique now becomes very useful as an estimating tool, as all the major factors affecting cost can be included in the formula, although it still suffers from some of the defects previously described for two-variable regression. The technique has been applied reasonably successfully to reinforced concrete frames, simple building types and even services installations. One word of warning is that the choice of variables is left to the user, and he may leave out a most important variable because he does not know of it, or does not consider it worthwhile. Trial and error is required to establish which variables are significant, and commonsense should be used to ensure that silly correlations are not included even if they improve the model. (For example the height of the architect may be correlated in the sample to the cost of the building.) When choosing variables it is preferable to choose those which are well correlated to cost, and avoid those that are simply correlated with each other, and it is sensible to keep the number of variables well below the number of data points.

Finally, until now, we have been talking about linear relationships between the variables. In a number of situations the plot of the relationship may be curved or even broken. In the former case a curve will need to be fitted to the data (computer programs are capable of doing this) and in the latter, the regression model will need to be split into sections to cope with the thresholds apparent in the sample data obtained.

A useful visual guide to see how well any regression model fits the data is a graphical analysis of the 'residuals'. Residuals are the difference between the estimated value of

Cost per square metre of glazed area

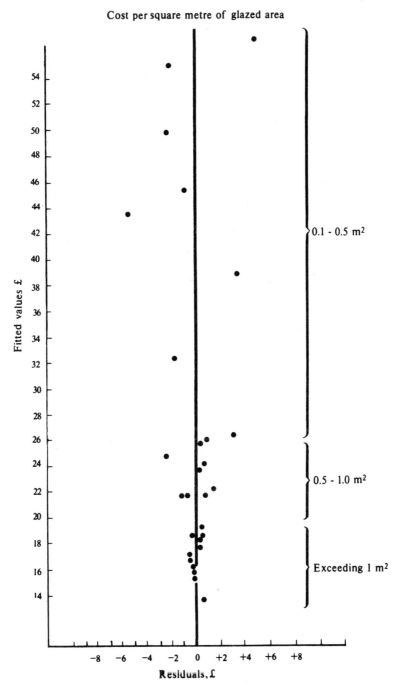

Fig. 20.7. Graphical analysis of residuals. Fixed light windows with mill finish.

the item using the regression model and its actual value taken from the sample. If the residuals are plotted against the fitted values and to the same scale, a very clear indication of the closeness of fit and any peculiarities in the model can be seen. Figure 20.7 shows a plot for fixed light aluminium windows where the variables chosen to predict the cost per square metre of glazed area were area and shape. There are in fact three models plotted, one for each area classification of window, and it can be seen from the sample data that the regression model is most reliable in the case of the larger windows.

Points to watch out for in graphical analysis are:

1 The existence of 'outliers', i.e. a few of the residuals much larger in magnitude than all the others. These should be investigated to discover the reason for the poor performance. In many cases it may be due to freak circumstances, in which case that particular item of data can be ignored.

2 The existence of a curved pattern of residuals which would suggest that a different model may be more beneficial.

3 A skew (or other non-normal) distribution. Non-normality can be detected more reliably by using normal probability paper. This concept can be followed up by studying most elementary statistical text books.

4 A progressive change in the variability of the residuals as the fitted values increase, indicating non-constant variance and a violation of one of the major assumptions in regression analysis.

This section on regression has been a very superficial introduction to a topic which will have its own section in any library of statistics. Before attempting to use the technique it is important that the model builder becomes full conversant with the method of construction, the tests to be applied to ensure reliability, and the limitations within the context of forecasting cost. A number of elementary statistical text books cover the basic principles in much more detail than it has been possible to achieve in this subsection and these should be read in detail.

Monte Carlo simulation

In the two previous techniques we have attempted to describe the cost of the building in algebraic or statistical terms so that we can estimate the cost effect of changing one or more of the design variables. We have recognised the problems associated with the variability of our basic data, and also the difficulties in predicting what will happen in the future. It is probable, however, that we could define a likely range of values for the unit cost of an item, and also the likely range of say the inflation rate over the next few years. What we need is a technique that will allow us to use this assumption of the probable range of values to explore the possible cost solutions available using the values within that range. If we repeat the exercise for two or more alternatives and obtain a distribution of solutions for each, we can compare the results and see which solution has the highest probability of being the most economic choice over the likely range of circumstances.

The Monte Carlo technique is one method by which we can simulate activities over time, of a kind which could not be represented adequately by any of the theoretical models. Simulation is simply a means of creating a typical life-history of the system (e.g.

the total building, the production process, maintenance, costs-in-use) and activities under given conditions, working out step-by-step what happens during each unit of life of the system. In order to do this we need to know the detailed characteristics and operation of the system and its relevant distributions.

Whilst the example chosen will be in the field of costs-in-use, the same principles can be applied to any other problem where values are not constant. The building of a brick wall with its variability of labour times for the various operatives involved is an example of another field, this time production, in which the technique can be used. In production there is often a sequence of events which are 'end on' to one another and which therefore form a 'queue' of activities. These events are very often dependent upon one another and, therefore, an irregular performance in one part of the system will probably have an impact on the total performance. For example, if the labourer supporting the bricklaying gang is not up to scratch then the overall performance of the gang will be affected. It is also unlikely that any one member of a gang will work at a constant rate and, therefore, variability in the team's performance is bound to occur for this reason. The Monte Carlo technique will allow simulation of that 'queue' of activities, or the special operations research technique of 'queuing' can be applied direct. Queuing as a technique is not discussed in this book but it is worth following up if your interest lies in this area.

Perhaps the most difficult forecast to make with regard to buildings is the cost of running and maintaining the property, not only because there are so many factors affecting the way the building performs (maintenance strategy, standard of use, design detailing, workmanship, etc.) but also because of the long period of time over which the prediction must be made. The problems of using 'costs-in-use studies' have been identified in Chapter 4, one of the major problems being the choice of assumptions that have to be made before a calculation takes place. The example of the 'factory cladding' calculation will be remembered in this connection.

To illustrate the ability of Monte Carlo simulation to assist us in decision-making by giving us more information on the range of possible future costs, a simple hypothetical example has been taken. This concerns a building component that requires no routine maintenance or repair but which normally needs to be replaced several times during the building life. Taxation, inflation, disruption costs, and secondary maintenance have not been included, thereby simplifying the calculation. The variables dealt with are therefore:

1 Component life.
2 Life of building.
3 Interest rate.

At the heart of the technique are the distributions from which the sampling of the simulations will take place. These distributions can only be determined from data collected over a long period of time and will relate to the variables previously mentioned. Figure 20.8 shows the histogram for building life in our hypothetical example. Each class interval is shown with its frequency of occurrence for this type of building on a probability scale between zero and one, with the value of one being the absolute certainty that the event will occur. Thus in this histogram there is a 50% chance that the life of the building will be between forty and sixty years, and only a 10% chance that it will be within zero and twenty years (for the purposes of the exercise it is assumed

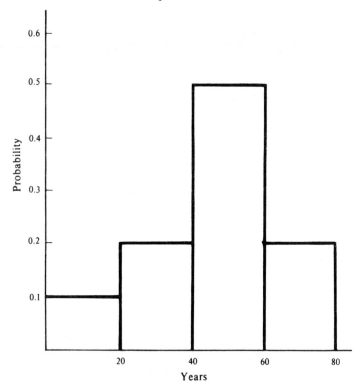

Fig. 20.8. Histogram of building life.

that there is no chance of it exceeding eighty years). It has been known for some buildings to be knocked down almost immediately after completion, for one reason or another, and therefore this possibility is included albeit at a very low probability.

The next step is to translate this histogram into a cumulative distribution curve or 'ogive' from which we can sample the distribution. This is achieved by adding each probability to the previous one, in order, starting with the lowest class interval (Table 13.)

These points are then plotted on a graph as shown in Fig. 20.9, which we can sample

Table 13.

Building life	Probability	Cumulative building life	Cumulative probability
0–20	0.1	0–20	0.1
20–40	0.2	0–40	0.3
40–60	0.5	0–60	0.8
60–80	0.2	0–80	1.0

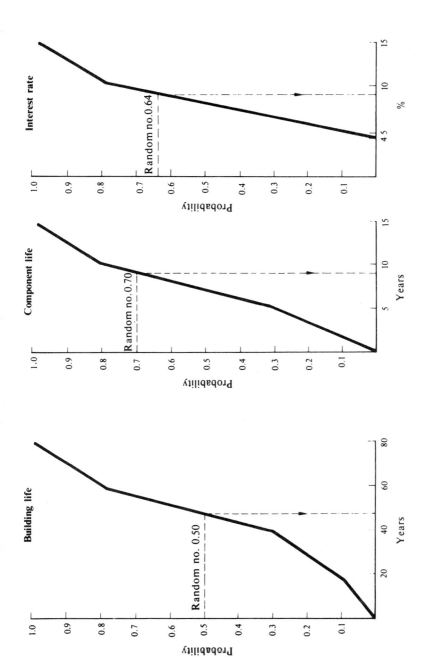

Fig. 20.9. Cumulative distribution curves.

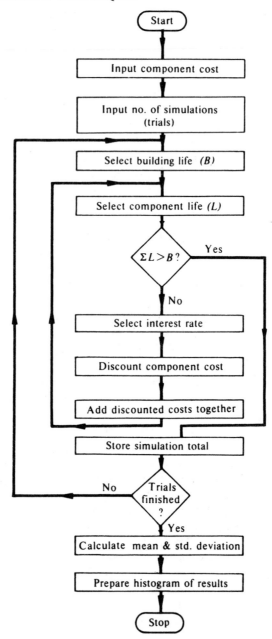

Fig. 20.10. Flow chart. Monte Carlo CIU model.

from the distribution. Sampling is a simple matter of selecting a random number (from random number tables or a random number generator) and reading off the resultant life of the building. For example, if the random number 0.30 came up then the life of building would be chosen as forty years. Random numbers are figures between set limits (usually zero to 0.99 or zero to 99) which occur entirely by chance. For example, they could be generated by a roulette wheel or by some sophisticated electronic process which ensures that all numbers have the same chance of being picked on the next selection. Most small calculators and micro-processors produce pseudo random numbers and if the operation of selection is performed too many times the same sequence of numbers will begin to appear again. Statistical tables containing a selection of random numbers are available for manual simulation and a sample sheet is included as Appendix B. If the other distributions for component life and interest rate are treated in a similar way (as shown in Fig. 20.9) then we are almost ready to commence the first simulation.

The only remaining consideration is the order in which the simulation should take place, and a useful way for demonstrating the sequence of events is by means of a flow chart as in Fig. 20.10. This is a short cut way of saying the following:

1 Determine the cost of the component.
2 Decide how many times you are going to do the simulation.
3 Sample from the distribution for building life by using random numbers. (This figure will remain fixed for the first simulation.)
4 Sample from the distribution for component life by using random numbers.
5 Check whether the sum of the lives of the components simulated to date exceeds the building life. If the building life has been exceeded then store result of total discounted values to date and check whether the required number of simulations has been made.
6 Sample from the distribution of interest rates by using random numbers.
7 Add the selected component life to any previous cumulative total of component lives for this simulation.
8 Calculate the discounted value of the component using the interest rate selected and the year when the component is replaced.
9 Add the resultant discounted value to previous total of discounted values (including initial cost) for this simulation.
10 Repeat items (4) to (9) until the check in item (5) tells you that the component has been replaced a sufficient number of times for the building life of this simulation.
11 Record the total discounted costs of the component for this simulation and store.
12 Repeat items (3) to (11) until the required number of simulations has been made.
13 Calculate the arithmetic mean and standard deviations of the stored results of the total number of simulations made.
14 Prepare distribution of results.

The above simulation would be very tedious if the calculations had to be done manually, but by using a computer the whole process is speeded up and requires no effort by the user other than inputting a small amount of data. Figure 20.11 shows the

```
****************DEMONSTRATION PROGRAM****************

THE PROGRAM DEMONSTRATES THE USE OF MONTE CARLO TECHS.
FOR CIU MODELS. IT SELECTS THE LIFE OF BUILDING,COMPONENT
LIFE,AND INTEREST RATE FROM THREE SEPARATE DISTRIBUTIONS
PREVIOUSLY DETERMINED. THE RESULTS OF THE SIMULATIONS ARE
STORED AND ARITH. MEAN,STD. DEVIATION AND HISTOGRAM
ARE CALCULATED FOR THE NUMBER OF TRIALS REQUESTED.

MONTE CARLO SIMULATION +CIU

WHAT IS INITIAL COST OF COMPONENT ?
+25
HOW MANY TRIALS DO YOU REQUIRE PER RUN ?
MAX. NUMBER = 999
+600
   ARITHMETIC MEAN      = £    65.0564
   STANDARD DEVIATION   = £    18.655
DO YOU WISH TO HAVE A HISTOGRAM OF RESULTS?
+YES


*********************************************************

       HISTOGRAM OF RESULTS FROM CIU SIMULATION

NOTE: EACH HORIZ'*'REPRESENTS A FREQUENCY OF FOUR

   >                        FREQUENCY

   0

  10

  20          ********
              ********
  30          *****
              *****
  40          ******************
              ******************
  50          ************************************
              ************************************
  60          *********************************
              *********************************
  70          ****************************
              ****************************
  80          *****************
              *****************
  90          *********
              *********
 100          *****
              *****
 110          **
              **
 120


   ARITH. MEAN        = £ 65.0564
   STD. DEVIATION     = £ 18.655
```

Fig. 20.11.

results of such a program run for six hundred simulations with a component costing £25.00 and the distributions shown in our cost-in-use example. It should be stressed that the above example has been over-simplified for the purposes of communication and in practice many more variables and a smoothing out of erratic movements in the interest rate may be required. If the process is now repeated using possible alternative components with different initial cost and different periods for replacement (i.e. the component life distribution is different) then a comparison can be made as to the probability of one component being a more economic solution than the other. Figure 20.12 shows this in diagrammatic form with the two distributions superimposed on each other. It can be seen immediately that component A is generally the better proposition but there is a point of overlap between the two. Within the range of values of the shaded

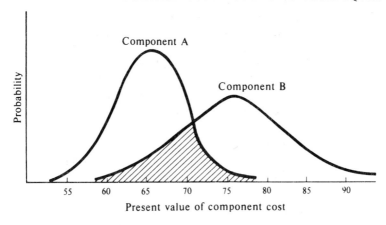

Fig. 20.12. Comparison of distributions of results of simulations for alternative components.

area, either A or B could prove to be the better proposition. However, the chances of finding component A to be the least expensive solution is much greater even within this range. For example, the area of the distribution of component A to the left of, or lower than, a present value of £70.00 is much greater than the area below the same present value in the distribution of component B. As the area under the curve represents the probability of the cost falling within a certain price range (in our case below £70.00) then it follows that it is much more likely component A will be a better choice than component B.

What the Monte Carlo technique has allowed us to do is to explore the future cost of a component within defined ranges which have been weighted from previous knowledge or experience by means of the distributions used. The steeper the slope in the ogive the greater the chance (class intervals remaining the same) that the values within a certain class will be selected by the random number. By repeating the exercise a number of times we have artificially 'lived through' or simulated the lifetime of the building a similar number of times. The distribution of results gives us a very good clue as to the likelihood of a particular cost arising, whereas to obtain this sort of information by more conventional mathematics would be extremely difficult and probably impossible.

Although the technique has overcome some of the problems of variability of data it has not helped us with problems such as technological change.

If a component were to come on to the market in future years which was a much more attractive proposition than the original component, then it is likely that this new material would be the replacement component. The model cannot predict this kind of variation and over the life of a building this type of change may occur several times – think back over the last sixty years!

Another problem lies in the distributions. One method of constructing them would be to look back into history to see what range occurred in the past. However, this may not be reasonable; a distribution for inflation constructed in, say, 1965 based on rates over the past one hundred years would certainly not have predicted inflation during the 1970s and probably not for the remainder of the century either. The same would apply to interest rates. Materials which have just been introduced would have no record to build upon, and hypothetical distributions would need to be produced.

The subjective building of distributions may not be too much of a burden however, provided that they can be justified by reason and sound professional judgement. After all, giving a range of likely possibilities is surely a better proposition than pulling a single figure out of mid-air with very little chance of it being correct. At least the risk in the decision of choosing between the alternatives is reasonably identified, and it is at least recognising that estimates cannot be rigid, accurate, predictions. Of course, only a very simple example has been demonstrated here which would need considerable refinement if it were to be included in a working model – if all the variables effecting a solution were required to be sampled the model would probably involve a score or more distributions for a single component. As with all cost prediction techniques it would be very difficult to find sufficient data to produce the distributions required. However, not all assumptions are cost significant and therefore a concentration on the more important distributions will probably be sufficient for this kind of predictive, exploratory exercise.

Monte Carlo simulation can be adapted for other uses such as undertaking a random search for the optimum solution to a design problem within a computer. Random numbers are used to select alternative shapes of buildings, structural frame types, specification, etc., to build up a causal model of the building cost. The machine searches for an improvement on its original selections and when it finds it, holds the new solution and aborts the previous one. This process continues until after a specified number of successively more attractive solutions, the program finishes, and the last solution is presumed to be the best or optimum solution. (In fact it rarely is the optimum as the number of alternatives is for practical purposes infinite). To assist in the search for the best alternative the computer program can guide the search by means of 'rules-of-thumb' established from the practical experience of the author of the program, and it can also incorporate short cut methods of rejecting obviously inappropriate solutions earlier in the routine to avoid unnecessary calculations. Even with these aids and a powerful computer it can take several hours to achieve a successful solution with a complex model.

Generally

There is no doubt that if the full potential of the computer is to be harnessed for the benefit of the cost adviser, the development of models must play a large part. In the past, the time-consuming task of calculating algebraic formula many times over has daunted even the keenest cost explorer, but with the computer and its ability to perform iterative processes without effort, a new generation of cost advice appears to be on the horizon. Unfortunately the construction of the models, the development of software for use by the computer, and the collection of the data to be applied, have not kept pace with the enormous advances in the electronic hardware of computing. In this chapter we have scratched the surface of an enormous area of potential knowledge which has yet to be developed for the benefit of the quantity surveyor. It has not been the intention to delve into the mathematical basis for these techniques but to show by simple illustration the opportunities available in just three model types. There are a very large number of books on operations research, statistics, and systems which the reader should follow up if his interest has been stimulated in this area of study.

Chapter 21
Models for Optimisation

Every day each member of the human race is engaged in one overriding common activity. That activity remains true, no matter what the geographical, political, religious, cultural and ideological climate in which the individual is found. It is common to all peoples and yet the choices made will vary almost as much as the number of individuals. Even when the same person is monitored, the activity remains the same, but the choices made will vary with time for the same individual making the same decision.

This common activity is the search for what the individual considers to be the 'best' solution. Whether it be selecting a chair in front of the television set or the quickest route from A to B; the choosing of the most compatible member of the opposite sex for a spouse or voting for a new government; the human mind is searching for what, in a given set of circumstances, is the most appropriate solution. The choice will be tempered by the mind's own perception of the problem, its own prejudices and preferences, and the many constraints, both physical and cultural, within which it has to work.

This search for the 'best' is known as the process of optimisation. It is task that is set the design team at the beginning of every building contract and yet it is an almost impossible goal to achieve. Some of the reasons for this difficulty are common to many optimisation problems, such as:

1. what criteria are to be used to judge the alternative choices? Is the utilitarian principle of 'what solution will bring the greatest happiness to the greatest number of people' to be applied? If so, this is an extremely difficult criterion to satisfy, particularly with a long-term and complex asset such as a building

2. what time is available to make the choice? We have already seen in Chapter 8 that one of the differences between design method and scientific method is the extra constraint of time. Thus we often have to be satisfied with optimal (i.e. close to the best) solutions rather than the optimum even when the criteria are clear

3. do objective measures exist for measuring the best solution? For a simple reinforced concrete framed column it may be possible to design the minimum size of cross section, but to determine the most pleasant internal arrangement of building space would be a problem of different magnitude altogether. The former has a well-established and formalised method of calculation and measurement. The latter will involve social issues, aesthetics, efficiency, lighting and so on, many of which have no means of accepted measurement

4. do we optimise for now or extend into the future? Buildings are not judged at a single point in time. Assessments are made by the owners, the occupants and the community throughout its life as to its quality and performance. Some decision as to what time scale it is reasonable to plan for, needs to be made

5. who will assess the optimum solution? In building design should it be the client,

the architect, society (through legislation on space, quality, cost, etc.) or the occupants? Naturally, the assessment of these members or groups will overlap, but who arbitrates in the case of conflict?

It is these issues and others like them which make it difficult to make judgements as to what is best. While it is quite possible to come to an apparent 'best' solution at sub-problem level, it is much more difficult to optimise between the results of the sub-optimisations. Very often there will be conflict between two optimum solutions at the sub-problem level, e.g. optimum cost and optimum comfort. There is, however, a danger that the hard tangible nature of a cost figure will carry more weight than the more subjective nature of a description of comfort. It is not until after the decision has been made that the error becomes apparent.

Those aspects of design which can be quantified often have an unfair advantage over those which rely on 'word models' such as a description or even visual models such as drawings. The advantage is even more unfair where the figures convey an apparent accuracy and confidence which conceal the assumptions and guesswork which lie behind the results presented. In life cycle costing we have already made the point in Chapter 4 that unless there is a strikingly obvious difference between alternatives in cost-in-use terms then the decision is probably best made on other grounds.

The same also applies to a lesser extent with initial cost. The use of unit prices for a data base abstracted from a model which does not always reflect cost occurrence on site, e.g. the bill of quantities under SMM.7 does not give confidence in establishing an optimum design which will set the parameters for the production process on site.

The counter argument to these concerns must be that it is an obligation placed on the cost adviser and the members of the design team to try and achieve the best solution. This 'ideal' must be tempered with a recognition that our processes are imperfect but capable of improvement. By seeking to use and develop models that provide a single best solution, we begin to establish a pattern of enquiry which will give us greater understanding of small parts of the building system and eventually the whole.

Whether it will ever be possible to provide the optimum total design, even for a simple building at a single point in time, is extremely unlikely. To optimise a spatial layout with twenty spaces with a different activity in each so that the spaces and activities are re-arranged to provide the best solution for internal circulation would take 6×10^{17} iterations if every possibility were tested. Even using a powerful computer which could make one total building evaluation of spatial arrangement every second would take over 2,000 years! Most clients are reluctant to wait that long!

If all the other design possibilities and options are included it can be seen that the lifespan of the earth is unlikely to be sufficient for an exhaustive evaluation of one building. Now this is a rather comforting thought for it means that perfection is still an ideal to be sought after, but with little chance of being achieved. We can however get somewhere near it through a combination of knowledge, experience and technique.

Over the years a conventional wisdom has developed which acts as a screening process for likely solutions. We do not consider some options because we 'know' (through education or experience) that they are generally unsatisfactory in this situation. The human brain develops its own search processes for seeking a solution and it is only in the detail (e.g. how much extra would an increase of 0.2 in the wall/floor ratio cost) that it needs the assistance of technique. Most evaluative computer programs are handling the detail rather than the major issues. Most designers will be able to provide a reasonable stab at a building form and specification that will satisfy the client's requirements for support and enclosure. it is the marginal improvement of that solution, but which may appear to the client to be quite large, to which most design aids are aimed.

One way of approaching the design problem could be to search for the minimum cost and then move away from this cost through the minimum acceptable level of design to the client towards the optimal solution. Figure 21.1 shows this possibility diagrammatically and is a development of the graph shown in Figure 8.2. Building 'B' is likely to be a simple building with a smaller range of options and therefore higher percentage commitment at the start (e.g. a warehouse). Building 'A' is likely to be a more complex building with few constraints and many more options.

The problem is posed of course as to how far we should go beyond the mere satisfying of the client's requirements before we stop 'improving' the design. A parallel exists in aircraft design where one approach is to find the 'least weight' solution and then add on the extra items which will add to passenger comfort and managerial efficiency until the anticipated benefits of these additions balance the extra cost both in initial and future (fuel, maintenance, etc). terms.

This balancing of the marginal benefit with the marginal unit cost introduces the concepts of utility and value. Utility is a concept which is well defined in economics and is sometimes defined as 'the amount of satisfaction a person derives from some thing or service without any reference to its usefulness'. This degree of satisfaction tends to decrease as the consumption of the good, benefit or service increases in amount. With successive new units of a good the total utility will grow at a slower and slower rate. The extra utility added by the addition of the last unit of good is known as marginal utility and the law of diminishing marginal utility states that this decreases as more units are added. This is because there is a tendency for our psychological ability to appreciate more of the good, to become less keen. Attempts have been made to measure utility but even when considering consumer demand theory for less complex situations than buildings the majority of economists would prefer to rank the comparative utility rather than measure it.

There is another law – the law of equal marginal utilities per pound or dollar, which says that each good is demanded up to the point where the marginal utility per pound spent on it is exactly the same as the marginal utility of a pound spent on any other good.

This equilibrium condition is a useful concept for it provides a clue to the theoretical upper limit for improvement which will be related to the resources at the client's disposal which he could use for other purposes should he so wish. At the point where the marginal improvement in benefit or utility equals the marginal increase in cost then a search for improvement should stop. In practical terms this benefit will be hard to measure and even the marginal increase in cost is by no means certain.

However, this balance of benefit to cost is close to a more general definition of value that is sometimes given. Value can be defined as the balance between the client's wants (i.e. shelter, security, income, etc.) and the resources at his disposal (i.e. finance, time, etc.) and the constraints within which he works (planning, site, technological, etc.) On some occasions the constraints on resources will be so great that the theoretical balance of utility and cost will not be able to be met. However, this will not affect an assessment

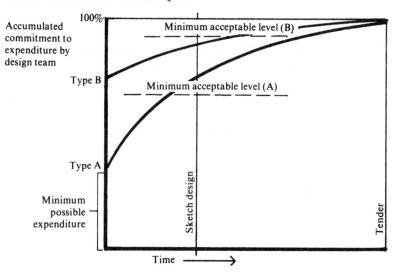

Fig. 21.1. Hypothetical cumulative expenditure over time for two building types A and B.

of value as constraints are accepted within the definition. If these concepts are accepted then it will be realised that it may be possible to have 'quality for money' but not 'value for money' as money is already part of the value assessment in projects where it is a criterion.

The purpose of the above discussion has been to lay the foundation for the introduction of optimising techniques. Our techniques should be searching for this theoretical balance between benefit and cost or where the constraints are tight should be attempting to provide the best satisfaction of wants (i.e. design quality) within the resources available. We must also realise that some, probably the majority, of benefits derived from a building are difficult to measure and as we have seen in Chapter 10 costs themselves are by no means certain. However, a step can be taken if we make some simplifying assumptions such as:

1 assume certainty (e.g. as to cost) when in reality there is variability.
2 insert human search processes to reduce the number of options considered (e.g. where the human brain can draw associations much more quickly than a computer – as in spatial layout).
3 assume simplified mathematical relationships (such as linear) between variables, providing there is a close approximation to reality.
4 assume 'rules-of-thumb', which may be based on experience, for the evaluation of the building. For example, the number of lifts may be related to the plan area of a building and its speed to the height. A large number of these empirical rules would allow the internal contents of a building to be costed with reasonable accuracy once a building shape had been determined.

In the techniques that follow simplifying assumptions such as these will need to be made and together with the qualifications already discussed, it is apparent that these techniques will only lead to optimal rather than optimum economic solutions. We can get within an approximation of the best solution but cannot guarantee to have found it. In choosing the techniques for this chapter the authors have concentrated on three processes which provide a representative sample of approaches and which have all been used in building cost experiments. it is too early to say whether one or more of the approaches is likely to become a lasting and common technique for the cost adviser. The development of computer systems will no doubt encourage the use of more mathematical techniques mainly because the chore of calculation will have been removed and data will be readily accessible. This does not mean however that the technique will be acceptable to the consultant. He may not trust the model because he cannot interact with it to use his own knowledge and experience. It has not been possible to go into detail within the space available, but it is hoped that sufficient information has been given for a clear appreciation of approach.

Structured search approach

This is perhaps the simplest of the possible optimisation methods. An outline of the procedure has already been given on page 102 related to the design process. In that procedure it was suggested that if a table of possible solutions were produced then one of these solutions could be chosen and form the basis of a design strategy. The problem is how to arrive at the table and what variables should the table consider.

An appreciation of the objective is given in Fig. 21.2. This simple diagram illustrates the point that there are a number of constraints acting on the solutions available and these form a boundary to what is possible. (Some of these boundaries will not be clear and will depend on other constraints.) Within the boundary will be found those solutions which are feasible, providing that the constraints do not rule out every

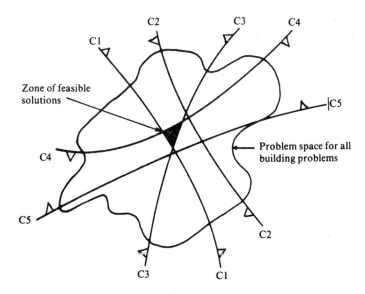

Fig. 21.2. Diagrammatic representation of how constraints reduce the number of solutions available.

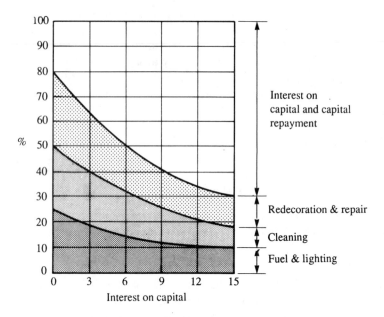

Fig. 21.3. Annual percentage spent on capital and future costs for a sample of Crown office buildings (DOE, R & D Bulletin Building Maintenance, Report of the Committee 1970).

possibility. Within this zone of feasible solutions, which in practice usually contains an infinite number of building shapes and combination of components, there will be 'better' economic solutions. To avoid trying to search through a wide range of possibilities in order to find a good solution the zone could be 'mapped' at regular intervals in order to establish a trend and if possible a least cost solution. From this 'least cost' reference point other alternatives can be considered which may be better from other design standpoints and their cost compared with the 'least cost' already calculated. It can then be seen whether it is worthwhile to pay this extra amount of money to achieve the improvement in design. If a similar table were produced for future costs then it might be possible to narrow down the number of solutions still further.

Since there are different objectives and constraints for initial and future finance, the separation of the two will enable the client to apply his own weighting to the importance of each. This will no doubt be influenced by a number of factors including the interest paid on capital as shown in Fig. 21.3.

Some of the building shape variables will be discrete (i.e. whole numbers) such as the number of storeys and therefore the chances of missing the 'best' solution in the 'mapping' process will be reduced for these variables. Others will be continuous and if these are used to generate other costs, e.g. plan area may be used to determine the number staircases, then there is always a danger that a cost threshold has been missed.

However, as has already been stated in this chapter, it is not possible to consider every possible combination of specification and shape. The opportunities are infinite and human experience must be brought to bear on the problem to act as a filter to avoid

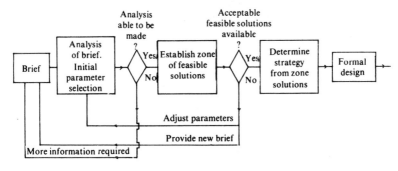

Fig. 21.4. Design model.

those options which are most likely to yield an unsatisfactory result. Even with this initial screening the possibilities are still enormous and some search strategy must be applied which will enable the information to be handled in a meaningful way. If a number of options are to be evaluated, it is essential that a computer is used. The danger here is that the operator will be swamped with paper, and confusion caused by over-kill of information will be the order of the day. Those projects which have the widest number of choices will be those found on 'green field' sites with no planning constrictions or heavily constrainted internal environments. An infill site between two adjacent buildings in the centre of the city will be severely constrained by shape and probably specification.

One possible procedure is that shown in Figs 21.4 and 21.5. It consists of a two-stage affair. The first stage involves the design team choosing ceretain specification options from an analysis of the brief for each of the major cost-significant components of the building. These are 'explored' from the point of view of cost as if they were independent of any building. Models such as the lift model in Chapter 20 or cost models for the comparison of different structural frames could be used for this purpose.

From this initial exploration, the specifrication and parameters of components for a building can be chosen which are then taken into the second stage. This next evaluation involves the use of these chosen specifications, which would be checked for compatibility, in another model which calculates the whole building cost for a number of building geometries.

It is anticipated that this kind of optimisation process would take place prior to design and therefore the level of detail required for the geometric model would be rather

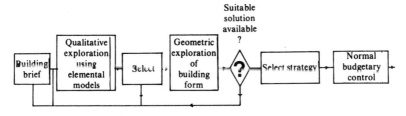

Fig. 21.5. Cost model related to above.

coarse. This would be essential in any case, if we were to allow the desinger to use his own particular skills of spatial organisation and aesthetics to full advantage. The geometric variables could be:

1 Number of storeys.
2 Accommodation area.
3 Plan shape index (see page 200).
4 Plan compactness ratio (see page 199).
5 Average storey height.
6 Glazed area.
7 Density of partitioning.

These are probably the minimum number to obtain some useful quantitative information for cost evaluation and at the same time 'trigger' the cost thresholds that exist, say, for the number and speed of lifts. There is no reason why these variables should not also interact within themselves. For example, the glazed area could vary with the assumed depth of plan. Alongside these parameters would be a series of 'rules-of-thumb' which would generate the requirements for other aspects of the building and these could include rules for circulation areas, staircases, toilet accommodation, service cores, etc. Constraints such as site area and shape, minimum plan area, etc. would also be included to avoid unnecessary evaluations.

The presentation of the final series of cost maps could be similar to that shown in Fig. 21.6. In this case the change in cost related to the plan shape index and number of storeys is shown for a commercial building. (Cost is in thousands of pounds.) All other variables and specifications have been kept constant and the 'map' contoured for ease of comprehension. It can be seen that for this set of constants the 'least cost' solution is at six storeys with a square plan shape. We can now ask the question 'how much extra will it cost to go for a four-storey building and more external modelling with, say, a plan shape index of 1:4' and then probably more importantly 'is it worth it?' The difference in cost must be balanced by the additional benefit to be derived from the improvement in the form of the building. The 'guide' costs (and the results should be put no higher than this) will give an indication of extra cost and the client will have a datum for measuring value. The argument has been put forward rather simplistically and there are all sorts of qualifications that would need to be made. There is a danger, particularly where public accountability is concerned, for the least cost to be chosen irrespective of the other benefits. It is incumbent on all members of the design team to strengthen the quality argument whilst at the same time developing more explicit quantitative models.

The cost maps are relatively easy to produce (although the models are more difficult) and it is therefore easy to loop back through the system and select a different specification for components, or undertake a different analysis of the brief, to see what impact this would have on the final output. (The maps of course would also look at other combinations of geometric variables and future costs as well.)

At the end of the day, from the wide selection of cost alternatives, a design strategy would be adopted which took into account cost, form and specification levels, albeit at a very coarse level. It may not be optimum cost but it may be a step towards optimum design. In essence the quantitative and qualitative issues are being made explicit to aid the approach to design. This is shown diagrammatically in Fig. 21.7. Once a strategy has been taken on board (the computer can reel off the assumptions made in the chosen

Plan shape	\multicolumn{10}{c}{Number of storeys}									
	1	2	3	4	5	6	7	8	9	10
1.0	2874	2031	1926	1804	1768	1720	1827	1865	1897	1933
1.5	2877	2035	1931	1810	1775	1727	1835	1874	1906	1942
2.0	2883	2043	1941	1822	1788	1742	1850	1890	1924	1961
2.5	2890	2053	1953	1836	1805	1756	1868	1909	1944	1982
3.0	2897	2063	1965	1850	1819	1775	1887	1929	1965	2004
3.5	2904	2073	1977	1864	1854	1795	1905	1949	1986	2026
4.0	2911	2083	1989	1877	1850	1810	1924	1969	2007	2048
4.5	2917	2092	2001	1891	1865	1826	1941	1988	2027	2070
5.0	2924	2102	2012	1904	1880	1842	1959	2006	2047	2090
5.5	2930	2111	2024	1917	1894	1858	1976	2025	2066	2111
6.0	2937	2120	2034	1930	1908	1874	1991	2043	2085	2131
6.5	2943	2128	2045	1942	1922	1889	2009	2060	2105	2150
7.0	2949	2137	2055	1954	1935	1903	2025	2077	2121	2169
7.5	2955	2145	2066	1966	1948	1918	2040	2093	2139	2188
8.0	2960	2156	2075	1977	1961	1932	2055	2110	2156	2206
8.5	2966	2166	2085	1988	1974	1945	2070	2125	2173	2223
9.0	2971	2169	2095	1999	1986	1959	2087	2141	2189	2241
9.5	2977	2176	2104	2010	1998	1972	2099	2156	2205	2258
10.0	2982	2184	2113	2020	2009	1985	2113	2171	2221	2274
10.5	2987	2191	2122	2031	2021	1996	2126	2185	2236	2290
11.0	2992	2198	2131	2041	2032	2010	2140	2200	2252	2306
11.5	2997	2205	2139	2051	2043	2022	2153	2214	2267	2322
12.0	3002	2212	2148	2060	2054	2034	2166	2227	2281	2338
12.5	3007	2219	2156	2070	2065	2045	2178	2241	2296	2353
13.0	3012	2225	2164	2079	2076	2057	2191	2254	2310	2368
13.5	3016	2232	2172	2089	2086	2068	2203	2267	2324	2382
14.0	3021	2238	2180	2098	2096	2080	2215	2280	2337	2397
14.5	3025	2245	2188	2107	2106	2091	2227	2293	2351	2411
15.0		2251	2196	2116	2116	2101	2239	2306	2364	
15.5		2257	2203	2124	2126	2112	2250	2318	2377	
16.0		2263	2211	2133	2135	2121	2262	2330	2390	

Fig. 21.6. Cost map of permissible solution space within natural and legal constraints (costs in thousands of pounds). Blank evaluations exceed constraints.

solution) then more formal budgetary control mechanisms such as elemental cost planning can be used as drawn information becomes available. There is in fact no reason why the computer should not produce the first broad outline cost plan for its evaluative routines. It is interesting to note that the cost patterns do not always conform to conventional ideas and indeed when 'future cost' contours are superimposed, may open the way to more adventurous design possibilities.

The procedure outlined above was developed into a pilot model by one of the authors some years ago. This is, however, only one approach to the structured search for optimisation and simpler or more complex models could be derived. In the above

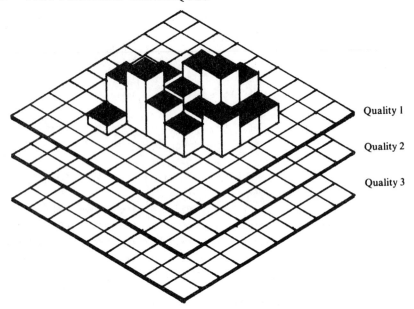

Quality 1

Quality 2

Quality 3

Fig. 21.7. Diagrammatic representation of cost map profile with different quality levels.

system most of the models were empirical although there is no reason why a variety of different types should not be used, particularly in the first stage of the process.

A simpler form of one-stage model could be developed from a regression equation where a table is produced from a systematic evaluation of the major variables. At the other extreme a multi-stage model could be devised which relates to the various 'entry points' into the analysis and synthesis stages of design. This more complex model would allow more flexibility for the designer and would make assumptions based on 'rules-of-thumb' with regard to information which is not available at the time. This is a small step towards the development of what is known as a 'knowledge base', necessary for the development of a computerised expert system. Currently the knowledge base is held in the human brain and accesses the data base to make a decision. As time goes by more of the experience of the cost adviser will be able to be made explicit in the machine and there will be less need for human intervention in the search. This subject is dealt with in Chapter 22.

Lastly, it should not be thought that the structured search approach will provide a panacea for all our problems, or that the theory behind the models is well developed. However, with the increase in computer power there is no reason why it should not provide a substantial step forward in providing better cost advice.

Monte Carlo 'hill climbing' approach

In the above approach to optimisation all the steps along the path to finding a solution were made explicit. At the end of the computation cost comparisons could be made alongside other design criteria. There was a good deal of human interaction to sieve through the wide range of possibilities and it was recognised that better solutions may lie between the points identified in the cost map.

Another approach might be to say – let the computer do all the work and select from the options that are available, trying out every combination that is possible. If this was done randomly rather than in a structured way as before, it would amount to a 'suck it and see' approach controlled by chance processes. The machine would select at random the geometry of the building comprising, say, a plan shape index, a certain number of storeys, an average storey height, a specification for each element, etc and then evaluate it from a cost file to obtain a total cost. It would hold this figure until it undertook an evaluation which was less than the previous lowest cost. The process would continue with all higher costs being discarded until no further improvement could take place. The final solution would then be considered to be the optimum.

Now it is obvious that even with the powerful computers we have today it would take several thousand years (as noted earlier in the chapter) to calculate a cost for every combination of shape and specification that is possible. Once again for large problems some sort of strategy must be employed to get to the end solution more quickly and it must be accepted that we will only get near to the optimum, not on it.

The technique for random sampling is quite straightforward and is identical to that outlined in Chapter 20 (page 270). Where discrete variables are concerned, e.g. the number of floors, then random numbers within a certain range can be allocated to each number in the range. Even with continuous variables these can be classified and included as discrete variables to speed up the operation. For example, supposing the machine had to select the span and bay size of a reinforced concrete framed floor, then the options could be structured as shown in Table 14.

Table 14. Cost per square metre

Span	Bay size	20 m²	25 m²	30 m²
3 m		48	36	35
5 m		40	38	39
7 m		44	45	57

To select a bay size and span, the discrete distributions would be set up as shown in Figure 21.8 in which a span of 5 m and bay area of 20 m² have been chosen. The cost of £40 per square metre would now be included in the cost evaluation. On the next selection a different set of random numbers are likely to be generated and therefore probably a new bay area and span.

This kind of sampling would be undertaken for all building components.

The important question now is, how do we impose a strategy on this process to avoid the selection of the expensive solutions? In other circumstances (e.g. an external wall component with U-value provided and the final cost represented in present value terms) it may be less self-evident which solution or combination would be the cheaper.

This is where the 'hill climbing' aspect comes in. Imagine an undulating plain dotted with hills. The high points are the least cost solutions and the plain represents the higher

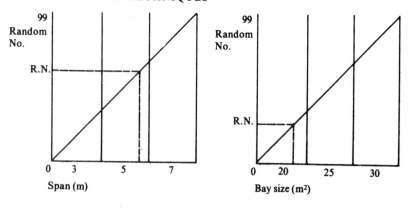

Fig. 21.8. Random selection of span and bay size from tables.

costs. The strategy to be adopted is metaphorically to find the highest mountain and climb it. One approach that could be taken is

1 Take the mean of all the options as a starting point for the first evaluation (i.e. the random number on the first round is therefore fixed at 50).
2 Assume that better solutions will be found near to previous good solutions.
3 Start with a fairly wide area of search around the 'good' point and as improvements continue to be found gradually narrow the search area (by restraining the band of random numbers available to be used).

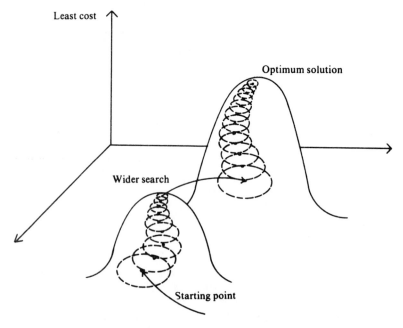

Fig. 21.9. Diagrammatic representation of 'hill climbing' approach.

4 When, after a specified certain number of evaluations or trials e.g. 200, no improvement has taken place, then the search area should be widened considerably to see if there is a higher mountain.

5 If no higher mountain is found after a specified number of trials assume that the optimum has been found. This is shown diagrammatically in Figure 21.9.

The process is shown in a procedural form in Fig. 21.10. Note that in this example the centre point for the new search radius is the value of the variable as found in the latest improvement.

A number of experiments have been undertaken with this method although once again it must be said that the complexity of a building makes it unlikely that complete structures can be effectively optimised by this technique. The closest to a useful model using this procedure for the cost of a building expressed in annual equivalent terms was that devised by Dr Clive Bastin of the Wyvern Partnership, Swindon. In his research at Bristol University he developed a model called AIM which moved accommodation units (e.g. rooms or compartments) around in three dimensions until the cost of movement between spaces, heating and ventilating and physical elements were in the

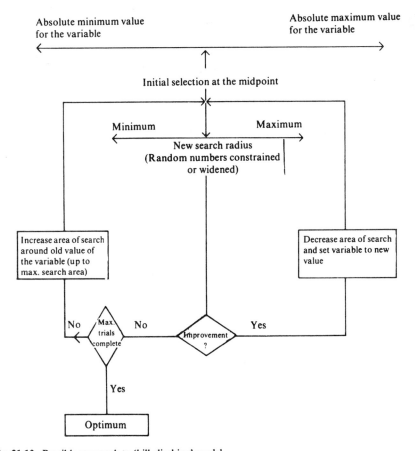

Fig. 21.10. Possible approach to 'hill climbing' model.

optimum relation to each other. The rooms were rectangular but were able to be expanded or contracted within certain constraints. Simplifying assumptions were made on structural requirements, costings, etc. and even so the optimisation of six or seven spaces took several hours starting from scratch with each space independent of the others. An additional search strategy was adopted to speed things up whereby an elemental cost profile was used to abort an evaluation in the early stages if it did not look like producing a good solution. If the element evaluation proved to be more than a certain percentage above the previous 'good' solution for the most important elements then that trial stopped. A sophisticated routine was included which moved the order of the element evaluation to make sure the critical costs were considered first – based on the machine's experience after the first few evaluations.

The process has been tested with students attempting to optimise the same number of spaces as the machine, using annual equivalent costs, but on a manual basis. In most cases the machine produced a better cost solution but it took appreciably longer. If the number of spaces is increased then the time requirements increase exponentially. The answer to an improvement in time probably lies in an improved search process and the anticipated developments in the speed of computing.

Figures 21.11 and 21.12 show the nature of the three-dimensional model together with the pattern of improvement in cost derived from one run of AIM. Despite the experiments it is unlikely that this technique will be used in practice in the near future except for fairly simple, well-structured and constrained problems. It has been used for optimisation of circulation space and for selection of cladding panels on an envelope but is rather cumbersome.

The advantages of the technique, such as the fact that it can cope with non-linear and discontinuous variables (i.e. those where there are one or more thresholds and costs do not change in direct proportion to quantity), are outweighed by the over-simplification of large problems to overcome the time problem. The two major criticisms voiced are, however, that the technique may miss the very highest 'mountain' altogether and consequently not find the best solution, and the considerable time involved in defining and structuring all the design possibilities.

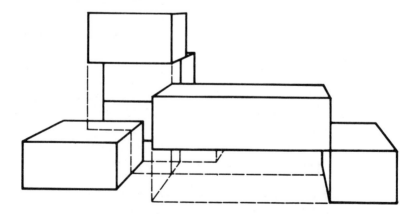

Fig. 21.11. AIM partial completion of optimisation process using 'hill climbing' model.

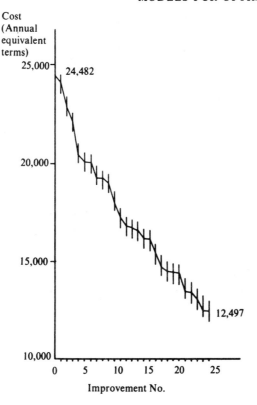

Fig. 21.12. Graph showing number of improvements and reduction in cost for one run of AIM.

Linear programming

The problem of how to utilise limited resources is a familar one to the cost adviser. He faces it in a variety of circumstances. For example, when attempting to get the best 'mix' of houses on a site, when trying to arrange the best use of plant and labour in construction or manufacture, and when trying to select the optimum arrangement of different cladding panels.

A range of mathematical techniques has been developed to tackle this particular type of problem and includes linear, non-linear and dynamic programming. Of these the simplest and most widely used is linear programming. In a book of this nature it is only possible to introduce the concept leaving the reader to tackle the mathematical basis of the technique and its development by further study.

To keep the approach simple we will deal only with linear programming and then only with a problem which has two unknowns so that it can be solved by graphical means.

To be able to use linear programming five conditions must be met:
    1 The problem must be capable of numerical formulation.
    2 A number of alternative courses of action (solutions) must be possible.

3 There must be one or more constraints on the alternative courses of action.

4 Some criterion must be available to optimise the various courses of action, e.g. least cost.

5 The relationship must be linear, although it is possible to deal with some non-linear relationships by simplifying or, for example, by transposing a curve into a straight line by taking the log.

The technique is used to maximise or minimise the objective of the exercise. For example, we may wish to maximise profit or minimise cost. A solution can only be found by graphical means for up to three variables; beyond this number mathematical routines are available such as the 'Simplex Method' involving a manual iterative process or more commonly computer programs are used to handle the complexity of the problem.

To introduce the technique a very simple example will be taken which first appeared in a similar form in *Building* magazine in March 1969. It nevertheless illustrates the basic principles very well and has a long history of use in some academic establishments!

Suppose the design team wish to clad a modular factory wall with two different cladding panels comprising different proportions of glazed to solid area. Each panel would have different *U*-values and costs. The problem is to determine the number of each type to provide an external wall of not more than 3.5 w/m²°C and a glazing area of at least 140 m² for a minimum cost. Both panels are 4 m high × 3 m wide and are able to be substituted by each other in every situation. The attributes of the two panels are given below:

Panel	Glazed area in m²	Solid area m²	U-value w/m²°C	Cost £	Number of each
A	7	5	4	500	x
B	2	10	2	400	y

The total area of cladding surface is to be 480 m² and therefore the total number of panels required is 480 ÷ 12 = 40 panels.

To solve this problem graphically it is necessary to go through a number of stages which identify the constraints and objective function and express them in a numerical form before plotting them on a graph and reading off the optimum figure.

Stage 1

Formulate the constraints and objective functions:

U-value constraint

$$U\text{-value} = \frac{U_A \cdot x \cdot 12 + U_B \cdot y \cdot 12}{x \cdot 12 + y \cdot 12} \leqslant 3.5$$

Transposing this equation and including the known *U*-values we get

$$\frac{12(4x + 2y)}{480} \leqslant 3.5$$

$$2x + y \qquad \leqslant 70 \qquad\qquad (1)$$

The glazing constraint is represented as

$$7x + 2y \qquad \geqslant 140 \qquad\qquad (2)$$

In addition the total number of panels must not exceed the cladding area and this can be expressed as:

$$x + y = 40 \qquad\qquad (3)$$

Since we cannot have a negative number of panels then it follows that

$$x \text{ and } y \geqslant 0 \qquad\qquad (4)$$

The 'objective function' is the expression which describes the line along which the maximum or minimum solution for which we are searching must lie. In our case we are trying to minimise cost and therefore the solution will lie along the line expressed by z.

$$500x + 400y = z \qquad\qquad (5)$$

Our objective is to find the minimum value of z within the constraints given. The formulation of these equations concludes the numerical formulation or mathematical model of the problem.

Stage 2

Plot the constraints on a graph.

As this problem only has two unknowns, i.e. the two different types of cladding panel, the constraints and objective function can be plotted on a two-dimensional graph. If there were three panels a three-dimensional graph would be necessary and beyond this a graphical solution would not be possible and other approaches would need to be taken. The object of the 'plot' is to identify the range of feasible solutions and then find the minimum (as in our case) or maximum according to the criterion chosen, e.g. least cost. The area of feasible solution before constraints are included is shown in Figure 21.13. As x and y are equal or greater than zero then any part of the shaded plane satisfies our constraint (4).

Thermal constraint

The first constraint to be plotted is the thermal constraint and the intercept with the two axes can be found by setting x and y to zero in the equation (1), thus:

when $x = 0$
$0 + y = 70$
$y = 70$

when $y = 0$
$2x + 0 = 70$
$x = 35$

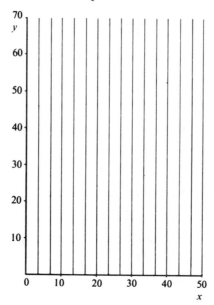

Fig. 21.13. Non-negative: $x,y>0$.

The fact that the constraint requires the total of x and y to be less than 70 means that all values below the line AB in Fig. 21.14 (i.e. the shaded area) satisfy this condition. All values above this line are not valid.

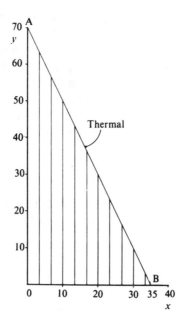

Fig. 21.14. Thermal constraint: $2x + y \leqslant 70$.

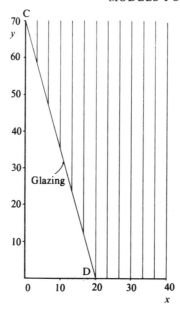

Fig. 21.15. Glazing constraint: $7x + 2y \geqslant 140$.

Glazing constraint

The equation for the glazing constraint was

$$7x + 2y \geqslant 140$$

Again we plot the equality condition to give the line CD shown in Figure 21.15.

when $x = \quad 0$
$0 + 2y = 140$
$\quad y = \quad 70$

when $y = \quad 0$
$7x + 0 = 140$
$\quad x = \quad 20$

In this case the 'greater than' condition means that all values above the line CD are feasible.

The combined effect of these two constraints is represented by the shaded area ACDB in Figure 21.16 which identifies the range of feasible solutions so far.

Number of panels

The final constraint specifies that the total number of panels must be equal to 40.

$$x + y = 40$$

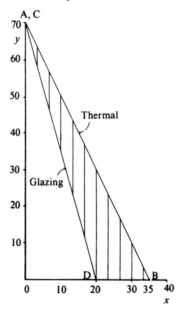

Fig. 21.16. Glazing and thermal constraints.

This condition is satisfied by the line FG in Figure 21.17. As this is an equality condition only those values of x and y which lie on FG satisfy the condition. However, only the part of this line crossing the area identified by lines ACDB will conform to all constraints. This is shown by the thicker line HJ.

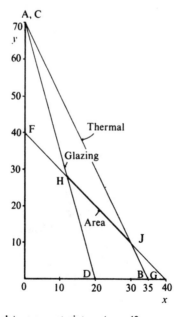

Fig. 21.17. Glazing + thermal + area constraints: $x + y = 40$.

Stage 3

Plot objective function.

The reader will recall that the objective function is represented by the line

$$500x + 400y = z$$

In order to find the minimum value of z it is necessary to plot the *slope* of the objective function and to move it parallel to itself away from the origin (intersection of x and y axis) until it meets the line HJ. The intersection with HJ will be the correct arrangement of x and y which will satisfy the constraints and provide minimum cost.

To plot the slope of the function

$$ax + by = z$$

it is necessary to take any point on the x axis and a point on the y axis so that

$$\frac{y}{x} = \frac{a}{b}$$

and join the two points.

The objective function is $500x + 400y = z$ so that points on the x and y axes must be chosen so that

$$\frac{y}{x} = \frac{500}{400}$$

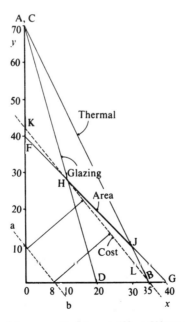

Fig. 21.18. Glazing + thermal + area constraints cost: $400x + 500y = z$

This is satisfied by taking for example $y = 10$ on the y axis and $x = 8$ on the x axis since

$$\frac{y}{x} = \frac{10}{8} = \frac{500}{400}$$

The line ab shown dotted in Fig. 21.18 is the plot of the slope of this line. We now move this line parallel to itself towards HJ until it touches it. (Note that if we were maximising cost we would be moving ab from above HJ towards the origin.)

Stage 4

Calculate results.

The point of intersection is with point H and this is therefore the optimum solutioni to the problem. Reading from the graph at point H we obtain the following values for x and y:

$$x = 12 \text{ representing 12 of panel A}$$
$$y = 28 \text{ representing 28 of panel B}$$

The shell U-value is

$$\frac{4 \times 12 \times 12 + 2 \times 28 \times 12}{12 \times 12 + 28 \times 12} = 2.6 \text{ w/m}^{20}\text{C}$$

which satisfies constraint (1).
    The glazing area is

$$7 \times 12 + 2 \times 28 = 140 \text{ m}^2$$

which satisfies constraint (2).
    The total number of panels is

$$12 + 28 = 40$$

which satisfies constraint (3).
    The total cost is

$$500 \times 12 + 400 \times 28 = £17,200$$

which is the minimum cost.

Summary

Although this example shows clearly the principle of the technique it will also be observed that the whole problem has been over-simplified in that many other design factors, e.g. aspect, noise, aesthetic values, would need to be taken into account. In addition not all problems could be specified so clearly in numerical and linear terms; other problems would arise if the result had not been an integer (whole number) as we would not have known instantly whether to round up or down. In more complex examples this can be a real problem and integer search techniques would need to be applied. Also many constraints are subject to thresholds, e.g. the number of lifts required for a given number of storeys or floor area. Providing the relationships still remain linear it is possible to overcome this problem although it complicates the process. Computer programs are available to undertake much of this work and this becomes almost essential when dealing with a large number of unknowns. Although manual iterations are possible (but beyond the scope of this book) it becomes a very time-consuming business. The stages of structuring the problem are however very similar, no matter what the method, and starting in a simple way will assist in developing the reader's knowledge for the complex problems.

Linear programming has been used for several different types of building optimisation problems including:

1 Hotel design (INVEST from ABACUS, University of Strathclyde).
2 Housing development mix. (See article by Terry Pitt in *Chartered Surveyor BQS Quarterly* Vol 4, No 3 Spring 1977, pp 40–42.)
3 Site management. (See R. Oxley and J. Poskitt, (1986) *Management Techniques Applied to the Construction Industry*, 4th edn. BSP Professional Books and J. F. Woodward, *Construction Management and Design*, Macmillan, 1975.)

The reader is recommended to follow up this simple introduction with some reading of introductory textbooks in Operations Research. If it is possible to gain access to one of the many computer programs available for the technique even greater speed of comprehension will result providing the underlying structure is understood.

Generally

The purpose of this chapter has been to introduce some concepts in optimisation which it is hoped will be of benefit to the reader. A number of warnings have already been given but a final parting shot may not go amiss. There is always the danger with any new process, and particularly those which are computerised, that the inexperienced user will become so engrossed with the technique and its methodology that he loses sight of the overall main objective and the limitations of the tool he is adopting. By all means encourage the use of numerical techniques where they will enlighten, improve and assist in the development of design. However, they are not an end in themselves and the danger is that used unwisely they will give too much weight to a part of the design problem, which though important may be less significant than is first supposed.

Chapter 22
Information Technology and Cost Planning

In Chapter 21 it was suggested that there were a number of ways in which cost information could be improved in order to provide better advice to construction clients. It was also suggested that computer technology could play a large part in making the improvement. The power of the machine to hold vast quantities of information and recall and manipulate that information must be of interest to all those charged with providing cost advice, the purpose of this chapter is to describe some key developments which exemplify the manner in which information technology is developing to support the subject area. Already in various chapters we have noted the need for computing power. Certainly the techniques described in Chapters 20 and 21 would be difficult to implement or would be uneconomical in practice unless computers were employed to undertake the calculations.

Information Technology has a wider remit than just the application of computer modelling. The SERC IT applications panel for the Environment Committee defined IT research as the application and development of technology which enables improved data capture, transmission, manipulation or interpretation of information in support of decision-makers or the manufacture of an improved product. If this definition is accepted then there are a large number of systems on the market which have arisen from IT research and which are now supporting commercial activity within the field of building economics and financial management. Almost without exception the supporting technology is computer based although strictly speaking IT encompasses a wider range of applications. For the purpose of this book we will look at four broad ranges of application namely:

1 Data retrieval systems
2 Computer aided design
3 Knowledge based systems
4 Integrated Project Management Systems

This classification is not meant to represent a robust mutually exclusive set of IT applications but merely a convenient way of discussing some useful application areas relevant to the subject of this book. However, before we look at these it is worth considering the difference between humans and computers. Table 15 summarises some of the major issues and is taken from (1983) *Microcomputers in Building Appraisal*, by P. S. Brandon and R. G. Moore, BSP Professional Books.

This summary gives a good indication of what machines should not do with the current state of the technology. They are good at recall and calculation but have difficulty with creative processes and handling information which is unstructured. Humans on the other hand tend to be slow at calculation and less reliable on recall. The object must be to harness the strengths of each in order to get the best system. It should also be an evolutionary process so that consultants and practitioners accept the

Table 15. Summary of comparison between humans and computer

Characteristics	Machine	Man
Speed of operation	Extremely fast and superior to man in most respects	Slower at physical and simple mental work
Stamina	Can sustain operation *ad infinitum* and limited only by reliability of mechanical and electrical devices	Unable to sustain long periods of work without rest
Accuracy and consistency	Extremely accurate, consistent and reliable. Poor at error correction	Unreliable and inconsistent particularly when dealing with repetitive work. Good at error correction
Senses	Consistent sensitivity. Finds difficulty in sensing from a variety of sources simultaneously. Wider range of sensitivity, dependent upon sensing device	Ability to combine sensory powers. Senses can be affected by environment, drugs etc.
Memory	Generally smaller (in micros, very much smaller) but very accurate storage and recall	Large memory but subject to memory loss and error in recall
Overload	Sudden breakdown	Gradual degradation
Cognitive ('knowing') processes	Follows instructions with complete accuracy, performing logical and/or arithmetical operations	May follow instructions precisely or in a haphazard fashion. May misunderstand them
	Logic largely a matter of determining whether a statement is true or false. Therefore mainly a process of comparison with given information	Processes may be interrupted by 'creative leaps' short circuiting a tedious procedure
	Man made criteria required for judging whether true or false	Perceiving, remembering, imagining, conceiving, judging, reasoning affected by feelings, emotions, motivation and so on.
Logic	Good at arithmetic operations but can only draw and compare simple analogies	Good at comparing and judging unlike things, at inductive logic and at drawing analogies/ metaphors directly relevant to the problem in hand
		Good at making complex decisions
Input	Machine input exceptionally fast but human input slow	Slow at character reading but fast at sensory recognition
Output	Very fast, neat and tidy. Output to backing store exceptionally fast. Lack of flexibility in manipulation	Slow output at all times either by speech or hand. Very flexible in manipulation.

technology as it develops, can trust the output and feel in control. For this reason the majority of computer systems have adopted manual approaches as a starting point in their development. Most manual processes provide a framework or reference point to which or from which the consultant adds his expertise. They provide structure and form but do not dictate a solution. Elemental cost planning for example uses a standard classification system and historic unit rates but the consultant can vary specifications, quantities, rates as he feels appropriate to the new situation. He can envisage the wider economic issues, the nature of the client, the features and location of the site etc. which it would be difficult to encompass within a machine. The computer has no knowledge of the world, other than the very limited amount given by the programmer or user, and has no means of self reference so it does not 'understand' why it is doing things. These issues will become very important as software develops and the machine appears to be replicating human decision making. It is always wise to keep in mind the limitations of the technology to avoid the creation of an oppressive tool.

At a more mundane level computers are not very economic where large amounts of detailed information needs to be input before a solution can be obtained. Measurement for bills or quantities is an example of such a problem. The process is similar to any estimating process as follows:

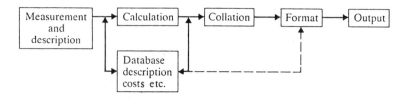

The problem is that of inputting the many thousands of dimensions and descriptions. Apart from measuring there is the problem of checking them and many would argue that a manual method such as 'cut and shuffle' is just as quick and cheaper. It is only when the same information is used and manipulated for other tasks that the real benefit of the machine can be seen. In the case of bills of quantities the same information can be used for cost analyses, valuations, final accounts, financial reports, cash flows etc. If repetition of input can be avoided at each stage then substantial savings can be made. Firms that have committed themselves to machine operation for the majority of their work usually get the largest return from their investment into computing power. This is partly because running manual and computing systems together inevitably means some overlap and therefore abortive work and partly because the computer literacy of employees obviously increases when there is not a manual alternative.

History of IT concepts related to cost planning

It has already been stated that the early use of computer power was to speed up manual processes. These manual processes were developed to compensate for human inadequacies such as slow limbs, poor memory and poor speed of calculation. The machine could undertake computation at enormous speed and this was seen as a major advantage. Unfortunately very little analysis of the manual processes was undertaken. If it had been investigated it would have been found that calculation and collationi represented a fairly small proportion of the total process. Much time is spent in discussion, in searching through literature and diagnosing drawings and other information. The benefits are limited therefore to greater efficiency in perhaps only 15% of the total time. It was only where the application was essentially all calculation that real return could be seen. In the UK the turning point was probably the NEDO Price Adjustment Formula in the early 1980s. Some firms and local authorities were able to claim a payback period of less than a year, on both hardware and software, just on this one operation.

Once firms had a machine in their possession their computer literacy improved and they began to look for other ways of harnessing the machines potential. Prior to 1980 there had been a continuous period of development of bill of quantity systems for over two decades. At one time over thirty five different systems were being used or developed. However it was not until the microcomputer arrived that a mass market was discovered. The reduction in hardware and software cost plus a shortage of labour to undertake routine work encouraged many firms to look again at the computer. The packages were still adopting manual concepts but as time went on the machine was used to integrate information and act as a data source.

By the mid-1980s a large number of firms had purchased machines and were using them for early cost forecasting and the BCIS introduced its 'On-line' service for access to its data bank. Accounts, estimating and other routine tasks were now undertaken commonly by computer. Computer aided draughting was also gaining in popularity and 3-D modelling of buildings also became more commonplace in larger practices.

Towards the end of the 1980s 'ELSIE' the first commercial expert system related to building economics and construction management became available. Developed by the RICS in collaboration with the University of Salford it pointed the way to the future potential of machines to provide consultancy advice in conjunction with humans. It is now widely used and is likely to be the forerunner of a large number of such systems. Ostensibly it follows traditional manual practices but at the same time is far more efficient, consistent and reliable.

In the 1990s we can expect a greater integration of software packages across disciplines. One requirement will be a common way of describing buildings in the machine so that all consultants can access a common data model for their requirements. It would appear at the moment that an 'object oriented' approach offers the best solution but only time will tell.

Research into cost planning techniques over the last two decades has largely centred on or relied upon computer techniques. In the beginning researchers helped transfer manual techniques to the machine. Subsequently they saw the potential for sensitivity analysis, repeating the calculations tirelessly, changing one or more variables at a time to see the effect. Techniques such as regression, which had its major interest in the 1970s, and simulation began to be developed. However few of them were adopted in practice on a large scale. A major problem with these techniques were that they were 'data hungry' and therefore required a high overhead in maintaining their reliability. More importantly they were 'black box' in nature i.e. information was put into the machine and information was output but the process in between was hidden from the user.

Consequently the user was being asked to accept that the computer model was perfect, (which it wasn't) and at the same time he was barred from bringing his own knowledge and expertise to bear on the problem. Very few practitioners were, or are, willing to place their faith in such models when their clients and their own livelihood depend on the result. Acceptability is a key factor in modelling which does not appear to have been given a high priority among researchers.

In the late 1980s emphasis swung towards the potential of knowledge based systems and in turn the potential for linking with conventional computer models to see whether 'intelligence' can be brought into these systems to relieve mundane work and improve consistency and performance.

The 1990s will no doubt see major developments in IT which will change the techniques employed, the nature of the professions and client expectations. It should be expected that greater standardisation and greater integration of systems will continue to erode the traditional boundaries between disciplines. The power of the machine may mean that one person or firm can control a wider sphere of activity beyond their normal discipline. This will signal a race between firms to control and manage the technology. It will be interesting to see which professions maintain their status and which firms survive.

In order to illustrate this wide subject four key applications impacting on building economics are described.

The BCIS on-line data base

As noted earlier in the book the RICS Building Cost Information Service has led the way in providing useful and pertinent cost information to practitioners in the UK. This service has been a mailing of hard copy information on economic indicators, indices, labour rates as well as the circulation of cost analyses supplied by subscribing members. It was a natural progression from paper format to supplying the information on a screen at the end of a telephone line.

Subscribers to the new 'On-Line' service can obtain most of the information described in chapter 10 from their own personal computer with modem. Since new data is being fed into the system all the time the 'On-Line' users have access to the latest information within the database.

At the time of writing the information includes:
detailed cost analyses
concise cost analyses
$£/M^2$ questionnaire returns
tender price indices
location factors
cost indices
NEDO formulae indices
daywork rates
construction output figures
construction new order figures
retail prices index
news (latest news on such matters as wage awards, plant hire costs)

briefing (text of current BCIS quarterly briefing)

average building prices

tender price studies

It is expected that even this list will be extended in the future. An overview of the system can be seen in Fig. 22.1.

In addition to fast access to data an operator can use the computing power of the machine to undertake manipulation of the data. Hence it is possible to call up one or

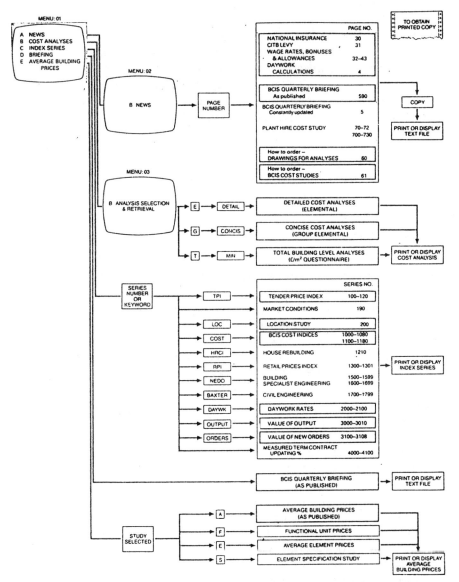

THE BUILDING COST INFORMATION SERVICE
THE ROYAL INSTITUTION OF CHARTERED SURVEYORS

Fig. 22.1. Overview of BCIS on-line system.

In addition to fast access to data an operator can use the computing power of the machine to undertake manipulation of the data. Hence it is possible to call up one or more analyses similar to a new project and go through a fairly conventional cost planning routine. The real advance however is being able to fine tune an estimate quickly and efficiently. Every subscriber to the 'On-line' system is provided with

a communication system to access the system

a data entry module for the user to include his own analyses

a report generator to display and update cost analyses

a package for providing an approximate estimate.

The latter feature is a suite of programs which can be used to manipulate cost analyses to form a budget estimate and cost plan. The approximate estimating package (AEP) is fully integrated with the rest of the on-line service and can draw on information from a user's own data as well as from the BCIS cost data banks.

In broad outline the initial procedure for operating the package is as follows:

Enter outline details of project

Select analyses to be manipulated

Define method of updating/adjusting analyses

Enter indices used for updating/adjusting analyses

Print indices used for updating/adjusting

Estimate either using analysis by analysis approach or element by element approach

Combine estimates, decide on total and calculate cost plan

Print Cost Plan

Manipulate cost plan

At the start of a new project the user must work through the logical sequence outlined above but subsequently the parts can be used in any order.

The system is now well established and other software houses are also developing packages which can access BCIS information and manipulate in different ways. For example ABS Oldacres are selling a product called BICEP developed by the University of Salford and Portsmouth City Council. This has additional features such as a wider choice of estimating method, statistical background to elemental costs (taken from the sample) and simple regression analyses for estimating. It also has more graphical output which can enable the user to assimilate the information more quickly. There is little doubt that databases of cost and their integrated programs will become a major feature of the cost consultants life in years to come.

Computer aided design

This topic is of course extremely wide ranging although it is often thought of in terms of drafting packages only. However it covers all those aspects of a building which can be modelled in the computer and which can assist the design team in their decision making process. It can, therefore, include energy evaluation, structural calculations and cost evaluation. The extent to which any particular system can support design varies considerably. Some systems found on small personal computers are very little more than two-dimensional drafting packages. Others found on larger workstations model the building in three dimensions with solid geometry and can undertake a number of different evaluations.

The three dimensional systems allow more information to be held with regard to the graphic information. For example walls will not be defined in plan as parallel lines, but rather as solids having height as well as width and depth. The user can then choose to look at an elevation, which merely requires the computer to work out the way the wall solids will appear from another angle. Perspectives can be produced in the same way. If further information is attached to the solid, such as a description or identification code, or a U-value or cost then other evaluations can take place. It is relatively easy to schedule like items, or to calculate the thermal efficiency of the external envelope, or cost the items that are listed (if of course the costs are proportional to quantity). In addition checks can be included in the system to ensure that such things as lines join up properly, the details conform to good practice, that building regulations are complied with and fittings are not forgotten. This extra discipline placed on the designer will in turn encourage the design to be complete. As each entry into the machine is retained in the memory, then short of a power failure, the item will be included in any schedule or evaluation without error. If it is on the 'drawings' it will be on the 'schedule'. Obviously the more the standard method of measurement is simplified and moves towards a 'counting' rather than analogue measurement process then the easier it will be to move to comprehensive automatic measurement.

If cost planning (i.e. forecasting and control) is seen as an aid to design then CAD packages are of considerable interest. In essence CAD systems are large databases of information with evaluative and manipulative routines attached. The database contains information on the co-ordinates of the building geometry, specification of items, location of components, spacial data, performance measures (e.g. U-values) and so forth. If these are available then it follows that quantitative information can be abstracted, costs can be attached to specification items and similar items can be collated and scheduled. Indeed there are several systems which allow this kind of useful information to be produced and therefore replicate the mechanical processes of measurement, collation, computation and pricing adopted when preparing a cost forecast. It is easier to derive the data required for early cost forecasts than for later estimates such as a bill of quantities. The detail of the design model at outline proposals and scheme design stages is still fairly coarse and the complexity of interaction between components is ignored. It is also easier for the machine to 'recognise' boundaries to spaces and walls at this early stage. Nevertheless there are some systems which claim that they can schedule up to 80% of the items for a full bill of quantities – although not necessarily complying with one of the standard method of measurements. Where the building is formed from a standard kit of parts, arranged in different ways then it is possible to produce a full bill of quantities with standard price rates attached. One example of this is the link between the McDonnell Douglas 'General Drafting System' (GDS) and ABS Oldacres BQ-Micro package applied to Mobil Oil petrol filling stations. Once design of a standard station is complete the machine will automatically schedule the component items with their quantities and generate a complete tendering document. However the number of components is severely limited, the specification standardised and the design conforms to a 'house style'. It is a much more difficult problem to take a non-standard high-tech laboratory, hospital or office block and generate a full BQ.

In general the power of CAD from the cost planners point of view is:

    1 All members of the design team are working off the same model and there is

therefore consistency across all disciplines and everyone should know the current state of the design.

2 Computation of quantities is largely automatic.

3 Specifications can be attached to quantities and all similar items collated and scheduled.

4 Costs can be attached to specifications/quantities to produce broad cost estimates.

5 The cost estimate can be altered automatically with changes in design.

This sounds very attractive but there are problems e.g.:

1 Even with three dimensional models it is not always easy to differentiate and classify items for scheduling. The technical problems will be overcome in time (thus posing a threat to all measurers) but it will require greater investment and possibly less attachment to standard methods of measurement.

2 In cost estimating there is an element of judgement and this tempers the mechanical application of standard unit rates. This moderation process is derived from the expertise of the consultant and should not be underestimated. It would require much more intelligence in the system to cope with this.

3 Scheduling systems do not easily cope with incomplete design. The computer can be used to check whether the design is complete and consistent but it is unusual even with CAD for designs to be complete at tender stage. This could result in omissions or failure of the scheduling system. Humans usually account for these problems by calling on their resource of expertise.

4 The time taken to input information can be extensive and the maintenance of the database of standard information, e.g. on costs, is not a trivial exercise. It is only when that data is used a number of times for different tasks or where it is likely that there is a high degree of modification will the overhead cost seem worthwhile.

However the advantages can outweigh the problems and particularly at the early stages of design. If several outline solutions can be generated by the designer and quite comparable cost estimates produced for each then the quality of decision making should be improved. Unfortunately most CAD systems were designed as draughting tools at working drawing stage. It was recognised that most of the architect's time was being spent on this time consuming task whereas the key decisions at the commencement of the design process had comparatively fewer hours allocated. Perhaps the ideal CAD tool would be one in which a simple model was created at the start, which could be manipulated and tested before moving to the next stage of refinement. As the design is refined through the three dimensional model the machine would be evaluating the building on-line. Ideally the concept of the 'design cockpit' would be introduced which would provide 'windows' to view the various 'instant' evaluations as design developed. This concept is not remote and various aspects of such a system have already been developed but not within one package. As standardisation of communication takes place and components are available on standard databases so it will become possible to integrate the current developments.

One requirement for increasing the speed of evolution of these packages will be that the industry accepts a standard description of the components of a building, probably in terms of 'objects' which have certain attributes. This will enable all disciplines to build the models they require based on a known set of parameters. There is much interest in this concept at the present time and it is likely that the research required to develop the idea will be undertaken in the 1990s. Inevitably it will take time to design and develop the standard objects and associated databases but this can be done in phases.

The technology is moving very fast but useful informaltion for cost planners can be obtained from 'CAD' Techniques – Opportunities for Chartered Quantity Surveyors' by Brian Atkin and published by The Royal Institution of Chartered Surveyors, 12 Great George Street, London SW1P 3AD. In addition the publication *Multidisciplinary CAD in the Building Industry* by Alex Faulkner and Alan Day and published by the CICA, Guildhall Place, Cambridge CB2 3QQ also contains some interesting commentary.

Intelligent Knowledge Based Systems (IKBS)

Whereas CAD has been available to the construction industry for the past two decades or more, knowledge-based systems or expert systems have only recently been made available commercially.

The British Computer Society specialist interest group in expert systems has offered the following definition:

'An expert system is regarded as the embodiment within a computer of a knowledge based component from an expert skill in such a form that the system can offer intelligent advice or take an intelligent decision about a processing function. A desirable characteristic, which many would consider fundamental, is the capability of the system, on demand, to justify its own line of reasoning in a manner directly intelligible to the enquirer'.

There are of course grave dangers in applying the term 'intelligent' to such systems at the present time because they only have certain of the characteristics of simple intelligence and do not begin to compare with that possessed by humans.

The key features of expert systems are the 'knowledge base' and what is usually called the 'inference engine'. The knowledge base is a collection of facts, beliefs and heuristic rules (i.e. rules of thumb) expressed in some form of logical, relationship in order to make sense of the individual items. The inference engine undertakes the task of reaching conclusions through the logic. If the knowledge base is sufficiently comprehensive then the system will mimic at least part of what is perceived to be the decision making process of a human being. Other parts which are now being seen as just as important are an explanation facility and a capacity to refine the knowledge quickly. In time natural language interfacing with the system, so that the user is not so restricted in the way he addresses the system will also be considered essential. The technology is at a very early stage. Even though it was developed in the 1960s it was not until the middle 1980s that the supporting tools became commercially available to develop the systems relatively quickly.

One of the first commercial expert systems available for the construction industry, worldwide, was developed by The Royal Institution of Chartered Surveyors (Quantity Surveying Division) in collaboration with the University of Salford, Department of Surveying. The system called ELSIE (L–C for lead consultant) provides consultancy advice at the strategic planning stage of the development process i.e. before design takes

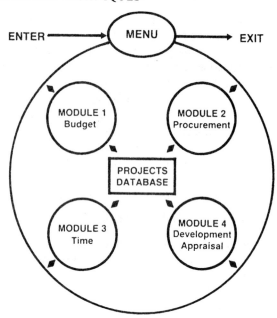

Fig. 22.2. ELSIE: Conceptual overview.

place. It takes the kind of information available in a clients brief and translates this information into a series of reports. These reports cover the following and are structured in the form of modules within the system:

Initial Budget:	How much will the development cost?
Time:	How long will the development take?
Procurement:	What is the contractual relationships between the parties to the development process.
Development Appraisal:	What is the profitability of the scheme.

The four modules are linked together see Fig. 22.2 and common information is transferred from one to the other through a database. At the end of a consultation between 20 and 30 pages of printed report are given which can be handed direct to the client if the operator so wishes. At the present time two broad categories of building type are covered by the consultation but others are being developed.

Because the system uses the knowledge of experts it operates in a similar way to an expert including the models he adopts. It has been noted before, that cost models, for example, are merely reference points or skeletons to which the consultant adds his own expertise. The expert system, therefore, uses the same approach. It adds to the cost model, in the case of ELSIE a form of coarse approximate quantities, the extra dimension of the consultants knowledge. At this early stage the consultant has to call on his experience of previous projects and put into the model what he considers to be appropriate and reasonable assumptions. These assumptions he derives by diagnosing the needs of the client from the brief or other method of communication. In reality much of his 'understanding' of the clients brief would come from conversations, visual comprehension of a location and even body language. The unstructured information that comes through the consultants senses 'triggers' thought patterns which allow him to arrive at a solution. At the present time, machines do not have the capability to sense and diagnose in the same way as humans. Consequently there will always be a need for consultants to check and possibly modify the solution suggested by a machine until the technology improves very substantially.

In the ELSIE initial budget system a solution is postulated by the machine from over 2000 'rules' and the user can then modify the answer by changing up to 150 variables which have been derived. To arrive at the first solution the machine asks between 24 and 30 questions depending on the answers given. It uses rules, logic and inference to generate variables of size, shape and specification and will instantly give a response (in terms of cost) of a change in any one of them.

The relationships between variables are often shown in a paper model called an inference net first so they can be checked by the knowledge provider before they are placed in the machine. An example of an inference net for external walls is shown in Fig. 22.3. Note that the final result to the right of the page is a cost based on quantity multiplied by rate. All the other variables are there to determine what these two values should be in order to derive the cost. In the system many of these variables are derived from each other and not requested from the operator. This explains why 150 variables can be determined from only asking around 24–30 questions. The ELSIE initial budget module had around 30 of these inference nets to plan and record the knowledge in the system.

Because the knowledge and relationships reside in the machine any changes to other variables are automatically altered should this be required and the machine is always consistent. In use on multi-million pound projects the results have proved to be within plus or minus 5% of the expected tender figure when operated by a knowledgeable consultant. This is rather better than could be expected if several estimators were asked to undertake the same task with the same information. In addition on more than one occasion the machine has discovered flaws in estimates prepared by skilled cost planners.

The advantages for cost modelling of this method are of major importance. Most of the research work established in recent years has resulted in black box techniques. Regression and simulation use statistical relationships, with comparatively few variables to derive a solution. Information is input into the model (the black box) and information comes out at the other and without any intervention or enhancement by the consultant. Nor is the consultant able to view or check, without considerable effort, what is happening. The assumption must be, therefore, that the model is perfect – which we know it isn't. Consequently acceptability of these new models has been low. The transparency and explanation facility of an expert system overcomes these problems to some extent and acceptability is higher, this explains the fast take up of the ELSIE system which now runs into hundreds of systems sold.

It is not possible to explain the nature of expert systems in a publication of this nature but if further information is required there is a hardback report published by the RICS entitled *Expert Systems, The Strategic Planning of Construction Projects* by P. S. Brandon, A. Basden, I. Hamilton, J. Stockley and is available from Imaginor Systems, 41 Leslie Hough Way, Salford University Business Park, Salford M6 6AJ. This volume provides a suitable introduction to expert systems and the ELSIE system in particular. The software is also available from this address.

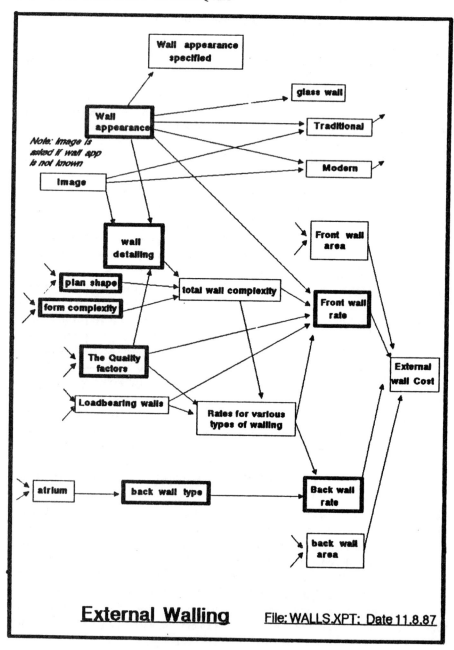

Fig. 22.3. ELSIE: Inference net diagram – external walls.

Integrated construction management systems

The developments in Information Technology have made it more difficult for a subject such as building economics, to see itself as a self-contained isolated discipline. It has always been dependant on the actions and views of others and its own knowledge and information is, in turn, used by other consultants and clients for their work. In the context of a building project the figures of the investment analyst or the design from the Architect are used as parameters and reference points for the cost planner. The estimates produced from the information provided are then used for assessing profitability levels or modifying the design. It is a quirk of history, and the limitations of human ability that have resulted in the diversification of professions into specialisms each with perceived boundaries to the part of the problem which they address.

However, much of the information that each profession uses is common. We have seen already that CAD could provide a common data base which all could access to avoid misunderstandings, incorrect measurement and bad communication. In cost planning it now seems rather outdated to consider merely the traditional view of the subject, in terms of estimating and control during design, when this is part of the total financial management of the project. This management covers the whole of the financial processes from the decision to invest to occupation and the choice of demolition or refurbishment. To isolate any particular part of the information flow is inefficient and possibly counter productive. If every piece of information gathered in the process can be collected and made easily available in a data base for use whenever it is required then a major advance has taken place. Instead of measuring the same item, say floor area, several times (by investor, architect, surveyor, contractors, sub-contractor, etc.) the information can be provided instantly for all to use.

The refinement process of information can also be enhanced by using coarse measurements and breaking them down into constituent parts as more knowledge becomes available. It may be possible to do some of this work automatically using knowledge based systems. For example, it is possible now to take an early estimate, refine it as working drawings became available, prepare tender documents and then allocate items to a network of activities to generate a cash flow using standard rates. The RIPAC system developed by Rider Hunts in Adelaide and being sold in the UK by CSSP of Market House, 12–13 Market Square, Bromley, Kent BC1 1NA is an example of a very extensive construction management system which uses the power of a relational data base to undertake this work. At the moment much manual input and classification of items is required depending on what output is needed. In time it is possible to foresee some of these processes being automated. After all, the human uses rules to undertake the task and it should be possible to capture at least some of these in the machine.

One of the biggest selling systems in the UK is the CATO (Computer Aided Taking Off) system sold by Elstree Computing. At the time of writing this system does not use a relational database but nevertheless covers a very comprehensive range of financial management activities related to a project. Fig. 22.4 shows the packages which are integrated together to allow estimating, cost monitoring, valuations and final accounts to be brought together.

As the computer technology develops so it will be possible to integrate and automate more of these processes. This will create its own problems. The larger the system the

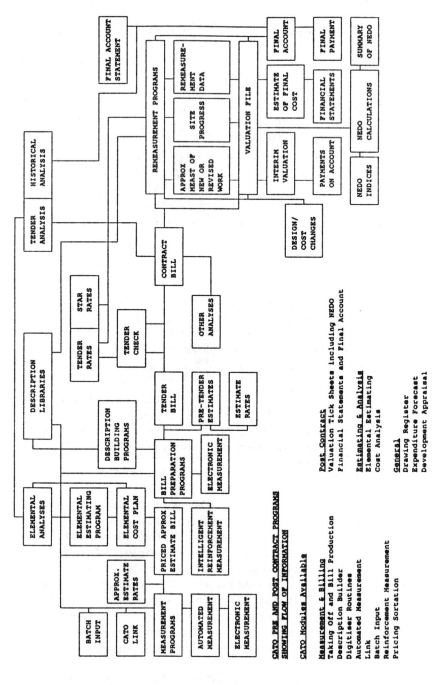

**CATO PRE AND POST CONTRACT PROGRAMS
SHOWING FLOW OF INFORMATION**

CATO Modules Available

Measurement & Billing
Taking Off and Bill Production
Description Builder
Digitiser Routines
Automated Measurement
Link
Batch Input
Reinforcement Measurement
Pricing Sortation

Post Contract
Valuation Tick Sheets including NEDO
Financial Statements and Final Account

Estimating & Analysis
Elemental Estimating
Cost Analysis

General
Drawing Register
Expenditure Forecast
Development Appraisal

Fig. 22.4. CATO: Pre and post contract programs showing flow of information.

more complex will be the relationships. This cannot be approached in the 'ad-hoc' manner which has often been the pattern of the past. There is a need to design the data base with its inter-relationships in a similar way to a manufacturing process or a building. A 'blue print' of the intended systems will provide a framework which would allow the development of databases across the industry in such a way that their data would be compatible. It is important to distinguish the 'design' from the 'physical' database. In a pilot study undertaken at the University of Salford for the RICS Quantity Surveyors Division, a technique called Entity-Relationship modelling was used to develop a high level model of some of the important relationships between data. Once the model was developed it was then possible to implement it in a physical database using packages such as DBase IV or Oracle without problems of communication. The technique provides a robust method of describing the relationships without ambiguity. This enables any firm to develop a system for its own requirements efficiently and effectively knowing that it will be possible to link with external systems in the future. The report entitled *Integrated Databases for Quantity Surveying* by P. Brandon and J. Kirkham is published by the RICS Quantity Surveyors Division, 12 Great George Street, London SW1 3AD.

Conclusion

Since financial management is largely the collection, manipulation and analysis of data, it follows that information technology has much to offer the discipline. Whether it be CAD, knowledge based systems or integrated data bases, each provides a major opportunity for advancement. As time goes on so these systems will be linked to other external systems and the information revolution will be well on its way. In a book of this nature it is impossible to do justice to the subject and in some areas the technology is still in its infancy – we do not know how far it can develop. Scientists are working on neural networks and molecular computers. It is sometimes difficult to believe that just over one working lifetime has passed since the first electronic computer. The development of the next fifty years will revolutionise what we do now in all aspects of our life and the subject of this book will be no exception.

Appendices

Appendix A
List of Elements from Standard Form
of Cost Analysis

This appendix contains the element definitions from the Standard Form of Cost Analysis, but not the full requirements for unit quantity costs, etc., for which the Form itself should be consulted. (Note: The amplified form (i.e. to the sub-element level) is no longer published by the BCIS but is provided here because some firms still use it within their offices.)

1 Sub-structure

All work below underside of screed or where no screed exists to underside of lowest floor finish including damp-proof membrane, together with relevant excavations and foundations.

Notes

 1 Where lowest floor construction does not otherwise provide a platform, the flooring surface shall be included with this element (e.g. if joisted floor, floor boarding would be included here).

 2 Stanchions and columns (with relevant casings) shall be included with 'Frame' (2.A).

 3 Cost of piling and driving shall be shown separately stating system, number and average length of pile.

 4 The cost of external enclosing walls to basements shall be included with 'External walls' (2.E) and stated separately for each form of construction.

2.A Frame

Loadbearing framework of concrete, steel, or timber. Main floor and roof beams, ties and roof trusses of framed buildings. Casing to stanchions and beams for structural or protective purposes.

Notes

 1 Structual walls which form an integral part of the loadbearing framework shall be included either with 'External walls' (2.E) or 'Internal walls and partitions' (2.G) as appropriate.

 2 Beams which form an integral part of a floor or roof which cannot be segregated therefrom shall be included in the appropriate element.

 3 In unframed buildings roof and floor beams shall be included with 'Upper floors' (2.B) or 'Roof structure' (2.C.1) as appropriate.

4 If the 'Stair structure' (2.D.1) has had to be included in this element it should be noted separately.

2.B Upper floors

Upper floors, continuous access floors, balconies and structural screeds (access and private balconies each stated separately), suspended floors over or in basements stated separately.

Notes

1 Where floor construction does not otherwise provide a platform the flooring surface shall be included with this element (e.g. if joisted floor, floor boarding would be included here).
2 Beams which form an integral part of a floor slab shall be included with this element.
3 If the 'Stair structure' (2.D.1) has had to be included in this element it should be noted separately.

2.C Roof

2.C.1 Roof structure

Construction, including eaves and verges, plates and ceiling joists, gable ends, internal walls and chimneys above plate level, parapet walls, and balustrades.

Notes

1 Beams which form an integral part of a roof shall be included with this element.
2 Roof housings (e.g. lift motor and plant rooms) shall be broken down into the appropriate constituent elements.

2.C.2 Roof coverings

Roof screeds and finishings. Battening, felt, slating, tiling, and the like.
Flashings and trims.
Insulation.
Eaves and verge treatment.

2.C.3 Roof drainage

Gutters where not integral with roof structure, rainwater heads, and roof outlets. (Rainwater downpipes to be included in 'Internal drainage' (5.C.1).)

2.C.4 Rooflights

Rooflights, opening gear, frame, kerbs, and glazing.
Pavement lights.

2.D Stairs

2.D.1 Stair structure

Construction of ramps, stairs, and landings other than at floor levels.
Ladders.
Escape staircases.

Notes
>   1 The cost of external escape staircases shall be shown separately.
>   2 If the staircase structure has had to be included in the elements 'Frame' (2.A) or 'Upper floors' (2.B) this should be stated.

2.D.2 Stair finishes

Finishes to treads, risers, landings (other than at floor levels), ramp surfaces, strings and soffits.

2.D.3 Stair balustrades and handrails

Balustrades and handrails to stairs, landings, and stairwells.

2.E External walls

External enclosing walls including that to basements but excluding items included with 'Roof structure' (2.C.1).
Chimneys forming part of external walls up to plate level.
Curtain walling, sheeting rails, and cladding.
Vertical tanking.
Insulation.
Applied external finishes.

Notes
>   1 The cost of structural walls which form an integral and important part of the loadbearing framework shall be shown separately.
>   2 Basement walls shall be shown separately and the quantity and cost given for each form of construction.
>   3 If walls are self-finished on internal face, this shall be stated.

2.F Windows and external doors

2.F.1 Windows

Sashes, frames, linings, and trims.
Ironmongery and glazing.
Shop fronts.
Lintels, sills, cavity damp-proof courses and work to reveals of openings.

2.F.2 External doors

Doors, fanlights, and sidelights.
Frames, linings, and trims.
Ironmongery and glazing.
Lintels, thresholds, cavity damp-proof courses and work to reveals of openings.

2.G Internal walls and partitions

Internal walls, partitions and insulation.
Chimneys forming part of internal walls up to plate level.
Screens, borrowed lights and glazing.
Movable space-dividing partitions.
Internal balustrades exluding items included with 'Stair balustrades and handrails' (2.D.3)

Notes

1 The cost of structural walls which form an integral and important part of the loadbearing framework shall be shown separately.
2 The cost of proprietary partitioning shall be shown separately stating if self-finished. Doors, etc., provided therein together with ironmongery, should be included stating the number of units installed.
3 The cost of proprietary WC cubicles shall be shown separately stating the number provided.
4 If design is cross-wall construction, the specification shall be stated and the cost shown separately.

2.H Internal doors

Doors, fanlights, and sidelights.
Sliding and folding doors.
Hatches.
Frames, linings, and trims.
Ironmongery and glazing.
Lintels, thresholds and work to reveals of openings.

3.A Wall finishes

Preparatory work and finishes to surfaces of walls internally.
Picture, dado and similar rails.

Notes

1 Surfaces which are self-finished (e.g. self-finished partitions, fair faced work) shall be included in the appropriate element.
2 Insulation which is a wall finishing shall be included here.
3 The cost of finishes applied to the inside face of external walls shall be shown separately.

3.B Floor finishes

Preparatory work, screeds, skirtings, and finishes to floor surfaces excluding items included with 'Stair finishes' (2.D.2), and structural screeds included with 'Upper floors' (2.B).

Note
Where the floor construction does not otherwise provide a platform the flooring surface will be included either in 'Sub-structure' (1) or 'Upper floors' (2.B) as appropriate.

3.C Ceiling finishes

3.C.1 Finishes to ceilings

Preparatory work and finishes to surfaces of soffits excluding items included with 'Stair finishes' (2.D.2) but including sides and soffits of beams not forming part of a wall surface.
Cornices, coves.

3.C.2 Suspended ceilings

Construction and finishes of suspended ceilings.

Notes
　　1　Where ceilings principally provide a source of heat, artificial lighting, or ventilation, they shall be included with the appropriate 'Services' element and the cost shall be stated separately.
　　2　The cost of finishes or suspended ceilings to soffits immediately below roofs shall be shown separately.

4.A Fittings and furnishings

4.A.1 Fittings, fixtures, and furniture

Fixed and loose fittings and furniture including shelving, cupboards, wardrobes, benches, seating, counters, and the like.
Blinds, blind boxes, curtain tracks, and pelmets.
Blackboards, pin-up boards, notice boards, signs, lettering, mirrors, and the like.
Ironmongery.

4.A.2 Soft furnishings

Curtains, loose carpets, or similar soft furnishing materials.

4.A.3 Works of art

Works of art if not included in a finishes element or elsewhere.

Note

Where items in this element have a significant effect on other elements a note should be included in the appropriate element.

4.A.4 Equipment

Non-mechanical and non-electrical equipment related to the function or need of the building (e.g. gymnasia equipment).

5.A Sanitary appliances

Baths, basins, sinks, etc.
WC's, slop sinks, urinals, and the like.
Toilet-roll holders, towel rails, etc.
Traps, waste fittings, overflows, and taps as appropriate.

5.B Services equipment

Kitchen, laundry, hospital and dental equipment, and other specialist mechanical and electrical equipment related to the function of the building.

Note

Local incinerators shall be included with 'Refuse disposal'.

5.C Disposal installations

5.C.1 Internal drainage

Waste pipes to 'Sanitary appliances' (5.A) and 'Services equipment' (5.B).
Soil, anti-syphonage, and ventilation pipes. Rainwater downpipes.
Floor channels and gratings and drains in ground within buildings up to external face of external walls.

Note

Rainwater gutters are included in 'Roof drainage' (2.C.3).

5.C.2 Refuse disposal

Refuse ducts, waste disposal (grinding) units, chutes, and bins.
Local incinerators and flues thereto.
Paper shredders and incinerators.

5.D Water installations

5.D.1 Mains supply

Incoming water main from external face of external wall at point of entry into building including valves, water meters, rising main to (but excluding) storage tanks, and main taps. Insulation.

5.D.2 Cold water service

Storage tanks, pumps, pressure boosters, distribution pipework to sanitary appliances and to services equipment.
Valves and taps not included with 'Sanitary appliances' (5.A) and/or 'Services equipment' (5.B). Insulation.

Note
Header tanks, cold water supplies, etc., for heating systems should be included in 'Heat source' (5.E).

5.D.3 Hot water service

Hot water and/or mixed water services.
Storage cylinders, pumps, calorifiers, instantaneous water heaters, distribution pipework to sanitary appliances and services equipment. Valves and taps not included with 'Sanitary appliances' (5.A) and/or 'Services equipment' (5.B). Insulation.

5.D.4 Steam and condensate

Steam distribution and condensate return pipework to and from services equipment within the building including all valves, fittings, etc. Insulation.

Note
Steam and condensate pipework installed in connection with space heating or the like shall be included as appropriate with 'Heat source' (5.E) or 'Space heating and air treatment' (5.F).

5.E Heat source

Boilers, mounting, firing equipment, pressurising equipment instrumentation and control, ID and FD fans, gantries, flues and chimneys, fuel conveyors, and calorifiers. Cold and treated water supplies and tanks, fuel oil and/or gas supplies, storage tanks, etc , pipework (water or steam mains) pumps, valves, and other equipment. Insulation.

Notes
1 Chimneys and flues which are an integral part of the structure shall be included with the appropriate structural element.

2 Local heat source shall be included with 'Local heating' (5.F.4).

3 Where more than one heat source is provided each shall be analysed separately.

5.F Space heating and air treatment

1 Heating only by

5.F.1 Water and/or steam

Heat emission units (radiators, pipe coils, etc.) valves and fittings, instrumentation and control, and distribution pipework from 'Heat source' (5.E).

5.F.2 Ducted warm air

Ductwork, grilles, fans, filters, etc., instrumentation and control.

5.F.3 Electricity

Cable heating systems, off-peak heating systems, including storage radiators.

Note
Electrically-operated heat emission units other than storage radiators should be included under 'Local heating' (5.F.4).

5.F.4 Local heating

Fireplaces (except flues), radiant heaters, small electrical or gas appliances, etc.

5.F.5 Other heating systems

2 Air treatment

Notes
  1 System described as having:
    'Air treated locally' shall be deemed to include all systems where air treatment (heating or cooling) is performed either in or adjacent to the space to be treated.
    'Air treated centrally' shall be deemed to include all systems where air treatment (heating or cooling) is performed at a central point and ducted to the space to being treated.
  2 The combination of treatments used shall be stated, i.e.:
    Heating   Dehumidification or drying
    Cooling   Filtration
    Humidification   Pressurisation
    and whether inlet extract or recirculation.
  3 High velocity system shall be identified as:
    Fan coil   Induction units 2 pipe

Dual duct	Induction units 3 pipe
Reheat	Induction units 4 pipe
Multi-zone	Any other system (state which).

5.F.6 Heating with ventilation (air treated locally)

Distribution pipework ducting, grilles, heat emission units including heating calorifiers except those which are part of 'Heat source' (5.E) instrumentation and control.

5.F.7 Heating with ventilation (air treated centrally)

All work as detailed under (5.F.6) for system where air treated centrally.

5.F.8 Heating with cooling (air treated locally)

All work as detailed under (5.F.6) including chilled water systems and/or cold or treated water feeds. The whole of the costs of the cooling plant and distribution pipework to local cooling units shall be shown separately.

5.F.9 Heating with cooling (air treated centrally)

All work detailed under (5.F.8) for system where air treated centrally.

Note
Where more than one system is used, design criteria, specification notes and costs should be given for each.

5.G Ventilating systems

Mechanical ventilating system not incorporating heating or cooling installations including dust and fume extraction and fresh air injection, unit extract fans, rotating ventilators and instrumentation and controls.

5.H Electrical installations

5.H.1 Electric source and mains

All work from external face of building up to and including local distribution boards including main switchgear, main and sub-main cables, control gear, power factor correction equipment, stand-by equipment, earthing, etc.

Notes
1 Installations for electric heating ('built-in' systems) shall be included with 'Space heating and air treatment' (5.F.3).
2 The cost of stand-by equipment shall be stated separately.

5.H.2 Electric power supplies

All wiring, cables, conduits, switches from local distribution boards, etc, to and including outlet points for the following:

General purpose socket outlets.
Services equipment.
Disposal installations.
Water installations.
Heat source.
Space heating and air treatment.
Gas installation.
Lift and conveyor installations.
Protective installations.
Communication installations.
Special installations.

Note
The cost of the power supply to these installations should, where possible, be shown separately.

5.H.3 Electric lighting

All wiring, cables, conduits, switches, etc., from local distribution boards and fittings to and including outlet points.

5.H.4 Electric lighting fittings

Lighting fittings including fixing.
Where lighting fittings supplied direct by client, this should be stated.

5.I Gas installation

Town and natural gas services from meter or from point of entry where there is no individual meter: distribution pipework to appliances and equipment.

5.J Lift and conveyor installations

5.J.1 Lifts and hoists

The complete installation including gantries, trolleys, blocks, hooks and ropes, downshop leads, pendant controls, and electrical work from and including isolator.

Notes
1 The cost of special structural work, e.g. lift walls, lift motor rooms, etc., shall be included in the appropriate structural elements.
2 Remaining electrical work shall be included with 'Electric power supplies' (5.H.2).

3 The cost of each type of lift or hoist shall be stated separately.

5.J.2 Escalators

As detailed under 5.J.1.

5.J.3 Conveyors

As detailed under 5.J.1.

5.K Protective installations

5.K.1 Sprinkler installations

The complete sprinkler installation and CO_2 extinguishing system including tanks, control mechanism, etc.

Note
Electrical work shall be included with 'Electrical power supplies' (5.H.2).

5.K.2 Fire-fighting installations

Hosereels, hand extinguishers, asbestos blankets, water and sand buckets, foam inlets, dry risers (and wet risers where only serving fire-fighting equipment).

5.K.3 Lightning protection

The complete lightning protection installation from finials conductor tapes, to and including earthing.

Note
The cost of lightning protection to boiler and vent stacks shall be stated separately.

5.L Communication installations

The following installations shall be included:
Warning installations (fire and theft)
    Burglar and security alarms.
    Fire alarms.
Visual and audio installations
    Door signals.
    Timed signals.
    Call signals.
    Clocks.
    Telephones.
    Public address.

Radio.
Television.
Pneumatic message systems.

Notes

    1 The cost of each installation shall be stated separately if possible along with an indication of the specification.

    2 The cost of the work in connection with electrical supply shall be included with 'Electric power supplies' (5.H.2).

5.M Special installations

All other mechanical and/or electrical installations (separately identifiable) which have not been included elsewhere, e.g. Chemical gases; Medical gases; Vacuum cleaning; Window cleaning equipment and cradles; Compressed air; Treated water; Refrigerated stores.

Notes

    1 The cost of each installation shall, where possible, be shown separately along with an indication of the specification.

    2 Items deemed to be included under 'Refrigerated stores' comprise all plant required to provide refrigerated conditions (i.e. cooling towers, compressors, instrumentation and controls, cold room thermal insulation and vapour sealing, cold room doors, etc.) for cold rooms, refrigerated stores and the like other than that required for 'Space heating and air treatment' (5.F.8 and 5.F.9).

5.N Builder's work in connection with services

Builder's work in connection with mechanical and electrical services.

Notes

    1 The cost of builder's work in connection with each of the services elements (5.A to 5.M) shall, where possible, be shown separately.

    2 Where tank rooms, housings, and the like are included in the gross floor area, their component parts shall be analysed in detail under the appropriate elements. Where this is not the case the cost of such items shall be included here.

5.O Builder's profit and attendance on services

Builder's profit and attendance in connection with mechanical and electrical services.

Note

The cost of profit and attendance in connection with each of the services elements (5.A to 5.M) shall, where possible, be shown separately.

6.A Site works

6.A.1 Site preparation

Clearance and demolitions.
Preparatory earth works to form new contours.

6.A.2 Surface treatment

The cost of the following items shall be stated separately if possible:
Roads and associated footways.
Vehicle parks.
Paths and paved areas.
Playing fields.
Playgrounds.
Games courts.
Retaining walls.
Land drainage.
Landscape work.

6.A.3 Site enclosure and division

Gates and entrance.
Fencing, walling, and hedges.

6.A.4 Fittings and furniture

Notice boards, flag poles, seats, signs.

6.B Drainage

Surface water drainage.
Foul drainage.
Sewage treatment.

Note
To include all drainage works (other than land drainage included with 'Surface treatment' (6.A.2)) outside the building to and including disposal point, connection to sewer or to treatment plant.

6.C External services

6.C.1 Water mains

Main from existing supply up to external face of building.

6.C.2 Fire mains

Main from existing supply up to external face of building; fire hydrants.

6.C.3 Heating mains

Main from existing supply or heat source up to external face of building.

6.C.4 Gas mains

Main from existing supply up to external face of building.

6.C.5 Electric mains

Main from existing supply up to external face of building.

6.C.6 Site lighting

Distribution, fittings, and equipment.

6.C.7 Other mains and services

Mains relating to other service installations (each shown separately).

6.C.8 Builder's work in connection with external services

Builder's work in connection with external mechanical and electrical services: e.g. pits, trenches, ducts, etc.

Note
The cost of builder's work shall be stated separately for each of the sub-sections (6.C.1) to (6.C.7).

6.C.9 Builder's profit and attendance on external mechanical and electrical services

Note
The cost of profit and attendances shall be stated separately for each of the sub-sections (6.C.1) to (6.C.7).

6.D Minor building work

6.D.1 Ancillary buildings

Separate minor buildings such as sub-stations, bicycle stores, horticultural buildings, and the like, inclusive of local engineering services.

6.D.2 Alterations to existing buildings

Alterations and minor additions, shoring, repair, and maintenance to existing buildings.

Preliminaries

Priced items in Preliminaries Bill and Summary but excluding contractors' price adjustments. Individual costs of the main preliminary items should be given.

Notes

1 Professional fees will not form part of the cost analysis.
2 Lump sum adjustments shall be spread pro rata amongst all elements of the building and external works based on all work excluding Prime Cost and Provisional Sums.

Appendix B
Table of Random Numbers

23	61	75	21	92	45	10	84	83	02	37	63	30	22
63	07	17	94	14	51	06	29	74	37	38	07	70	33
30	00	47	21	50	65	79	34	69	98	85	57	29	33
62	08	86	93	47	31	63	36	73	91	74	24	83	72
58	71	31	21	27	47	67	68	58	44	32	76	42	08
10	95	59	69	21	19	48	35	91	72	71	26	07	96
97	92	03	46	34	51	01	91	77	28	57	61	96	76
55	04	42	32	74	70	00	64	97	01	76	89	43	56
16	33	76	33	61	62	03	89	25	02	63	16	45	56
28	09	81	53	27	74	71	89	20	56	00	68	73	24
73	34	51	61	94	55	44	87	63	00	05	41	40	57
68	14	43	19	08	98	69	19	02	69	46	27	37	99
58	55	54	94	74	50	48	58	95	22	42	81	85	51
42	67	37	72	62	15	60	96	78	60	66	92	12	03
60	86	00	89	92	50	77	65	01	97	59	45	35	30
98	57	04	16	45	85	43	96	86	60	67	45	73	30
70	94	73	75	58	05	21	60	76	31	64	87	75	16
27	14	30	92	14	56	03	23	40	17	36	23	77	01
20	01	18	69	70	75	52	41	73	11	28	56	68	71
47	40	09	73	13	01	33	32	86	82	78	26	79	70
35	98	39	86	77	27	32	07	45	87	80	37	58	64
92	16	16	89	54	10	35	19	35	32	70	48	95	13
86	99	97	52	77	10	86	46	71	34	64	29	70	17
41	77	60	89	81	47	66	18	96	82	93	63	26	07
74	73	29	59	80	96	03	08	04	04	80	31	33	57
27	18	60	83	77	28	85	37	55	75	15	92	44	04
98	99	15	37	28	84	06	98	61	81	68	45	46	06
09	95	85	97	82	98	12	88	63	32	39	63	30	73
20	89	27	55	20	66	67	55	32	18	96	23	59	56
95	43	31	90	54	25	06	28	19	31	82	49	69	80
11	07	21	09	98	61	54	29	96	75	18	59	04	72
60	91	95	50	85	60	12	85	20	27	96	37	93	22
57	80	32	41	10	01	71	92	60	73	50	61	15	43
98	67	62	69	47	71	67	17	48	74	64	64	30	60
13	78	72	88	76	51	89	46	03	93	65	18	69	72
27	97	36	55	67	89	21	39	25	93	99	96	82	18
31	59	28	01	51	40	88	17	31	89	02	62	17	74
90	00	45	19	80	87	58	51	69	08	74	69	53	01
65	09	67	96	52	30	51	97	54	29	16	15	76	40
05	13	79	46	87	08	91	99	16	88	30	53	78	14

Appendix C
Definition of Statistical Terms

Generally

Variable

This is a quantitative measurement to which a numerical value is assigned, e.g. area of a building or number of bed spaces in a building.

Variables can be discrete, i.e. they can only take integer values, or continuous, i.e. they can take any value including fractional values between two limits. In the example above 'area' would be a continuous variable and 'number of bed spaces' a discrete variable.

Parameter

This is a quantity which is constant in a particular case considered, but which varies in different cases. The parameters of a rectangular building may be length, width and height. Although all rectangular buildings can be described in this manner, for any specific rectangular building these quantities will be fixed.

Abbreviations

Arithmetic mean   = $\bar{x}$ or μ
Standard deviation = s or σ
To sum a series   = Σ

Averages

These are measures of location or measures of central tendency and are single figures which sum up certain characteristics of a whole group of figures.

Arithmetic mean

This is the sum (Σ) of all the variables divided by the number of members in the series:

$$\bar{x} = \frac{1}{n} \Sigma x$$

where $\bar{x}$ = arithmetic mean
        n = number of members in the sample
        x = variable.

Geometric mean

This is the nth root of the product of the values of the variable in question from the 1st to the nth:

$$\text{Geometric mean} = \sqrt[n]{(X_1 \times X_2 \times X_3 \times X_4 \ldots \times X_n)}$$

where n = number of the variables in the sample
            X_1 to X_n = the individual values of each variable.

It is preferable to use this mean when *relative* changes in a variable quantity are averaged, e.g. the average annual rate of increase in building costs over the last n years where each year's increase will be relative to the previous year's value.

Median

The median of a data is the value of the middle observation when the items have been arranged in either ascending or descending order. If there are an even number of observations the median is taken as the mean of the middle two observations. The median is only affected by the middle or two observations; the remaining observations do not have any direct influence on its size.

Mode

The mode is the value in the data set that occurs most frequently. Some sets of data have no modal value and others have two or more. By definition there can be no mode in a simple series, where no value occurs more than once.

Measures of dispersion

Averages measure the position or location of a series of observations. They do not give any indication of the total spread of values that go to make up the 'average' figure. To help us understand the sample set of data better, some measures of dispersion are available.

Standard deviation

This is sometimes known as the root mean squared deviation and this name is a description of the method of calculation. The deviations (i.e. the difference between *an* observation and the arithmetic mean calculated for each item of data) is squared to get rid of negative values. The arithmetic mean of these squared deviations is then calculated and the square root obtained.

The resultant figure is the standard deviation and it is expressed in absolute terms and is given in the same unit of measurement as the variable itself.

The formula is:

$$\text{standard deviation } (s) = \sqrt{\left[\frac{\Sigma(x - \bar{x})^2}{n}\right]}$$

where x = observed value
   $\bar{x}$ = arithmetic mean of the data
   n = number of observations.

For manual calculation purposes the formula is normally given in another form:

$$s = \sqrt{\left[\frac{\Sigma x^2}{n} - \left(\frac{\Sigma x}{n}\right)^2\right]} .$$

This formula requires only the aggregate of the series (Σx) and the aggregate of the squares of the individual values in the series (Σx^2) for calculation purposes.

The standard deviation is a useful guide to the amount of dispersion in the data. It can be used with, for example, the arithmetic mean to describe a complete distribution such as the normal distribution from which further knowledge of the data set can be obtained.

Coefficient of variation

There are occasions when the standard deviation, being an absolute measure of dispersion, is inadequate and a relative form becomes preferable, e.g. if a comparison between the variability of distributions of say, plaster and concrete prices is required, the two distributions will have different arithmetic means. Even if they had the same amount of variability their standard deviations would therefore be different.

Where a relative measure of variability is required the coefficient of variation should be used.

$$\text{Coefficient of variation} = \frac{100s}{\bar{x}} \%$$

where s = standard deviation of the data
   $\bar{x}$ = arithmetic mean of the data.

The formula simply expresses the standard deviation as a percentage of the arithmetic mean.

Distributions

Frequency distribution

This is the manner in which the frequency of occurrence of a variable is spread over its range of values.

It can be expressed in numerical form or by a histogram or curve.

Normal distribution

This is continuous symmetrical distribution in the form of a bell shaped curve which slopes downwards on both sides from the maximum value towards the x-axis but never touches the x-axis. A normal distribution is completely specified by two parameters, the mean and the standard deviation.

It can be shown mathematically that whenever a measurement is affected by a large number of small independent factors, no one of which predominates, the distribution of observations will be 'normal'.

Because of its characteristics it is possible, knowing the arithmetic mean and standard deviation, to assess the proportion of a variable under a portion of the curve. This assists in determining the probability of an event occurring.

Skewed distribution

This is a distribution which is not symmetrical but has its maximum value to one side or other of the arithmetic mean.

Appendix D
Discounting and Interest Formulae and Tables

The interest and discount formulae and tables in this appendix may be used in respect of weekly, monthly or yearly periods.

$$0.096\% \text{ per week} = 0.417\% \text{ per month} = 5\% \text{ per annum}$$
$$0.192\% \text{ per week} = 0.833\% \text{ per month} = 10\% \text{ per annum}$$
$$0.289\% \text{ per week} = 1.25\% \text{ per month} = 15\% \text{ per annum}$$
$$0.385\% \text{ per week} = 1.667\% \text{ per month} = 20\% \text{ per annum}$$

In the following formulae n represents the number of periods and i the interest rate expressed as a decimal fraction of the principal, e.g. $5\% = 0.05$.

Formula 1: Compound interest $(1 + i)^n$

Formula 2: Future value of £1 invested at regular intervals $\dfrac{(1 + i)^n - 1}{i}$

Formula 3: Present value of £1 $\dfrac{1}{(1 + i)^n}$

Formula 4: Present value of £1 payable at regular intervals ('years purchase')

$$\frac{(1 + i)^n - 1}{i(1 + i)^n}$$

Formula 5: Annuity purchased by £1 $\dfrac{i(1 + i)^n}{(1 + i)^n - 1}$

Formula 6: Sinking fund $\dfrac{i}{(1 + i)^n - 1}$

Note: No provision is made in Table 3 and in those that follow for repayment of the original sum by means of a sinking fund. When this is required the relevant value from Table 6 should be added.

Formulae 3, 5 and 6 are the reciprocals of formulae 1, 4 and 2 respectively.

TABLE 1—COMPOUND INTEREST

Value at end of each period of £1 invested at beginning of period 1 and accumulating at compound interest from 1% to 30% per period

Period	1% £	1.5% £	2% £	2.5% £	3% £	4% £	5% £	6% £	7% £	8% £
1	1.01	1.02	1.02	1.03	1.03	1.04	1.05	1.06	1.07	1.08
2	1.02	1.03	1.04	1.05	1.06	1.08	1.10	1.12	1.14	1.17
3	1.03	1.05	1.06	1.08	1.09	1.12	1.16	1.19	1.23	1.26
4	1.04	1.06	1.08	1.10	1.13	1.17	1.22	1.26	1.31	1.36
5	1.05	1.08	1.10	1.13	1.16	1.22	1.28	1.34	1.40	1.47
6	1.06	1.09	1.13	1.16	1.19	1.27	1.34	1.42	1.50	1.59
7	1.07	1.11	1.15	1.19	1.23	1.32	1.41	1.50	1.61	1.71
8	1.08	1.13	1.17	1.22	1.27	1.37	1.48	1.59	1.72	1.85
9	1.09	1.14	1.20	1.25	1.30	1.42	1.55	1.69	1.84	2.00
10	1.10	1.16	1.22	1.28	1.34	1.48	1.63	1.79	1.97	2.16
11	1.12	1.18	1.24	1.31	1.38	1.54	1.71	1.90	2.10	2.33
12	1.13	1.20	1.27	1.34	1.43	1.60	1.80	2.01	2.25	2.52
13	1.14	1.21	1.29	1.38	1.47	1.67	1.89	2.13	2.41	2.72
14	1.15	1.23	1.32	1.41	1.51	1.73	1.98	2.26	2.58	2.94
15	1.16	1.25	1.35	1.45	1.56	1.80	2.08	2.40	2.76	3.17
16	1.17	1.27	1.37	1.48	1.60	1.87	2.18	2.54	2.95	3.43
17	1.18	1.29	1.40	1.52	1.65	1.95	2.29	2.69	3.16	3.70
18	1.20	1.31	1.43	1.56	1.70	2.03	2.41	2.85	3.38	4.00
19	1.21	1.33	1.46	1.60	1.75	2.11	2.53	3.03	3.62	4.32
20	1.22	1.35	1.49	1.64	1.81	2.19	2.65	3.21	3.87	4.66

Period	1% £	1.5% £	2% £	2.5% £	3% £	4% £	5% £	6% £	7% £	8% £
21	1.23	1.37	1.52	1.68	1.86	2.28	2.79	3.40	4.14	5.03
22	1.24	1.39	1.55	1.72	1.92	2.37	2.93	3.60	4.43	5.44
23	1.26	1.41	1.58	1.76	1.97	2.46	3.07	3.82	4.74	5.87
24	1.27	1.43	1.61	1.81	2.03	2.56	3.23	4.05	5.07	6.34
25	1.28	1.45	1.64	1.85	2.09	2.67	3.39	4.29	5.43	6.85
26	1.30	1.47	1.67	1.90	2.16	2.77	3.56	4.55	5.81	7.40
27	1.31	1.49	1.71	1.95	2.22	2.88	3.73	4.82	6.21	7.99
28	1.32	1.52	1.74	2.00	2.29	3.00	3.92	5.11	6.65	8.63
29	1.33	1.54	1.78	2.05	2.36	3.12	4.12	5.42	7.11	9.32
30	1.35	1.56	1.81	2.10	2.43	3.24	4.32	5.74	7.61	10.06
31	1.36	1.59	1.85	2.15	2.50	3.37	4.54	6.09	8.15	10.87
32	1.37	1.61	1.88	2.20	2.58	3.51	4.76	6.45	8.72	11.74
33	1.39	1.63	1.92	2.26	2.65	3.65	5.00	6.84	9.33	12.68
34	1.40	1.66	1.96	2.32	2.73	3.79	5.25	7.25	9.98	13.69
35	1.42	1.68	2.00	2.37	2.81	3.95	5.52	7.69	10.68	14.79
36	1.43	1.71	2.04	2.43	2.90	4.10	5.79	8.15	11.42	15.97
37	1.45	1.73	2.08	2.49	2.99	4.27	6.08	8.64	12.22	17.25
38	1.46	1.76	2.12	2.56	3.07	4.44	6.39	9.15	13.08	18.63
39	1.47	1.79	2.16	2.62	3.17	4.62	6.70	9.70	13.99	20.12
40	1.49	1.81	2.21	2.69	3.26	4.80	7.04	10.29	14.97	21.72

TABLE 1—COMPOUND INTEREST

Value at end of each period of £1 invested at beginning of period 1 and accumulating at compound interest from 1% to 30% per period

Period	9% £	10% £	11% £	12% £	13% £	14% £	15% £	20% £	25% £	30% £
1	1.09	1.10	1.11	1.12	1.13	1.14	1.15	1.20	1.25	1.30
2	1.19	1.21	1.23	1.25	1.28	1.30	1.32	1.44	1.56	1.69
3	1.30	1.33	1.37	1.40	1.44	1.48	1.52	1.73	1.95	2.20
4	1.41	1.46	1.52	1.57	1.63	1.69	1.75	2.07	2.44	2.86
5	1.54	1.61	1.69	1.76	1.84	1.93	2.01	2.49	3.05	3.71
6	1.68	1.77	1.87	1.97	2.08	2.19	2.31	2.99	3.81	4.83
7	1.83	1.95	2.08	2.21	2.35	2.50	2.66	3.58	4.77	6.27
8	1.99	2.14	2.30	2.48	2.66	2.85	3.06	4.30	5.96	8.16
9	2.17	2.36	2.56	2.77	3.00	3.25	3.52	5.16	7.45	10.60
10	2.37	2.59	2.84	3.11	3.39	3.71	4.05	6.19	9.31	13.79
11	2.58	2.85	3.15	3.48	3.84	4.23	4.65	7.43	11.64	17.92
12	2.81	3.14	3.50	3.90	4.33	4.82	5.35	8.92	14.55	23.30
13	3.07	3.45	3.88	4.36	4.90	5.49	6.15	10.70	18.19	30.29
14	3.34	3.80	4.31	4.89	5.53	6.26	7.08	12.84	22.74	39.37
15	3.64	4.18	4.78	5.47	6.25	7.14	8.14	15.41	28.42	51.19
16	3.97	4.59	5.31	6.13	7.07	8.14	9.36	18.49	35.53	66.54
17	4.33	5.05	5.90	6.87	7.99	9.28	10.76	22.19	44.41	86.50
18	4.72	5.56	6.54	7.69	9.02	10.58	12.38	26.62	55.51	112.46
19	5.14	6.12	7.26	8.61	10.20	12.06	14.23	31.95	69.39	146.19
20	5.60	6.73	8.06	9.65	11.52	13.74	16.37	38.34	86.74	190.05

Period	9% £	10% £	11% £	12% £	13% £	14% £	15% £	20% £	25% £	30% £
21	6.11	7.40	8.95	10.80	13.02	15.67	18.82	46.01	108.42	247.06
22	6.66	8.14	9.93	12.10	14.71	17.86	21.64	55.21	135.53	321.18
23	7.26	8.95	11.03	13.55	16.63	20.36	24.89	66.25	169.41	417.54
24	7.91	9.85	12.24	15.18	18.79	23.21	28.63	79.50	211.76	542.80
25	8.62	10.83	13.59	17.00	21.23	26.46	32.92	95.40	264.70	705.64
26	9.40	11.92	15.08	19.04	23.99	30.17	37.86	114.48	330.87	917.33
27	10.25	13.11	16.74	21.32	27.11	34.39	43.54	137.37	413.59	1192.53
28	11.17	14.42	18.58	23.88	30.63	39.20	50.07	164.84	516.99	1550.29
29	12.17	15.86	20.62	26.75	34.62	44.69	57.58	197.81	646.23	2015.38
30	13.27	17.45	22.89	29.86	39.12	50.95	66.21	237.38	807.79	2620.00
31	14.46	19.19	25.41	33.56	44.20	58.08	76.14	284.85	1009.74	3405.99
32	15.76	21.11	28.21	37.58	49.95	66.21	87.57	341.82	1262.18	4427.79
33	17.18	23.23	31.31	42.09	56.44	75.48	100.70	410.19	1577.72	5756.13
34	18.73	25.55	34.75	47.14	63.78	86.05	115.80	492.22	1972.15	7482.97
35	20.41	28.10	38.57	52.80	72.07	98.10	133.18	590.67	2465.19	9727.86
36	22.25	30.91	42.82	59.14	81.44	111.83	153.15	708.80	3081.49	12646.22
37	24.25	34.00	47.53	66.23	92.02	127.49	176.12	850.56	3851.86	16440.08
38	26.44	37.40	52.76	74.18	103.99	145.34	202.54	1020.67	4814.82	21372.11
39	28.82	41.14	58.56	83.08	117.51	165.69	232.92	1224.81	6018.53	27783.74
40	31.41	45.26	65.00	93.05	132.78	188.88	267.86	1469.77	7523.16	36118.86

TABLE 2—FUTURE VALUE OF £1 INVESTED AT REGULAR INTERVALS

Value of £1 invested regularly at end of each period (i.e. weekly, monthly, or yearly) accumulating at compound interest from 1% to 30% per period

Period	1% £	1.5% £	2% £	2.5% £	3% £	4% £	5% £	6% £	7% £	8% £
1	1.00	1.00	1.00	1.00	1.00	1.00	1.00	1.00	1.00	1.00
2	2.01	2.02	2.02	2.03	2.03	2.04	2.05	2.06	2.07	2.08
3	3.03	3.05	3.06	3.08	3.09	3.12	3.15	3.18	3.21	3.25
4	4.06	4.09	4.12	4.15	4.18	4.25	4.31	4.37	4.44	4.51
5	5.10	5.15	5.20	5.26	5.31	5.42	5.53	5.64	5.75	5.87
6	6.15	6.23	6.31	6.39	6.47	6.63	6.80	6.98	7.15	7.34
7	7.21	7.32	7.43	7.55	7.66	7.90	8.14	8.39	8.65	8.92
8	8.29	8.43	8.58	8.74	8.89	9.21	9.55	9.90	10.26	10.64
9	9.37	9.56	9.75	9.95	10.16	10.58	11.03	11.49	11.98	12.49
10	10.46	10.70	10.95	11.20	11.46	12.01	12.58	13.18	13.82	14.49
11	11.57	11.86	12.17	12.48	12.81	13.49	14.21	14.97	15.78	16.65
12	12.68	13.04	13.41	13.80	14.19	15.03	15.92	16.87	17.89	18.98
13	13.81	14.24	14.68	15.14	15.62	16.63	17.71	18.88	20.14	21.50
14	14.95	15.45	15.97	16.52	17.09	18.29	19.60	21.02	22.55	24.21
15	16.10	16.68	17.29	17.93	18.60	20.02	21.58	23.28	25.13	27.15
16	17.26	17.93	18.64	19.38	20.16	21.82	23.66	25.67	27.89	30.32
17	18.43	19.20	20.01	20.86	21.76	23.70	25.84	28.21	30.84	33.75
18	19.61	20.49	21.41	22.39	23.41	25.65	28.13	30.91	34.00	37.45
19	20.81	21.80	22.84	23.95	25.12	27.67	30.54	33.76	37.38	41.45
20	22.02	23.12	24.30	25.54	26.87	29.78	33.07	36.79	41.00	45.76

Period	1% £	1.5% £	2% £	2.5% £	3% £	4% £	5% £	6% £	7% £	8% £
21	23.24	24.47	25.78	27.18	28.68	31.97	35.72	39.99	44.87	50.42
22	24.47	25.84	27.30	28.86	30.54	34.25	38.51	43.39	49.01	55.46
23	25.72	27.23	28.84	30.58	32.45	36.62	41.43	47.00	53.44	60.89
24	26.97	28.63	30.42	32.35	34.43	39.08	44.50	50.82	58.18	66.76
25	28.24	30.06	32.03	34.16	36.46	41.65	47.73	54.86	63.25	73.11
26	29.53	31.51	33.67	36.01	38.55	44.31	51.11	59.16	68.68	79.95
27	30.82	32.99	35.34	37.91	40.71	47.08	54.67	63.71	74.48	87.35
28	32.13	34.48	37.05	39.86	42.93	49.97	58.40	68.53	80.70	95.34
29	33.45	36.00	38.79	41.86	45.22	52.97	62.32	73.64	87.35	103.97
30	34.78	37.54	40.57	43.90	47.58	56.08	66.44	79.06	94.46	113.28
31	36.13	39.10	42.38	46.00	50.00	59.33	70.76	84.80	102.07	123.35
32	37.49	40.69	44.23	48.15	52.50	62.70	75.30	90.89	110.22	134.21
33	38.87	42.30	46.11	50.35	55.08	66.21	80.06	97.34	118.93	145.95
34	40.26	43.93	48.03	52.61	57.73	69.86	85.07	104.18	128.26	158.63
35	41.66	45.59	49.99	54.93	60.46	73.65	90.32	111.43	138.24	172.32
36	43.08	47.28	51.99	57.30	63.28	77.60	95.84	119.12	148.91	187.10
37	44.51	48.99	54.03	59.73	66.17	81.70	101.63	127.27	160.34	203.07
38	45.95	50.72	56.11	62.23	69.16	85.97	107.71	135.90	172.56	220.32
39	47.41	52.48	58.24	64.78	72.23	90.41	114.10	145.06	185.64	238.94
40	48.89	54.27	60.40	67.40	75.40	95.03	120.80	154.76	199.64	259.06

TABLE 2—FUTURE VALUE OF £1 INVESTED AT REGULAR INTERVALS

Value of £1 invested regularly at end of each period (i.e. weekly, monthly, or yearly) accumulating at compound interest from 1% to 30% per period

Period	9% £	10% £	11% £	12% £	13% £	14% £	15% £	20% £	25% £	30% £
1	1.00	1.00	1.00	1.00	1.00	1.00	1.00	1.00	1.00	1.00
2	2.09	2.10	2.11	2.12	2.13	2.14	2.15	2.20	2.25	2.30
3	3.28	3.31	3.34	3.37	3.41	3.44	3.47	3.64	3.81	3.99
4	4.57	4.64	4.71	4.78	4.85	4.92	4.99	5.37	5.77	6.19
5	5.98	6.11	6.23	6.35	6.48	6.61	6.74	7.44	8.21	9.04
6	7.52	7.72	7.91	8.12	8.32	8.54	8.75	9.93	11.26	12.76
7	9.20	9.49	9.78	10.09	10.40	10.73	11.07	12.92	15.07	17.58
8	11.03	11.44	11.86	12.30	12.76	13.23	13.73	16.50	19.84	23.86
9	13.02	13.58	14.16	14.78	15.42	16.09	16.79	20.80	25.80	32.01
10	15.19	15.94	16.72	17.55	18.42	19.34	20.30	25.96	33.25	42.62
11	17.56	18.53	19.56	20.65	21.81	23.04	24.35	32.15	42.57	56.41
12	20.14	21.38	22.71	24.13	25.65	27.27	29.00	39.58	54.21	74.33
13	22.95	24.52	26.21	28.03	29.98	32.09	34.35	48.50	68.76	97.63
14	26.02	27.97	30.09	32.39	34.88	37.58	40.50	59.20	86.95	127.91
15	29.36	31.77	34.41	37.28	40.42	43.84	47.58	72.04	109.69	167.29
16	33.00	35.95	39.19	42.75	46.67	50.98	55.72	87.44	138.11	218.47
17	36.97	40.54	44.50	48.88	53.74	59.12	65.08	105.93	173.64	285.01
18	41.30	45.60	50.40	55.75	61.73	68.39	75.84	128.12	218.04	371.52
19	46.02	51.16	56.94	63.44	70.75	78.97	88.21	154.74	273.56	483.97
20	51.16	57.27	64.20	72.05	80.95	91.02	102.44	186.69	342.94	630.17

Periods	9% £	10% £	11% £	12% £	13% £	14% £	15% £	20% £	25% £	30% £
21	56.76	64.00	72.27	81.70	92.47	104.77	118.81	225.03	429.68	820.22
22	62.87	71.40	81.21	92.50	105.49	120.44	137.63	271.03	538.10	1067.28
23	69.53	79.54	91.15	104.60	120.20	138.30	159.28	326.24	673.63	1388.46
24	76.79	88.50	102.17	118.16	136.83	158.66	184.17	392.48	843.03	1806.00
25	84.70	98.35	114.41	133.33	155.62	181.87	212.79	471.98	1054.79	2348.80
26	93.32	109.18	128.00	150.33	176.85	208.33	245.71	567.38	1319.49	3054.44
27	102.72	121.10	143.08	169.37	200.84	238.50	283.57	681.85	1650.36	3971.78
28	112.97	134.21	159.82	190.70	227.95	272.89	327.10	819.22	2063.95	5164.31
29	124.14	148.63	178.40	214.58	258.58	312.09	377.17	984.07	2580.94	6714.60
30	136.31	164.49	199.02	241.33	293.20	356.79	434.75	1181.88	3227.17	8729.99
31	149.58	181.94	221.91	271.29	332.32	407.74	500.96	1419.26	4034.97	11349.98
32	164.04	201.14	247.32	304.85	376.52	465.82	577.10	1704.11	5044.71	14755.98
33	179.80	222.25	275.53	342.43	426.46	532.04	664.67	2045.93	6306.89	19183.77
34	196.98	245.48	306.84	384.52	482.90	607.52	765.37	2456.12	7884.61	24939.90
35	215.71	271.02	341.59	431.66	546.68	693.57	881.17	2948.34	9856.76	32422.87
36	236.12	299.13	380.16	484.46	618.75	791.67	1014.35	3539.01	12321.95	42150.73
37	258.38	330.04	422.98	543.60	700.19	903.51	1167.50	4247.81	15403.44	54796.95
38	282.63	364.04	470.51	609.83	792.21	1031.00	1343.62	5098.37	19255.30	71237.03
39	309.07	401.45	523.27	684.01	896.20	1176.34	1546.17	6119.05	24070.12	92609.14
40	337.88	442.59	581.83	767.09	1013.70	1342.03	1779.09	7343.86	30088.66	120392.88

TABLE 3—PRESENT VALUE OF £1

Present value of £1 payable (or receivable) at end of any period 1 to 40, discounted at interest rates from 1% to 30% per period. Values shown in pence.

Period	1%	1.5%	2%	2.5%	3%	4%	5%	6%	7%	8%
	p	p	p	p	p	p	p	p	p	p
1	99.0	98.5	98.0	97.6	97.1	96.2	95.2	94.3	93.5	92.6
2	98.0	97.1	96.1	95.2	94.3	92.5	90.7	89.0	87.3	85.7
3	97.1	95.6	94.2	92.9	91.5	88.9	86.4	84.0	81.6	79.4
4	96.1	94.2	92.4	90.6	88.8	85.5	82.3	79.2	76.3	73.5
5	95.1	92.8	90.6	88.4	86.3	82.2	78.4	74.7	71.3	68.1
6	94.2	91.5	88.8	86.2	83.7	79.0	74.6	70.5	66.6	63.0
7	93.3	90.1	87.1	84.1	81.3	76.0	71.1	66.5	62.3	58.3
8	92.3	88.8	85.3	82.1	78.9	73.1	67.7	62.7	58.2	54.0
9	91.4	87.5	83.7	80.1	76.6	70.3	64.5	59.2	54.4	50.0
10	90.5	86.2	82.0	78.1	74.4	67.6	61.4	55.8	50.8	46.3
11	89.6	84.9	80.4	76.2	72.2	65.0	58.5	52.7	47.5	42.9
12	88.7	83.6	78.8	74.4	70.1	62.5	55.7	49.7	44.4	39.7
13	87.9	82.4	77.3	72.5	68.1	60.1	53.0	46.9	41.5	36.8
14	87.0	81.2	75.8	70.8	66.1	57.7	50.5	44.2	38.8	34.0
15	86.1	80.0	74.3	69.0	64.2	55.5	48.1	41.7	36.2	31.5
16	85.3	78.8	72.8	67.4	62.3	53.4	45.8	39.4	33.9	29.2
17	84.4	77.6	71.4	65.7	60.5	51.3	43.6	37.1	31.7	27.0
18	83.6	76.5	70.0	64.1	58.7	49.4	41.6	35.0	29.6	25.0
19	82.8	75.4	68.6	62.6	57.0	47.5	39.6	33.1	27.7	23.2
20	82.0	74.2	67.3	61.0	55.4	45.6	37.7	31.2	25.8	21.5

Period	1% *p*	1.5% *p*	2% *p*	2.5% *p*	3% *p*	4% *p*	5% *p*	6% *p*	7% *p*	8% *p*
21	81.1	73.1	66.0	59.5	53.8	43.9	35.9	29.4	24.2	19.9
22	80.3	72.1	64.7	58.1	52.2	42.2	34.2	27.8	22.6	18.4
23	79.5	71.0	63.4	56.7	50.7	40.6	32.6	26.2	21.1	17.0
24	78.8	70.0	62.2	55.3	49.2	39.0	31.0	24.7	19.7	15.8
25	78.0	68.9	61.0	53.9	47.8	37.5	29.5	23.3	18.4	14.6
26	77.2	67.9	59.8	52.6	46.4	36.1	28.1	22.0	17.2	13.5
27	76.4	66.9	58.6	51.3	45.0	34.7	26.8	20.7	16.1	12.5
28	75.7	65.9	57.4	50.1	43.7	33.3	25.5	19.6	15.0	11.6
29	74.9	64.9	56.3	48.9	42.4	32.1	24.3	18.5	14.1	10.7
30	74.2	64.0	55.2	47.7	41.2	30.8	23.1	17.4	13.1	9.9
31	73.5	63.0	54.1	46.5	40.0	29.6	22.0	16.4	12.3	9.2
32	72.7	62.1	53.1	45.4	38.8	28.5	22.0	15.5	11.5	8.5
33	72.0	61.2	52.0	44.3	37.7	27.4	20.0	14.6	10.7	7.9
34	71.3	60.3	51.0	43.2	36.6	26.4	19.0	13.8	10.0	7.3
35	70.6	59.4	50.0	42.1	35.5	25.3	18.1	13.0	9.4	6.8
36	69.9	58.5	49.0	41.1	34.5	24.4	17.3	12.3	8.8	6.3
37	69.2	57.6	48.1	40.1	33.5	23.4	16.4	11.6	8.2	5.8
38	68.5	56.8	47.1	39.1	32.5	22.5	15.7	10.9	7.6	5.4
39	67.8	56.0	46.2	38.2	31.6	21.7	14.9	10.3	7.1	5.0
40	67.2	55.1	45.3	37.2	30.7	20.8	14.2	9.7	6.7	4.6

TABLE 3—PRESENT VALUE OF £1

Present value of £1 payable (or receivable) at end of any period 1 to 40, discounted at interest rates from 1% to 30% per period. Values shown in pence.

Period	9%	10%	11%	12%	13%	14%	15%	20%	25%	30%
	p	p	p	p	p	p	p	p	p	p
1	91.7	90.9	90.1	89.3	88.5	87.7	87.0	83.3	80.0	76.9
2	84.2	82.6	81.2	79.7	78.3	76.9	75.6	69.4	64.0	59.2
3	77.2	75.1	73.1	71.2	69.3	67.5	65.8	57.9	51.2	45.5
4	70.8	68.3	65.9	63.6	61.3	59.2	57.2	48.2	41.0	35.0
5	65.0	62.1	59.3	56.7	54.3	51.9	49.7	40.2	32.8	26.9
6	59.6	56.4	53.5	50.7	48.0	45.6	43.2	33.5	26.2	20.7
7	54.7	51.3	48.2	45.2	42.5	40.0	37.6	27.9	21.0	15.9
8	50.2	46.7	43.4	40.4	37.6	35.1	32.7	23.3	16.8	12.3
9	46.0	42.4	39.1	36.1	33.3	30.8	28.4	19.4	13.4	9.4
10	42.2	38.6	35.2	32.2	29.5	27.0	24.7	16.2	10.7	7.3
11	38.8	35.0	31.7	28.7	26.1	23.7	21.5	13.5	8.6	5.6
12	35.6	31.9	28.6	25.7	23.1	20.8	18.7	11.2	6.9	4.3
13	32.6	29.0	25.8	22.9	20.4	18.2	16.3	9.3	5.5	3.3
14	29.9	26.3	23.2	20.5	18.1	16.0	14.1	7.8	4.4	2.5
15	27.5	23.9	20.9	18.3	16.0	14.0	12.3	6.5	3.5	2.0
16	25.2	21.8	18.8	16.3	14.1	12.3	10.7	5.4	2.8	1.5
17	23.1	19.8	17.0	14.6	12.5	10.8	9.3	4.5	2.3	1.2
18	21.2	18.0	15.3	13.0	11.1	9.5	8.1	3.8	1.8	0.9
19	19.4	16.4	13.8	11.6	9.8	8.3	7.0	3.1	1.4	0.7
20	17.8	14.9	12.4	10.4	8.7	7.3	6.1	2.6	1.2	0.5

Period	9% p	10% p	11% p	12% p	13% p	14% p	15% p	20% p	25% p	30% p
21	16.4	13.5	11.2	9.3	7.7	6.4	5.3	2.2	0.9	0.4
22	15.0	12.3	10.1	8.3	6.8	5.6	4.6	1.8	0.7	0.3
23	13.8	11.2	9.1	7.4	6.0	4.9	4.0	1.5	0.6	0.2
24	12.6	10.2	8.2	6.6	5.3	4.3	3.5	1.3	0.5	0.2
25	11.6	9.2	7.4	5.9	4.7	3.8	3.0	1.0	0.4	0.1
26	10.6	8.4	6.6	5.3	4.2	3.3	2.6	0.9	0.3	0.1
27	9.8	7.6	6.0	4.7	3.7	2.9	2.3	0.7	0.2	0.1
28	9.0	6.9	5.4	4.2	3.3	2.6	2.0	0.6	0.2	0.1
29	8.2	6.3	4.8	3.7	2.9	2.2	1.7	0.5	0.2	0.1
30	7.5	5.7	4.4	3.3	2.6	2.0	1.5	0.4	0.1	
31	6.9	5.2	3.9	3.0	2.3	1.7	1.3	0.4	0.1	
32	6.3	4.7	3.5	2.7	2.0	1.5	1.1	0.3	0.1	
33	5.8	4.3	3.2	2.4	1.8	1.3	1.0	0.2	0.1	
34	5.3	3.9	2.9	2.1	1.6	1.2	1.0	0.2	0.1	
35	4.9	3.6	2.6	1.9	1.4	1.0	0.9	0.2		
36	4.5	3.2	2.3	1.7	1.2	0.9	0.8	0.1		
37	4.1	2.9	2.1	1.5	1.1	0.8	0.7	0.1		
38	3.8	2.7	1.9	1.3	1.0	0.7	0.6	0.1		
39	3.5	2.4	1.7	1.2	0.9	0.6	0.5	0.1		
40	3.2	2.2	1.5	1.1	0.8	0.5	0.4	0.1		

TABLE 4—PRESENT VALUE OF £1 PAYABLE AT REGULAR INTERVALS ("YEARS PURCHASE")

Present value of £1 payable (or receivable) regularly at end of each period (i.e. weekly, monthly, or yearly) discounted at interest rates from 1% to 30% per period.

Period	1% £	1.5% £	2% £	2.5% £	3% £	4% £	5% £	6% £	7% £	8% £
1	0.99	0.99	0.98	0.98	0.97	0.96	0.95	0.94	0.93	0.93
2	1.97	1.96	1.94	1.93	1.91	1.89	1.86	1.83	1.81	1.78
3	2.94	2.91	2.88	2.86	2.83	2.78	2.72	2.67	2.62	2.58
4	3.90	3.85	3.81	3.76	3.72	3.63	3.55	3.47	3.39	3.31
5	4.85	4.78	4.71	4.65	4.58	4.45	4.33	4.21	4.10	3.99
6	5.80	5.70	5.60	5.51	5.42	5.24	5.08	4.92	4.77	4.62
7	6.73	6.60	6.47	6.35	6.23	6.00	5.79	5.58	5.39	5.21
8	7.65	7.49	7.33	7.17	7.02	6.73	6.46	6.21	5.97	5.75
9	8.57	8.36	8.16	7.97	7.79	7.44	7.11	6.80	6.52	6.25
10	9.47	9.22	8.98	8.75	8.53	8.11	7.72	7.36	7.02	6.71
11	10.37	10.07	9.79	9.51	9.25	8.76	8.31	7.89	7.50	7.14
12	11.26	10.91	10.58	10.26	9.95	9.39	8.86	8.38	7.94	7.54
13	12.13	11.73	11.35	10.98	10.63	9.99	9.39	8.85	8.36	7.90
14	13.00	12.54	12.11	11.69	11.30	10.56	9.90	9.29	8.75	8.24
15	13.87	13.34	12.85	12.38	11.94	11.12	10.38	9.71	9.11	8.56
16	14.72	14.13	13.58	13.06	12.56	11.65	10.84	10.11	9.45	8.85
17	15.56	14.91	14.29	13.71	13.17	12.17	11.27	10.48	9.76	9.12
18	16.40	15.67	14.99	14.35	13.75	12.66	11.69	10.83	10.06	9.37
19	17.23	16.43	15.68	14.98	14.32	13.13	12.09	11.16	10.34	9.60
20	18.05	17.17	16.35	15.59	14.88	13.59	12.46	11.47	10.59	9.82

Period	1% £	1.5% £	2% £	2.5% £	3% £	4% £	5% £	6% £	7% £	8% £
21	18.86	17.90	17.01	16.18	15.42	14.03	12.82	11.76	10.84	10.02
22	19.66	18.62	17.66	16.77	15.94	14.45	13.16	12.04	11.06	10.20
23	20.46	19.33	18.29	17.33	16.44	14.86	13.49	12.30	11.27	10.37
24	21.24	20.03	18.91	17.88	16.94	15.25	13.80	12.55	11.47	10.53
25	22.02	20.72	19.52	18.42	17.41	15.62	14.09	12.78	11.65	10.67
26	22.80	21.40	20.12	18.95	17.88	15.98	14.38	13.00	11.83	10.81
27	23.56	22.07	20.71	19.46	18.33	16.33	14.64	13.21	11.99	10.94
28	24.32	22.73	21.28	19.96	18.76	16.66	14.90	13.41	12.14	11.05
29	25.07	23.38	21.84	20.45	19.19	16.98	15.14	13.59	12.28	11.16
30	25.81	24.02	22.40	20.93	19.60	17.29	15.37	13.76	12.41	11.26
31	26.54	24.65	22.94	21.40	20.00	17.59	15.59	13.93	12.53	11.35
32	27.27	25.27	23.47	21.85	20.39	17.87	15.80	14.08	12.65	11.43
33	27.99	25.88	23.99	22.29	20.77	18.15	16.00	14.23	12.75	11.51
34	28.70	26.48	24.50	22.72	21.13	18.41	16.19	14.37	12.85	11.59
35	29.41	27.08	25.00	23.15	21.49	18.66	16.37	14.50	12.95	11.65
36	30.11	27.66	25.49	23.56	21.83	18.91	16.55	14.62	13.04	11.72
37	30.80	28.24	25.97	23.96	22.17	19.14	16.71	14.74	13.12	11.78
38	31.48	28.81	26.44	24.35	22.49	19.37	16.87	14.85	13.19	11.83
39	32.16	29.36	26.90	24.73	22.81	19.58	17.02	14.95	13.26	11.88
40	32.83	29.92	27.36	25.10	23.11	19.79	17.16	15.05	13.33	11.92

TABLE 4—PRESENT VALUE OF £1 PAYABLE AT REGULAR INTERVALS ("YEARS PURCHASE")

Present value of £1 payable (or receivable) regularly at end of each period (i.e. weekly, monthly, or yearly) discounted at interest rates from 1% to 30% per period.

Period	9% £	10% £	11% £	12% £	13% £	14% £	15% £	20% £	25% £	30% £
1	0.92	0.91	0.90	0.89	0.88	0.88	0.87	0.83	0.80	0.77
2	1.76	1.74	1.71	1.69	1.67	1.65	1.63	1.53	1.44	1.36
3	2.53	2.49	2.44	2.40	2.36	2.32	2.28	2.11	1.95	1.82
4	3.24	3.17	3.10	3.04	2.97	2.91	2.85	2.59	2.36	2.17
5	3.89	3.79	3.70	3.60	3.52	3.43	3.35	2.99	2.69	2.44
6	4.49	4.36	4.23	4.11	4.00	3.89	3.78	3.33	2.95	2.64
7	5.03	4.87	4.71	4.56	4.42	4.29	4.16	3.60	3.16	2.80
8	5.53	5.33	5.15	4.97	4.80	4.64	4.49	3.84	3.33	2.92
9	6.00	5.76	5.54	5.33	5.13	4.95	4.77	4.03	3.46	3.02
10	6.42	6.14	5.89	5.65	5.43	5.22	5.02	4.19	3.57	3.09
11	6.81	6.50	6.21	5.94	5.69	5.45	5.23	4.33	3.66	3.15
12	7.16	6.81	6.49	6.19	5.92	5.66	5.42	4.44	3.73	3.19
13	7.49	7.10	6.75	6.42	6.12	5.84	5.58	4.53	3.78	3.22
14	7.79	7.37	6.98	6.63	6.30	6.00	5.72	4.61	3.82	3.25
15	8.06	7.61	7.19	6.81	6.46	6.14	5.85	4.68	3.86	3.27
16	8.31	7.82	7.38	6.97	6.60	6.27	5.95	4.73	3.89	3.28
17	8.54	8.02	7.55	7.12	6.73	6.37	6.05	4.77	3.91	3.29
18	8.76	8.20	7.70	7.25	6.84	6.47	6.13	4.81	3.93	3.30
19	8.95	8.36	7.84	7.37	6.94	6.55	6.20	4.84	3.94	3.31
20	9.13	8.51	7.96	7.47	7.02	6.62	6.26	4.87	3.95	3.32

Period	9% £	10% £	11% £	12% £	13% £	14% £	15% £	20% £	25% £	30% £
21	9.29	8.65	8.08	7.56	7.10	6.69	6.31	4.89	3.96	3.32
22	9.44	8.77	8.18	7.64	7.17	6.74	6.36	4.91	3.97	3.32
23	9.58	8.88	8.27	7.72	7.23	6.79	6.40	4.92	3.98	3.33
24	9.71	8.98	8.35	7.78	7.28	6.84	6.43	4.94	3.98	3.33
25	9.82	9.08	8.42	7.84	7.33	6.87	6.46	4.95	3.98	3.33
26	9.93	9.16	8.49	7.90	7.37	6.91	6.49	4.96	3.99	3.33
27	10.03	9.24	8.55	7.94	7.41	6.94	6.51	4.96	3.99	3.33
28	10.12	9.31	8.60	7.98	7.44	6.96	6.53	4.97	3.99	3.33
29	10.20	9.37	8.65	8.02	7.47	6.98	6.55	4.97	3.99	3.33
30	10.27	9.43	8.69	8.06	7.50	7.00	6.57	4.98	4.00	3.33
31	10.34	9.48	8.73	8.08	7.52	7.02	6.58	4.98	4.00	3.33
32	10.41	9.53	8.77	8.11	7.54	7.03	6.59	4.99	4.00	3.33
33	10.46	9.57	8.80	8.14	7.56	7.05	6.60	4.99	4.00	
34	10.52	9.61	8.83	8.16	7.57	7.06	6.61	4.99	4.00	
35	10.57	9.64	8.86	8.18	7.59	7.07	6.62	4.99	4.00	
36	10.61	9.68	8.88	8.19	7.60	7.08	6.62	4.99		
37	10.65	9.71	8.90	8.21	7.61	7.09	6.63	4.99		
38	10.69	9.73	8.92	8.22	7.62	7.09	6.63	5.00		
39	10.73	9.76	8.94	8.23	7.63	7.10	6.64	5.00		
40	10.76	9.78	8.95	8.24	7.63	7.11	6.64	5.00		

TABLE 5—ANNUITY £1 WILL PURCHASE (ANNUAL EQUIVALENT)

Annual equivalent of £1 invested at the beginning of the period or the annuity purchased by a lump sum payment of £1 at rates of interest from 1% to 30% per period.

Period	1% £	1.5% £	2% £	2.5% £	3% £	4% £	5% £	6% £	7% £	8% £
1	1.010	1.015	1.020	1.025	1.030	1.040	1.050	1.060	1.070	1.080
2	0.508	0.511	0.515	0.519	0.523	0.530	0.538	0.545	0.553	0.561
3	0.340	0.343	0.347	0.350	0.354	0.360	0.367	0.374	0.381	0.388
4	0.256	0.259	0.263	0.266	0.269	0.275	0.282	0.289	0.295	0.302
5	0.206	0.209	0.212	0.215	0.218	0.225	0.231	0.237	0.244	0.250
6	0.173	0.176	0.179	0.182	0.185	0.191	0.197	0.203	0.210	0.216
7	0.149	0.152	0.155	0.157	0.161	0.167	0.173	0.179	0.186	0.192
8	0.131	0.134	0.137	0.139	0.142	0.149	0.155	0.161	0.167	0.174
9	0.117	0.120	0.123	0.125	0.128	0.134	0.141	0.147	0.153	0.160
10	0.106	0.108	0.111	0.114	0.117	0.123	0.130	0.136	0.142	0.149
11	0.096	0.099	0.102	0.105	0.108	0.114	0.120	0.127	0.133	0.140
12	0.089	0.092	0.095	0.097	0.100	0.107	0.113	0.119	0.126	0.133
13	0.082	0.085	0.088	0.091	0.094	0.100	0.106	0.113	0.120	0.127
14	0.077	0.080	0.083	0.086	0.089	0.095	0.101	0.108	0.114	0.121
15	0.072	0.075	0.078	0.081	0.084	0.090	0.096	0.103	0.110	0.117
16	0.068	0.071	0.074	0.077	0.080	0.086	0.092	0.099	0.106	0.113
17	0.064	0.067	0.070	0.073	0.076	0.082	0.089	0.095	0.102	0.110
18	0.061	0.064	0.067	0.070	0.073	0.079	0.086	0.092	0.099	0.107
19	0.058	0.061	0.064	0.067	0.070	0.076	0.083	0.090	0.097	0.104
20	0.055	0.058	0.061	0.064	0.067	0.074	0.080	0.087	0.094	0.102

Period	1% £	1.5% £	2% £	2.5% £	3% £	4% £	5% £	6% £	7% £	8% £
21	0.053	0.056	0.059	0.062	0.065	0.071	0.078	0.085	0.092	0.100
22	0.051	0.054	0.057	0.060	0.063	0.069	0.076	0.083	0.090	0.098
23	0.049	0.052	0.055	0.058	0.061	0.067	0.074	0.081	0.089	0.096
24	0.047	0.050	0.053	0.056	0.059	0.066	0.072	0.080	0.087	0.095
25	0.045	0.048	0.051	0.054	0.057	0.064	0.071	0.078	0.086	0.094
26	0.044	0.047	0.050	0.053	0.056	0.063	0.070	0.077	0.085	0.093
27	0.042	0.045	0.048	0.051	0.055	0.061	0.068	0.076	0.083	0.091
28	0.041	0.044	0.047	0.050	0.053	0.060	0.067	0.075	0.082	0.090
29	0.040	0.043	0.046	0.049	0.052	0.059	0.066	0.074	0.081	0.090
30	0.039	0.042	0.045	0.048	0.051	0.058	0.065	0.073	0.081	0.089
31	0.038	0.041	0.044	0.047	0.050	0.057	0.064	0.072	0.080	0.088
32	0.037	0.040	0.043	0.046	0.049	0.056	0.063	0.071	0.079	0.087
33	0.036	0.039	0.042	0.045	0.048	0.055	0.062	0.070	0.078	0.087
34	0.035	0.038	0.041	0.044	0.047	0.054	0.062	0.070	0.078	0.086
35	0.034	0.037	0.040	0.043	0.047	0.054	0.061	0.069	0.077	0.086
36	0.033	0.036	0.039	0.042	0.046	0.053	0.060	0.068	0.077	0.085
37	0.032	0.035	0.039	0.042	0.045	0.052	0.060	0.068	0.076	0.085
38	0.032	0.035	0.038	0.041	0.044	0.052	0.059	0.067	0.076	0.085
39	0.031	0.034	0.037	0.040	0.044	0.051	0.059	0.067	0.075	0.084
40	0.030	0.033	0.037	0.040	0.043	0.051	0.058	0.066	0.075	0.084

TABLE 5—ANNUITY £1 WILL PURCHASE (ANNUAL EQUIVALENT)

Annual equivalent of £1 invested at the beginning of the period or the annuity purchased by a lump sum payment of £1 at rates of interest from 1% to 30% per period.

Period	9% £	10% £	11% £	12% £	13% £	14% £	15% £	20% £	25% £	30% £
1	1.090	1.100	1.110	1.120	1.130	1.140	1.150	1.200	1.250	1.300
2	0.568	0.576	0.584	0.592	0.599	0.607	0.615	0.655	0.694	0.735
3	0.395	0.402	0.409	0.416	0.424	0.431	0.438	0.475	0.512	0.551
4	0.309	0.315	0.322	0.329	0.336	0.343	0.350	0.386	0.423	0.462
5	0.257	0.264	0.271	0.277	0.284	0.291	0.298	0.334	0.372	0.411
6	0.233	0.230	0.236	0.243	0.250	0.257	0.264	0.301	0.339	0.378
7	0.199	0.205	0.212	0.219	0.226	0.233	0.240	0.277	0.316	0.357
8	0.181	0.187	0.194	0.201	0.208	0.216	0.223	0.261	0.300	0.342
9	0.167	0.174	0.181	0.188	0.195	0.202	0.210	0.248	0.289	0.331
10	0.156	0.163	0.170	0.177	0.184	0.192	0.199	0.239	0.280	0.323
11	0.147	0.154	0.161	0.168	0.176	0.183	0.191	0.231	0.273	0.318
12	0.140	0.147	0.154	0.161	0.169	0.177	0.184	0.255	0.268	0.313
13	0.134	0.141	0.148	0.156	0.163	0.171	0.179	0.221	0.265	0.310
14	0.128	0.136	0.143	0.151	0.159	0.167	0.175	0.217	0.262	0.308
15	0.124	0.131	0.139	0.147	0.155	0.163	0.171	0.214	0.259	0.306
16	0.120	0.128	0.136	0.143	0.151	0.160	0.168	0.211	0.257	0.305
17	0.117	0.125	0.132	0.140	0.149	0.157	0.165	0.209	0.256	0.304
18	0.114	0.122	0.130	0.138	0.146	0.155	0.163	0.208	0.255	0.303
19	0.112	0.120	0.128	0.136	0.144	0.153	0.161	0.206	0.254	0.302
20	0.110	0.117	0.126	0.134	0.142	0.151	0.160	0.205	0.253	0.302

Period	9% £	10% £	11% £	12% £	13% £	14% £	15% £	20% £	25% £	30% £
21	0.108	0.116	0.124	0.132	0.141	0.150	0.158	0.204	0.252	0.301
22	0.106	0.114	0.122	0.131	0.139	0.148	0.157	0.204	0.252	0.301
23	0.104	0.113	0.121	0.130	0.138	0.147	0.156	0.203	0.251	0.301
24	0.103	0.111	0.128	0.128	0.137	0.146	0.155	0.203	0.251	0.301
25	0.102	0.110	0.119	0.127	0.136	0.145	0.155	0.202	0.251	0.300
26	0.101	0.109	0.118	0.127	0.136	0.145	0.154	0.202	0.251	0.300
27	0.100	0.108	0.117	0.126	0.135	0.144	0.154	0.201	0.250	0.300
28	0.099	0.107	0.116	0.125	0.134	0.144	0.153	0.201	0.250	0.300
29	0.098	0.107	0.116	0.125	0.134	0.143	0.153	0.201	0.250	0.300
30	0.097	0.106	0.115	0.124	0.133	0.143	0.152	0.201	0.250	0.300
31	0.097	0.105	0.115	0.124	0.133	0.142	0.152	0.201	0.250	0.300
32	0.096	0.105	0.114	0.123	0.133	0.142	0.152	0.200	0.250	0.300
33	0.096	0.104	0.114	0.123	0.132	0.142	0.152	0.200	0.250	0.300
34	0.095	0.104	0.113	0.123	0.132	0.142	0.151	0.200	0.250	0.300
35	0.095	0.104	0.113	0.122	0.132	0.141	0.151	0.200	0.250	0.300
36	0.094	0.103	0.113	0.122	0.132	0.141	0.151	0.200	0.250	0.300
37	0.094	0.103	0.112	0.122	0.131	0.141	0.151	0.200	0.250	0.300
38	0.94	0.103	0.112	0.122	0.131	0.141	0.151	0.200	0.250	0.300
39	0.093	0.102	0.112	0.121	0.131	0.141	0.151	0.200	0.250	0.300
40	0.093	0.102	0.112	0.121	0.131	0.141	0.151	0.200	0.250	0.300

TABLE 6—SINKING FUND

Amount (in pence) which has to be invested regularly at the end of each period (i.e. weekly, monthly, or yearly) in order to accumulate to £1 by the end of the chosen term; at compound interest of from 0.096% to 12% per period.

Period	1%	1.5%	2%	3%	4%	5%	6%	8%	10%	12%
	p	p	p	p	p	p	p	p	p	p
2	49.8	49.6	49.5	49.3	49.0	48.8	48.5	48.1	47.6	47.2
3	33.0	32.8	32.7	32.4	32.0	31.7	31.4	30.8	30.2	29.6
4	24.6	24.4	24.3	23.9	23.5	23.2	22.9	22.2	21.5	20.9
5	19.6	19.4	19.2	18.8	18.5	18.1	17.7	17.0	16.4	15.7
6	16.3	16.1	15.9	15.5	15.1	14.7	14.3	13.6	13.0	12.3
7	13.9	13.7	13.5	13.1	12.7	12.3	11.9	11.2	10.5	9.9
8	12.1	11.9	11.7	11.2	10.9	10.5	10.1	9.4	8.7	8.1
9	10.7	10.5	10.3	9.8	9.4	9.1	8.7	8.0	7.4	6.8
10	9.6	9.3	9.1	8.7	8.3	8.0	7.6	6.9	6.3	5.7
11	8.6	8.4	8.2	7.8	7.4	7.0	6.7	6.0	5.4	4.8
12	7.9	7.7	7.5	7.0	6.7	6.3	5.9	5.3	4.7	4.1
13	7.2	7.0	6.8	6.4	6.0	5.6	5.3	4.7	4.1	3.6
14	6.7	6.5	6.3	5.9	5.5	5.1	4.8	4.1	3.6	3.1
15	6.2	6.0	5.8	5.4	5.0	4.6	4.3	3.7	3.1	2.7
16	5.8	5.6	5.4	5.0	4.6	4.2	3.9	3.3	2.8	2.3
17	5.4	5.2	5.0	4.6	4.2	3.9	3.5	3.0	2.5	2.0
18	5.1	4.9	4.7	4.3	3.9	3.6	3.2	2.7	2.2	1.8
19	4.8	4.6	4.4	4.0	3.6	3.3	3.0	2.4	2.0	1.6
20	4.5	4.3	4.1	3.7	3.4	3.0	2.7	2.2	1.7	1.4

| Period | 1% | 1.5% | 2% | 3% | 4% | 5% | 6% | 8% | 10% | 12% |
	p	p	p	p	p	p	p	p	p	p
21	4.3	4.1	3.9	3.5	3.1	2.8	2.5	2.0	1.6	1.2
22	4.1	3.9	3.7	3.3	2.9	2.6	2.3	1.8	1.4	1.1
23	3.9	3.7	3.5	3.1	2.7	2.4	2.1	1.6	1.3	1.0
24	3.7	3.5	3.3	2.9	2.6	2.2	2.0	1.5	1.1	0.8
25	3.5	3.3	3.1	2.7	2.4	2.1	1.8	1.4	1.0	0.7
26	3.4	3.2	3.0	2.6	2.3	2.0	1.7	1.3	0.9	0.7
27	3.2	3.0	2.8	2.5	2.1	1.8	1.6	1.1	0.8	0.6
28	3.1	2.9	2.7	2.3	2.0	1.7	1.5	1.0	0.7	0.5
29	3.0	2.8	2.6	2.2	1.9	1.6	1.4	1.0	0.7	0.5
30	2.9	2.7	2.5	2.1	1.8	1.5	1.3	0.9	0.6	0.4
31	2.8	2.6	2.4	2.0	1.7	1.4	1.2	0.8	0.5	0.4
32	2.7	2.5	2.3	1.9	1.6	1.3	1.1	0.7	0.5	0.3
33	2.6	2.4	2.2	1.8	1.5	1.2	1.0	0.7	0.4	0.3
34	2.5	2.3	2.1	1.7	1.4	1.2	1.0	0.6	0.4	0.3
35	2.4	2.2	2.0	1.7	1.4	1.1	0.9	0.6	0.4	0.3
36	2.3	2.1	1.9	1.6	1.3	1.0	0.8	0.5	0.3	0.2
37	2.2	2.0	1.9	1.5	1.2	1.0	0.8	0.5	0.3	0.2
38	2.2	2.0	1.8	1.4	1.2	0.9	0.7	0.5	0.3	0.2
39	2.1	1.9	1.7	1.4	1.1	0.9	0.7	0.4	0.2	0.1
40	2.0	1.8	1.7	1.3	1.1	0.8	0.6	0.4	0.2	0.1

Index